Reliability Engineering in Systems Design and Operation

Balbir S. Dhillon

Associate Professor
Department of Mechanical Engineering
University of Ottawa

VNR VAN NOSTRAND REINHOLD COMPANY
NEW YORK   CINCINNATI   TORONTO   LONDON   MELBOURNE

Copyright © 1983 by Van Nostrand Reinhold Company Inc.

Library of Congress Catalog Card Number: 82–8355
ISBN: 0–442–27213–8

Manufactured in the United States of America

Published by Van Nostrand Reinhold Company Inc.
135 West 50th Street, New York, N.Y. 10020

Van Nostrand Reinhold Publishing
1410 Birchmount Road
Scarborough, Ontario M1P 2E7, Canada

Van Nostrand Reinhold
480 Latrobe Street
Melbourne, Victoria 3000, Australia

Van Nostrand Reinhold Company Limited
Molly Millars Lane
Wokingham, Berkshire, England

15 14 13 12 11 10 9 8 7 6 5 4 3 2 1

Library of Congress Cataloging in Publication Data
Dhillon, B. S.
   Reliability engineering in systems design and operation.

   Includes bibliographical references and index.
   1. Reliability (Engineering)   2. System design.
I. Title.
TA169.D47   1982   620'.00452   82–8355
ISBN 0–442–27213–8   AACR2

This book is affectionately dedicated to
THE MEMORY OF MY LATE BELOVED FATHER
KARNAIL S. DHILLON.

Preface

The field of system reliability engineering has advanced up to a level where it is beginning to be separated into various specialized areas such as reliability optimization, life cycle costing, and reliability growth modeling. In systems design and operation maximizing system reliability at a minimum cost, knowledge of these and various other reliability related areas is vital to engineers and reliability analysts. The author's intent in this book is to briefly update the commonly used conventional and advanced reliability and reliability related areas.

For example, areas directly related to reliability engineering are maintenance and safety engineering. Knowledge of these subjects is essential to engineers when designing and operating systems. An engineer faces inconvenience in obtaining information on the diverse areas of reliability and related matters because the information on these topics is available either in various texts or in technical reviews but not in a single volume. This book was written to fulfil the vital need for a single reference work. The book is intended for such readers as practicing engineers and undergraduate and graduate students with no previous knowledge of the subject. The emphasis is on the structure of concepts rather than on mathematical rigor and minute details. The source of most of the material presented in the book is given in references if a reader wishes to delve deeper into a specific area. The book contains over 1400 references which are listed at the end of each chapter. These references will provide readers with further information on each topic. In addition, the book contains several examples and their corresponding solutions.

Chapter 1 briefly discusses reliability engineering history, reliability publications, the need for the reliability engineering program, systems engineering, basic definitions, and the scope of the book. A list of military and other important publications on reliability is presented at the end of the chapter.

A review of the mathematics and probability theory required to understand subsequent chapters of the book is contained in Chapter 2.

Chapter 3 consists of basic reliability concepts. Both Chapters 2 and 3 will be beneficial to those readers not familiar with the field of reliability engineering. The topics such as supplementary variables method, interference theory, human reliability, common-cause failures, fault trees, software reliability, and renewal theory are presented in Chapter 4.

The reliability optimization and engineering reliability growth models, two topics of current interest, are presented in Chapters 5 and 6, respectively. These two topics are very important in the system design phase to make useful reliability decisions. Therefore, the material in both of these chapters is presented in considerable depth.

Chapter 7 discusses the subject of system safety engineering. This subject is closely related to reliability, especially in the system design phase. Various important areas such as system safety management, system safety analytical techniques, and the legal aspects of safety are discussed in the chapter.

Failure data analysis techniques are discussed in Chapter 8. An important subject of current interest, "Life Cycle Costing," is presented in Chapter 9. In addition, to the economic analysis techniques, the chapter contains several life cycle cost and system operation cost models. Chapter 10 presents the pertinent system operation phase discipline, maintenance engineering. The chapter discusses the maintenance management methods, equipment maintenance models and maintainability. The contents of Chapters 8 to 10 will be useful both at the system design and system operation phases.

The remaining two chapters, 11 and 12, are concerned with the application of reliability theory to medical and power equipment, respectively. These chapters are presented here because of the considerable interest in these topics in recent years.

In addition to system design and operation engineers, the book will also be useful to students of reliability, managers, reliability, maintainability, and safety engineers, maintenance engineers, and so on. In addition to being valuable for a long course on system reliability, the book can be used for various short courses such as introduction to system reliability (Chapters 1–3, 8), advanced reliability (Chapters 4–6, 7), reliability optimization (Chapters 1–3, 5), reliability growth modeling (Chapters 1–3, 6), life cycle costing (Chapters 1–3, 9), medical system reliability (Chapters 1–3, 11), Power System Reliability (Chapters 1–3, 8, 12), system safety (Chapters 1–3, 7), and maintenance engineering (Chapters 1–3, 8, 10).

I wish to thank the Department of Mechanical Engineering for assistance in typing a number of chapters of the book. I am particularly grateful to Mrs. D. Champion-Demers for her excellent typing. The author acknowledges the contributions of former colleagues and professionals at Ontario Hydro and his present colleagues at the University of Ottawa for their interest in the project.

I wish to thank my parents, relatives, and many friends for their constant encouragement throughout. Last but not least, I thank my wife Rosy for typing major portions of the manuscript, preparing all the diagrams, and so on. During the preparation of the manuscript her patience and tolerance were also appreciated.

Balbir S. Dhillon
Ottawa, Ontario

Contents

3.   INTRODUCTORY CONCEPTS IN ENGINEERING RELIABILITY/23

4. ADVANCED RELIABILITY EVALUATION CONCEPTS/50

5. RELIABILITY OPTIMIZATION/108

11. MEDICAL EQUIPMENT RELIABILITY/274

12. POWER EQUIPMENT RELIABILITY/289

1
Introduction

1.1 RELIABILITY ENGINEERING HISTORY

The field of reliability engineering is not new. Its history goes back to World War II, when the Germans reported to have first introduced the reliability concept to improve reliability of their V-1 and V-2 rockets.

In the United States during the period between 1945–1950 various Navy, Army, and Air Force studies were conducted on equipment repair and maintenance cost, failure of electronic equipment, and so on. As the result of these studies an ad hoc committee on reliability was established by the Department of Defense in 1950. This committee was transformed to a permanent group in 1952 and became known as the Advisory Group on the Reliability of Electronic Equipment (AGREE).

In the early fifties the IEEE Transactions on Reliability and the *Proceedings of the National Symposium on Reliability and Quality Control* came into being as a result of reliability awareness in the United States. In 1957 AGREE published a report that led directly to a specification on the reliability of military electronic equipment.

The early sixties witnessed the publication of many books on reliability. In addition, an international journal entitled *Microelectronics and Reliability,* published by the Pergamon Press, also had its origin during this period.

So far several books and other publications on reliability and maintainability engineering have appeared. The important books and the other publications on the subject are listed at the ends of Chapter 3 and Chapter 1, respectively.

Today many industries, government agencies, and others employ specialists known as reliability engineers, reliability group leaders, and reliability managers. The field of reliability has grown into many subbranches, some of which are: software reliability, mechanical reliability, human reliability, power systems reliability, maintainability engineering, life cycle costing, etc. Finally, it may be said the field of reliability is here for the good. Because of its many benefits, the importance of the field is increasing.

1.2 RELIABILITY PUBLICATIONS

Throughout the world there are various journals and conference proceedings which are specifically devoted to the reliability field or at least publish some articles on the subject annually. It is not the intention of the author to list all of them here. The following three journals and conference proceedings specifically publish papers on reliability engineering and are particularly important:

1. *IEEE Transactions on Reliability,* jointly published by the IEEE and the American Society for Quality Control (ASQC). It is published five times a year.
2. *Microelectronics and Reliability: An International Journal,* published by the Pergamon Press, Oxford, England. It is published bimonthly.
3. *Proceedings of the Annual Reliability and Maintainability Symposium* sponsored by nine American professional societies including the IEEE and the American Society of Mechanical Engineers. The symposium is held annually in the United States.

As its title suggests *Microelectronics and Reliability* is partially devoted to the reliability field. However, usually it publishes more than half of its articles in each issue on reliability. These three research publications are well known to most reliability specialists throughout the world. Sometimes papers are published on reliability in the following journals and conference proceedings:

1. *Naval Research Logistics Quarterly.*
2. *Technometrics.*
3. *Proceedings of the Annual Technical Meeting of the Institute of Environmental Sciences, U.S.A.*
4. *AIIE (American Institute of Industrial Engineers) Transactions.*
5. *Proceedings of Annual Pittsburgh Conference on Modeling and Simulation.*
6. *Journal of the Operations Research Society of America.*
7. Various conference proceedings of the American Society of Mechanical Engineers.
8. *Journal of Applied Probability.*
9. *Journal of Applied Mechanics.*
10. *IEEE Transactions on Power Apparatus and Systems,* published by IEEE.

11. *Proceedings of the Reliability Engineering Conference for the Electric Power Industry.*

As mentioned earlier, it is not the intent of the author to list all of the journals and conference proceedings which publish papers intermittently on the subject of reliability. However, the major ones are listed above. Furthermore, the important U.S. military standards and handbooks and other documents related to the reliability field are listed at the end of this chapter in references 1 to 54.

1.3 THE NEED FOR A RELIABILITY ASSURANCE PROGRAM

In order to minimize the overall costs of a system the high reliability is vital. In some U.S. military studies it was found that for some systems the yearly cost to maintain them in their operational state has been as high as 10 times the acquisition cost. Other factors which emphasize the need of high system reliability are the sophistication, high acquisition cost, inconvenience, loss of prestige, loss of national security, competitiveness, etc. Due to these reasons reliability consideration becomes very important in the planning, design and operation of systems. Quantitatively defined reliability plays an important role in the system planning, design, and operation because it can be traded off against cost and performance parameters at an early stage of the system development.

1.4 SYSTEMS ENGINEERING

Systems engineering is the science of designing major systems in their totality to make sure that the subsystems or components forming the system are designed, integrated, and operated optimally. The application of systems engineering is usually carried out in four phases: system analysis, system design, implementation, and operation.

System analysis comprises items such as problem formulation, project organization, system definition, and so on. The system design phase is made up of the forecasting, model building and simulation, reliability analysis, optimization, and so on. During the implementation stage the construction and documentation are the major items to be considered. Finally, the operation phase is concerned with initial operation, retrospective appraisal, and so on.

1.5 SYSTEMS RELIABILITY

Systems reliability may be divided into two broad categories, i.e., design reliability and operations reliability. Design reliability covers functions such as reliability analysis, design reviews, trade-off studies, reliability test analysis, and so on. Operations reliability is concerned with failure analysis, field operations reports, corrective action, and so on.

1.6 BRANCHES OF RELIABILITY ENGINEERING AND THEIR PRESENT DAY TRENDS

Today reliability engineering is broad in scope and is applied across various areas such as aerospace, transportation, electric power generation, and national security. Furthermore, it has developed into various specialized branches such as software reliability, human reliability, mechanical reliability, life cycle costing, power system reliability, and so on. In the last few years text books have been written on each of these areas. This is a clear sign that the field of reliability is developing at a significant rate. Currently the topics such as software reliability and power system reliability have been receiving somewhat more attention from the researchers in the field than the other branches of reliability engineering.

1.7 DEFINITIONS

Some commonly used definitions associated with reliability engineering are

1. Reliability: Reliability is the probability that an item will perform its function adequately for the desired period of time when operated according to specified conditions.
2. Hazard Rate (instantaneous failure rate): Hazard rate is defined as the rate of change of the failed components quantity divided by number of survived components at time t.
3. Active redundancy: Active redundancy is the term used when all redundant units are functioning simultaneously.
4. Maintainability: Maintainability is the probability that a failed item, will be restored to its satisfactory operational state.
5. Downtime: This is the total time during which the item is in an unsatisfactory operating mode.

Other reliability definitions may be found in references 1, 55, and 56.

1.8 SCOPE OF THE BOOK

Today many engineers are designing large, complex, and sophisticated systems and require knowledge of many specific areas of reliability. Furthermore, the operations engineers have to maintain such systems in the field with the minimum maintenance costs and maximum availability. Therefore, the knowledge of basic reliability and various specific areas of reliability are essential to these engineers and other personnel closely related to systems design and their operation in the field. By keeping this in mind, this book is an attempt to fulfil this specific need because there has been a considerable growth of knowledge in several specific areas of reliability and its applications. Most of the reliability and closely related areas essential to designing and operating large and complex systems are presented in this book. Prior knowledge of basic reliability mathematics, and basic reliability is not necessary because this material is covered in Chapters 2 and 3. This book is concerned with design and operations reliability, applications of reliability, and other areas, such as maintenance and safety engineering, which are closely related to reliability engineering. For example, a system may fail if the maintenance is performed incorrectly. Furthermore, an unreliable system may also be unsafe, because a loss of life or injury to users could be the result of system failures.

This book will be useful to reliability engineers, maintenance engineers, systems engineers, safety engineers, students of reliability, and others.

1.9 SUMMARY

The chapter briefly discusses reliability engineering history, reliability publications, the need for a reliability engineering program, systems engineering, systems reliability, branches of reliability engineering and their present day trends, definitions, and the scope of the book. A list of military standards and other publications on reliability is presented.

1.10 REFERENCES

Military Standard

1. Definitions of Effectiveness Terms for Reliability, Maintainability, Human Factors and Safety, MIL-STD-721B, 22 August 1966. Available from the Naval Publications and Forms Center, 5801 Tabor Ave., Philadelphia, PA, 19120.
2. Engineering Management, MIL-STD-499A, 1 May 1974. Available from the Naval Publications and Forms Center, 5801 Tabor Ave., Philadelphia, PA, 19120.
3. "List of Quality Standards, Specifications and Related Documents," Quality Progress, September 1976, pp. 30–35.
4. Maintainability Demonstration, MIL-STD-471A, 27 March 1973. Available from the Naval Publications and Forms Center, 5801 Tabor Ave., Philadelphia, PA, 19120.
5. Maintainability Program Requirements for Systems and Equipment, MIL-STD-470, 21 March, 1966. Available from the Naval Publications and Forms Center, 5801 Tabor Ave., Philadelphia, PA, 19120.
6. Procedures for Performing A Failure Mode and Effect Analysis for Shipboard Equipment,

MIL-STD-1629 (SHIPS), 1 November 1974. Available from the Naval Publications and Forms Center, 5801 Tabor Ave., Philadelphia, PA, 19120.

7. Reliability Assurance Program for Electronic Parts Specifications, MIL-STD-790C, 18 April 1968. Available from the Naval Publications and Forms Center, 5801 Tabor Ave., Philadelphia, PA, 19120.

8. Reliability Evaluation from Demonstration Data, MIL-STD-757, 19 June 1964. Available from the Naval Publications and Forms Center, 5801 Tabor Ave., Philadelphia, PA, 19120.

9. Reliability Prediction, MIL-STD-756A, 15 May 1963. Available from the Naval Publications and Forms Center, 5801 Tabor Ave., Philadelphia, PA, 19120.

10. Reliability Program for Systems and Equipment Development and Production, MIL-STD-785A, 28 March 1969. Available from the Naval Publications and Forms Center, 5801 Tabor Ave., Philadelphia, PA, 19120.

11. Reliability Tests: Exponential Distribution, MIL-STD-781B, 15 November 1967. Available from the Naval Publications and Forms Center, 5801 Tabor Ave., Philadelphia, PA, 19120.

12. System Safety Program for Systems and Associated Subsystems and Equipment, Requirements for, MIL-STD-882, 15 July 1969. Available from the Naval Publications and Forms Center, 5801 Tabor Ave., Philadelphia, PA, 19120.

Military Specifications

13. Airplane Strength and Rigidity Reliability Requirements, Repeated Loads and Fatigue," MIL-A-008866A, 21 January 1974. Available from the Naval Publications and Forms Center, 5801 Tabor Ave., Philadelphia, PA, 19120.

14. "Software Quality Assurance Program Requirements," MIL-S-52779, 5 April 1974. Available from the Naval Publications and Forms Center, 5801 Tabor Ave., Philadelphia, PA, 19120.

Military Handbooks

15. *Design for Reliability,* AMCP 706-196, ADA-027 370, January 1976. Available from the NTIS, Springfield, VA, 22161.

16. *Handbook of Reliability Engineering,* NAVWEPS 00-65-502, 1 June 1964. Available from the Superintendent of Documents, U.S. Government Printing Office, Washington, D.C., 20402.

17. *Handbook of Product Maintainability,* Reliability Division, American Society for Quality Control, August 1973.

18. *Life Cycle Environments,* AMCP 706-118, AD A-015 179, March 1975. Available from the NTIS, Springfield, VA, 22161.

19. *Maintenance Engineering Techniques,* AMCP 706-132, AD A-021 390, June 1975. Available from the NTIS, Springfield, VA, 22161.

20. *Maintainability Engineering, Theory and Practice,* AMCP 706-133, ADA-026 006, January 1976. Available from the NTIS, Springfield, VA, 22161.

21. *Maintainability Guide for Design,* AMCP 706-134, AD-754 202, October 1972. Available from the NTIS, Springfield, VA, 22161.

22. *Maintainability Prediction,* MIL-HDBK-472, 24 May 1966. Available from the Naval Publications and Forms Center, 5801 Tabor Ave., Philadelphia, PA, 19120.

23. *Mathematical Appendix and Glossary,* AMCP 706-200, ADA-027 372, January 1976. Available from the NTIS, Springfield, VA, 22161.

24. *Quality Assurance, Reliability Handbook,* AMCP 702-3, October 1968. Available from the Headquarters, United States Army Material Command, Washington, D.C., 20315.

25. *Reliability Design Handbook,* RDH-376, March 1976. Available from RADC, Griffiss Air Force Base, New York, 13441.
26. *Reliability Design Thermal Applications,* MIL-HDBK-251, 19 January 1978. Available from the Naval Publications and Forms Center, 5801 Tabor Ave., Philadelphia, PA, 19120.
27. *Reliability Engineering Design Handbook,* PB-121839 (NAVY). Available from the Naval Publications and Forms Center, 5801 Tabor Ave., Philadelphia, PA, 19120.
28. *Reliability Measurement,* AMCP 706-198, ADA-027 371, January 1976. Available from the NTIS, Springfield, VA, 22161.
29. *Reliability Prediction,* AMCP 706-197, ADA-032 105, January 1976. Available from the NTIS, Springfield, VA, 22161.
30. *Reliability Stress and Failure Rate Data for Electronic Equipment,* MIL-HDBK 217 C, 20 September 1974. Available from the Naval Publications and Forms Center, 5801 Tabor Ave., Philadelphia, PA, 19120.
31. *System Safety, Design Handbook,* 1967. Available from Wright-Patterson AFB, OH, 45433.
32. T. E. Paquette, *Microcircuit Device Reliability Linear Interface Data,* 1975. Available from RADC, Griffiss Air Force Base, New York, 13441.
33. "U.S. DOD Reliability Standardization Document Program," *IEEE Trans. on Reliability,* Vol. R-28, No. 3, August 1979, pp. 254–258.
34. *Value Engineering,* AMCP 706-104, AD-894 478, July 1971. Available from the NTIS, Springfield, VA, 22161.

NASA Reliability and Quality Publications

35. "An Introduction to the Assurance of Human Performance in Space Systems," NASA SP-6505. Available from the Naval Publications and Forms Center, 5801 Tabor Ave., Philadelphia, PA, 19120.
36. "Applicability of NASA Contract Quality Management and Failure Mode Effect Analysis Procedures to the USGS Outer Continental Shelf Oil and Gas Lease Management Program," NASA-TM X-2567. Available from the Naval Publications and Forms Center, 5801 Tabor Ave., Philadelphia, PA, 19120.
37. "Evaluation of Reliability Programs," NASA SP-6501, Available from the Naval Publications and Forms Center, 5801 Tabor Ave., Philadelphia, PA, 19120.
38. "Failure Analysis of Electronic Parts: Laboratory Methods," NASA SP-6508. Available from the Naval Publications and Forms Center, 5801 Tabor Ave., Philadelphia, PA, 19120.
39. "Failure Reporting and Management Techniques in the Surveyor Program," NASA-SP-6504. Available from the Naval Publications and Forms Center, 5801 Tabor Ave., Philadelphia, PA, 19120.
40. "Guide for Collection of Reliability, Availability and Maintainability Data from Field Performance of Electronic Items," 382, 1971. Available from the Naval Publications and Forms Center, 5801 Tabor Ave., Philadelphia, PA, 19120.
41. "Guide for the Inclusion of Reliability Clauses into Specifications for Components (or Parts) for Electronic Equipment," 409, 1973. Available from the Naval Publications and Forms Center, 5801 Tabor Ave., Philadelphia, PA, 19120.
42. "List of Basic Terms, Definitions and Related Mathematics for Reliability," 271, 1974. Available from the Naval Publications and Forms Center, 5801 Tabor Ave., Philadelphia, PA, 19120.
43. "Managerial Aspects of Reliability," 300, 1974. Available from the Naval Publications and Forms Center, 5801 Tabor Ave., Philadelphia, PA, 19120.
44. "Preliminary Reliability Considerations," 272, 1968. Available from the Naval Publications and Forms Center, 5801 Tabor Ave., Philadelphia, PA, 19120.

45. "Presentation of Reliability Data on Electronic Components (or Parts), Including Amendment 2 and Supplement 319 A," 319, 1970. Available from the Naval Publications and Forms Center, 5801 Tabor Ave., Philadelphia, PA, 19120.
46. "Quality Assurance Provisions for Government Agencies," NHB 5300 4 (2B), November 1971. Available from the Naval Publications and Forms Center, 5801 Tabor Ave., Philadelphia, PA, 19120.
47. "Reliability Program Provisions for Aeronautical and Space System Contracters," NHB 5300 4 (1A) April 1, 1970. Available from the Naval Publications and Forms Center, 5801 Tabor Ave., Philadelphia, PA, 19120.
48. "Safety, Reliability, Maintainability and Quality Provisions for the Space Shuttle Program," NHB 5300 4 (1D-1), August 1974. Available from the Naval Publications and Forms Center, 5801 Tabor Ave., Philadelphia, PA, 19120.

ISO Standards

49. A Guide to Quality Assurance. BS 4891 1972. Available from the American National Standards Institute, 1430 Broadway, New York, NY, 10018.

ANSI Standards

50. Draft Standard, General Principles for Reliability Analysis of Nuclear Power Generating Station Protection Systems, Trial-Use Guide (IEEE Std 352–1972), Issued for trial use and comment. N41.4. Available from the American National Standards Institute, 1430 Broadway, New York, NY, 10018.
51. Quality Assurance Program Requirements for Nuclear Power Plants. N45.2. Available from the American National Standards Institute, 1430 Broadway, New York, NY, 10018.
52. Quality Assurance Terms and Definitions N45.2.10. Available from the American National Standards Institute, 1430 Broadway, New York, NY, 10018.

ASQC Documents

53. A. J. McElroy, "IEEE Project 500 — Reliability Data Manual for Nuclear Power Generating Stations," *Proceedings of the 1976 Annual Reliability and Maintainability Symposium,* pp. 277–281.
54. *Handbook of Product Maintainability,* Reliability Division, August 1973. Available from the American Society for Quality Control, 161 West Wisconsin Ave., Milwaukee, WI, 53203.

Miscellaneous

55. W. H. Von Alven, *Reliability Engineering,* Prentice-Hall, Englewood Cliffs, N.J. 1964.
56. J. J. Naresky, "Reliability Definitions," *IEEE Trans. on Reliability,* Vol. R-19, November 1970, pp. 198–200.

2
Engineering Reliability Mathematics

2.1 INTRODUCTION

This chapter presents the basic mathematical concepts required to understand reliability engineering. These concepts such as statistical distribution, Laplace transform method, special functions, and so on are presented briefly. The depth of the topics presented should be sufficient for the reader to understand subsequent chapters of this book. However, if further reading is necessary, the reader should consult references 1 to 20.

2.2 BASIC PROBABILITY

This section presents some of the important properties of probability theory. These properties are as follows:

2.2.1 Statistical Independent Events Probability

If and only if

$$P(Y_1 \cdot Y_2 \cdot Y_3 \cdot \ldots Y_m) = P(Y_1) \cdot P(Y_2) \cdot P(Y_3) \ldots P(Y_m) \qquad (2.1)$$

then the events Y_1, Y_2, Y_3, ..., Y_m are said to be independent. $P(Y_1 \cdot Y_2 \cdot Y_3 \cdot \ldots Y_m)$ denotes the intersection probability of m events. And $P(Y_i)$ denotes the i_{th} Y event probability; for $i = 1, 2, 3 \ldots m$.

2.2.2 Union of *m* Events Probability

Probability of union of '*m*' events is

$$P(Y_1 + Y_2 + Y_3 + \ldots + Y_m) = \{P(Y_1) + P(Y_2) + P(Y_3) + \ldots$$
$$+ P(Y_m)\} - \{P(Y_1 Y_2) + P(Y_1 Y_3)$$
$$+ \ldots + P(Y_i Y_j)\} \ldots + (-1)^{m-1} \qquad (2.2)$$
$$\underset{i=j}{}$$
$$\{P(Y_1 Y_2 Y_3 \ldots Y_m)\}$$

where $(Y_1 + Y_2 + Y_3 + \ldots + Y_m)$ denotes the Y_m events union.

2.3 PROBABILITY DISTRIBUTIONS

This section presents distributions which have applications in reliability engineering. The probability density and cumulative probability distribution functions are as follows:

2.3.1 Exponential Distribution

This is one of the simplest distributions and is widely used in the field of reliability engineering [3]. The probability density function, $f(t)$, is defined as follows:

$$f(t) = \lambda\, e^{-\lambda t} \qquad \lambda > 0, \qquad t \geq 0 \qquad (2.3)$$

where t is time
λ is the constant failure rate

The cumulative probability distribution function, $F(t)$, is defined by

$$F(t) = \int_{-\infty}^{t} f(x) \cdot dx \qquad (2.4)$$

Thus, substituting (2.3) in (2.4) and integrating we get

$$F(t) = \int_{-\infty}^{t} \lambda\, e^{-\lambda x} dx = 1 - e^{-\lambda t} \qquad (2.5)$$

The derivative of (2.4) with respect to t yields the probability density function

$$f(t) = \frac{dF(t)}{dt} \qquad (2.6)$$

The area under the probability density curve is always equal to unity. This can be easily verified by letting $t \to \infty$ in Eq. (2.5). Similarly, for the remaining time continuous probability distributions, the cumulative probability distribution functions can be developed.

EXAMPLE 1

The exponential distribution cumulative distribution function is given by Eq. (2.5). Obtain the distribution probability density function.

Thus substituting (2.5) in (2.6) and differentiating we get

$$f(t) = \frac{d(1 - e^{-\lambda t})}{dt} = \lambda e^{-\lambda t} \tag{2.7}$$

2.3.2 Bathtub Hazard Rate Distribution

This distribution can represent increasing, decreasing and bathtub hazard rates as shown in Figure 3.1. The probability density function, $f(t)$, of the distribution is defined [4, 5] as follows:

$$f(t) = b\beta(\beta t)^{b-1} e^{-[e^{(\beta t)^b} - (\beta t)^b - 1]} \tag{2.8}$$

for β, $b > 0$, and $t \geq 0$

where b is the shape parameter
  β is the scale parameter
  t is time

The cumulative distribution function, $F(t)$, is obtained by substituting Eq. (2.8) in Eq. (2.4) and then integrating, as shown below.

$$F(t) = 1 - e^{-\{e^{(\beta t)^b} - 1\}} \tag{2.9}$$

The distribution parameter estimating procedures are given in Chapter 8. The special cases of this distribution are the extreme value distribution (at $b = 1$) and the bathtub curve (at $b = 0.5$).

2.3.3 Extreme Value Distribution

Probability density, $f(t)$, is

$$f(t) = \beta e^{\beta t} e^{-(e^{\beta t} - 1)} \qquad 0 < t < \infty \tag{2.10}$$

where t is time
  β is the scale parameter

This distribution is the special case of the bathtub distribution (at $b = 1$) and is used in the reliability field to represent the failure times of mechanical components.

2.3.4 Uniform Distribution

The distribution probability density function, $f(t)$, is defined by

$$f(t) = \frac{1}{\beta - \alpha} \qquad \alpha < t < \beta \tag{2.11}$$

where β, α are constants
    t is time

2.3.5 Normal Distribution

This two-parameter distribution has some applications in reliability engineering. The following is the probability density function of the distribution.

$$f(t) = \frac{1}{\sigma\sqrt{2\pi}} e^{-(t-\mu)^2/2\sigma^2} \qquad \sigma > 0, \ -\infty < t < \infty \tag{2.12}$$

where σ is the standard deviation
    μ is the mean
    t is time

2.3.6 Weibull Distribution

This is one of the most flexible distributions, and it is used to represent various types of physical phenomena. It is a three parameter distribution due to Weibull [9]. The probability density function, $f(t)$, of the distribution is defined by

$$f(t) = \frac{b}{\beta} \ (t - \alpha)^{b-1} \ e^{-\{(t-\alpha)^b/\beta\}} \tag{2.13}$$

for $t > \alpha$ and b, β, $\alpha > 0$

Where α, β and b are location, scale and shape parameters, respectively
    t is time

One should note that the Rayleigh and exponential distributions are special cases of the Weibull distribution at $b = 2$, $\alpha = 0$ and $b = 1$, $\alpha = 0$ respectively.

2.3.7 Log Normal Distribution

This distribution has its applications in Maintainability engineering in being able to represent failed systems repair times. The probability density is given below:

$$f(t) = \frac{1}{(t - \alpha)\sigma \sqrt{2\pi}} \; e^{-\{\log\,(t-\alpha)-\mu\}^2/2\sigma^2} \qquad (2.14)$$

for $t > \alpha > 0$, $\sigma > 0$

where α is a constant

μ is the mean log time to failure

σ is the standard deviation of log time to failure

t   is time

2.3.8 Beta Distribution

This is another two parameter distribution which also has applications in reliability engineering.

The probability density, $f(t)$, is defined as follows:

$$f(t) = \frac{(\alpha + \beta + 1)\,!}{\alpha!\;\beta!} \; t^\alpha (1 - t)^\beta \qquad (2.15)$$

for $0 < t < 1$, $\alpha > -1$, $\beta > -1$

where α and β are the distribution parameters

t   is time

2.3.9 Hazard Rate Model I

This is a two parameter distribution which can have increasing and decreasing hazard rates [5]. The hazard rate, $h(t)$, and the survival function $R(t)$, are defined as follows:

$$h(t) = \frac{\lambda(b + 1)\,[\ln(\lambda t + \alpha)]^b}{(\lambda t + \alpha)} \qquad (2.16)$$

for $b \geq 0$, $\lambda > 0$, $\alpha \geq 1$, $t \geq 0$

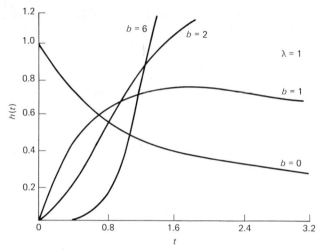

Figure 2.1. Hazard rate function plot.

where b is the shape parameter
λ is the scale parameter
α is the third parameter
t is time

Figure 2.1 shows hazard rate curve for several values of b. The distribution reliability function, $R(t)$, for $\alpha = 1$ is

$$R(t) = e^{-\{\ln(\lambda t + 1)\}^{b+1}} \tag{2.17}$$

The probability density, $f(t)$, is the product of Eqs. 2.16 and 2.17:

$$f(t) = h(t) \cdot R(t) \tag{2.18}$$

and the cumulative distribution function, $F(t)$, from Eq. 2.17 is

$$F(t) = 1 - R(t) \tag{2.19}$$

2.3.10 Hazard Rate Model II

This two parameter distribution is taken from reference 6. The hazard rate, $h(t)$, and reliability, $R(t)$, functions of the distribution are as follows:

$$h(t) = \frac{n\lambda t^{n-1}}{\lambda t^n + 1} \qquad \text{for } n \geq 1,\ \lambda > 0,\ t \geq 0 \tag{2.20}$$

where n is the shape parameter
λ is the scale parameter
t is time

Figure 2.2 shows increasing and decreasing hazard rates and

$$R(t) = e^{-\ln(\lambda\, t^n + 1)} \tag{2.21}$$

The probability density, $f(t)$, is given by the product of Eqs. 2.20 and 2.21:

$$f(t) = h(t) \cdot R(t) \tag{2.22}$$

and the cumulative distribution function, $F(t)$, from Eq. (2.21) is

$$F(t) = 1 - R(t) \tag{2.23}$$

2.3.11 The Mixed Weibull Distribution

This distribution is from Kao [15], who applied it to evaluate reliability of electron tubes. The probability density, $f(t)$, and cumulative distribution function, $F(t)$, are written as follows:

$$f(t) = \frac{c\gamma_1}{\mu_1}\left(\frac{t}{\mu_1}\right)^{\gamma_1-1}\exp\left(-\frac{t}{\mu_1}\right)^{\gamma_1} + \frac{(1-c)}{\mu_2}\gamma_2\left(\frac{t-\theta}{\mu_2}\right)^{\gamma_2-1} \\ \exp\left\{-\left(\frac{t-\theta}{\mu_2}\right)^{\gamma_2-1}\right\} \tag{2.24}$$

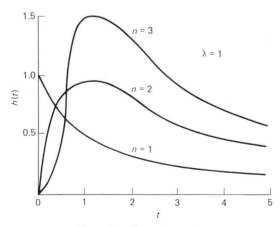

Figure 2.2. Hazard rate curves.

for $\mu_1, \mu_2 > 0, 0 < \gamma_1 < 1, \gamma_2 > 1, \theta > 0, 0 \le c \le 1$

where $\mu_1, \mu_2, \gamma_1, \gamma_2, \theta,$ and c   are the distribution parameters
    t is time

and

$$F(t) = c\left[1 - \exp\left\{-\left(\frac{t}{\mu_1}\right)^{\gamma_1}\right\}\right] - (1 - c)\left[1 - \exp\left\{-\left(\frac{t - \theta}{\mu_2}\right)^{\gamma_2}\right\}\right] \quad (2.25)$$

2.3.12 Fatigue-Life Distribution

This distribution is from Birnbaum and Saunders [2,15] and is mainly used to characterize failures caused by fatigue. The probability density function of the distribution is given by

$$f(t) = \frac{(t^2 - \mu^2)}{2\sqrt{2\pi}\,\beta^2\mu t^2\,(t/\mu)^{1/2} - (\mu/t)^{1/2}}\,\exp\left[-\frac{1}{2\beta^2}\left(\frac{t}{\mu} + \frac{\mu}{t} - 2\right)\right] \quad (2.26)$$

for $t > 0, \beta, \mu > 0$

where t is time
      β and μ are the shape and scale parameters, respectively

2.3.13 Rayleigh Distribution

This statistical function has applications in areas like reliability and the theory of sound. As mentioned in Section 2.3.6, it is the special case of the Weibull distribution when $\alpha = 0$ and $b = 2$. Therefore, from Eq. (2.13) the Rayleigh probability density is:

$$f(t) = \frac{2}{\beta}\,t\,e^{-t^2/\beta} \quad (2.27)$$

for $\beta > 0, t \ge 0$

2.3.14 Gamma Distribution

The gamma distribution is used in life test problems. The probability density of the distribution is

$$f(t) = \frac{\lambda(\lambda t)^{\beta-1}}{\Gamma(\beta)} e^{-\lambda t} \tag{2.28}$$

for λ, $\beta > 0$ and $t \geq 0$

where λ and β are the scale and shape parameters
   t is time

At $\beta = 1$, the gamma distribution becomes the exponential distribution.

2.3.15 Poisson Distribution

This is a discrete random variable distribution whose probability density function, $f(k)$, is defined as follows:

$$f(k) = \frac{(\lambda t)^k e^{-\lambda t}}{k!}, \text{ for } k = 0,1,2,3,4, \ldots. \tag{2.29}$$

where t is time
   λ is the constant rate
   k is the number of same kind of events

When one is interested in the probability of the occurrence of the same kind of events, k, one uses this distribution.

2.3.16 Binomial Distribution

This is another commonly used discrete random variable distribution.
   The probability density function, $f(x)$, of the distribution is given by

$$f(x) = \frac{k!}{x!\,(k-x)!} p^x q^{k-x} \qquad x = 0,1,2, \ldots, k \tag{2.30}$$

where k is the total number of trials
   x is the number of failures
   p is the single trial probability of success
   q is the single trial probability of failure

It has applications in reliability engineering, for example, when one is dealing with a situation in which an event is either a success or a failure. In other words either the event occurs or it does not occur.

2.4 OPTIMIZATION MATHEMATICS

This section presents the Lagrange multiplier method [18] to obtain maxima and minima of a function subject to constraint conditions. In order to find the relative maxima or minima of a function

$$f(t_1, t_2, t_3, t_4, \ldots, t_m) \tag{2.31}$$

where t_i is the ith time variable; for $i = 1, 2, 3 \ldots m$
  m is the number of variables

Subject to constraints

$$L_1(t_1, t_2, t_3, \ldots, t_m) = 0 \tag{2.32}$$

$$L_2(t_1, t_2, t_3, \ldots, t_m) = 0 \tag{2.33}$$

$$\vdots$$

$$L_i(t_1, t_2, t_3, \ldots, t_m) = 0 \tag{2.34}$$

We formulate

$$\begin{aligned} P(t_1, t_2, t_3, \ldots t_m) = f(t_1, t_2, \ldots, t_m) + \lambda_1 L_1(\cdot) + \lambda_2 L_2(\cdot) \\ + \lambda_3 L_3(\cdot) + \ldots + \lambda_i L_i(\cdot) \end{aligned} \tag{2.35}$$

where $\lambda_1, \lambda_2, \ldots, \lambda_i$ are Lagrange multipliers

The necessary conditions for maxima or minima are

$$\frac{\partial P(\cdot)}{\partial t_1} = 0 \qquad \frac{\partial P(\cdot)}{\partial t_2} = 0 \qquad \frac{\partial P(\cdot)}{\partial t_3} = 0 \ldots \frac{\partial P(\cdot)}{\partial t_m} = 0 \tag{2.36}$$

2.5 INTEGRALS OF SPECIAL FUNCTIONS

Integrals of special exponential functions relevant to the subsequent chapters of this book are presented in this section and listed in Table 2.1. In the table, t denotes the time variable. Other special integrals may be found in references 12 to 14.

2.6 LAPLACE TRANSFORMS METHOD

Most of the differential equations occurring in the subsequent chapters of this book can be solved by using the Laplace transforms technique [1,17]. Before we proceed to demonstrate this technique by solving one example, the Laplace transform, $F(s)$, of a function is defined as follows:

$$F(s) = \int_0^\infty e^{-st} f(t) \cdot dt \qquad (2.37)$$

where s is the Laplace transform variable

t is time

$f(t)$ is a function of t

Table 2.1. Integral of Special Functions

NUMBER	INTEGRAL
1.	$\int t\,e^{-t^2}\,dt = -\dfrac{1}{2} e^{-t^2}$
2.	$\int t^k e^{-t^2}\,dt = -\dfrac{1}{2} t^{k-1} e^{-t^2} + \dfrac{k-1}{2} \int t^{k-2} e^{-t^2}\,dt$
3.	$\int f(t) e^{-(\alpha t^2 + 2\beta t + \gamma)}\,dt = \dfrac{1}{\sqrt{\alpha}}\, e^{(\beta^2 - \alpha\gamma)/\alpha}\, f\!\left(\dfrac{\sqrt{\alpha}\,x - \beta}{\alpha}\right) e^{-x^2}\,dx$
	where $x = \sqrt{\alpha}\left(t + \dfrac{\beta}{\alpha}\right)$ for $\alpha > 0$
4.	$\displaystyle\int_0^\infty e^{-a^2 t^2}\,dt = \dfrac{\sqrt{2\pi}}{2a}$ for $a > 0$
5.	$\displaystyle\int_0^\infty e^{-\alpha t^b}\,dt = \dfrac{1}{b\alpha^{1/b}}\,\Gamma\!\left(\dfrac{1}{b}\right)$ for $b > 0,\ \alpha > 0$
6.	$\displaystyle\int_0^\infty t^n e^{-\alpha t^b}\,dt = \dfrac{1}{b}\dfrac{1}{\alpha^{(n+1)/b}}\,\Gamma\!\left(\dfrac{n+1}{b}\right)$ for $b,\ \alpha > 0$ and $n > -1$
7.	$\displaystyle\int_0^\infty e^{-(t^2/4b + at)}\,dt = \sqrt{\pi b}\, e^{ba^2}\, [1 - \Phi(a\sqrt{b})]$ for $b > 0$
	where $\Phi(a\sqrt{b}) = (2/\sqrt{\pi}) \displaystyle\int_0^{a\sqrt{b}} e^{-t^2}\,dt$
8.	$\displaystyle\int_0^\infty f(t) e^{-\alpha t^b}\,dt = \dfrac{1}{b(\alpha)^{1/b}} \int_0^\infty e^{-x} f\!\left(b\sqrt{\dfrac{x}{\alpha}}\right) x^{-(1-1/b)}\,dx$ for $\alpha,\ b > 0$
9.	$\displaystyle\int_0^\infty e^{(-be^{-t} - mt)}\,dt = b^{-m}\,\gamma(m,b)$ for $m \geq 0$
	where $\gamma(m,b) = \displaystyle\int_0^b e^{-t} t^{m-1}\,dt$ for $m > 0$
10.	$\displaystyle\int_0^\infty e^{-\alpha e^{mt}}\,dt = -\dfrac{1}{m} E_i(-\alpha)$ for $m \geq 1,\ \alpha > 0$
	where $E_i(-\alpha) = e^{-\alpha} - \left[\dfrac{1}{\alpha} + \displaystyle\int_0^\infty \dfrac{e^{-t}}{(\alpha + t)^2}\,dt\right]$
11.	$\Gamma(\alpha) = \displaystyle\int_0^\infty t^{\alpha-1} e^{-t}\,dt = (\alpha - 1)!$

By utilizing Eq. (2.37), the Laplace transform of $e^{-\alpha t}$, k, $t^{n-1}/(n-1)!$, $\dfrac{df(t)}{dt}$, $\dfrac{f(t)}{t}$, $\displaystyle\int_0^t f(x)\, G(t-x)\, dx$, respectively, are as follows:

$$F(s) = (s+\alpha)^{-1} \tag{2.38}$$

$$F(s) = k/s \tag{2.39}$$

$$F(s) = s^{-n},\ n>0 \tag{2.40}$$

$$s\,F(s) - f(0) \tag{2.41}$$

$$F(s) = \int_s^\infty F(u)\cdot du \tag{2.42}$$

and

$$F(s)\cdot g(s) \tag{2.43}$$

In the following example, the solution of a differential equation is obtained by utilizing the Laplace transform method.

EXAMPLE 2

Find the solution for the following differential equation for $t = 0$, $P(0) = 1$.

$$\frac{dP(t)}{dt} + \left(\sum_{i=1}^{n}\lambda_i\right)P(t) = 0 \tag{2.44}$$

where t is time
  n is the number of failure modes
  λ_i is the ith constant failure rate
  $P(t)$ is the probability at time t

Using the Laplace transform of Eq. (2.41) and the basic Laplace transform definition in Eq. (2.44) we get:

$$s\,P(s) - P(0) + \left(\sum_{i=1}^{n}\lambda_i\right)P(s) = 0 \tag{2.45}$$

For known initial condition, $P(0) = 1$, the equation (2.45) becomes

$$s\,P(s) - 1 + \left(\sum_{i=1}^{n}\lambda_i\right)P(s) = 0 \tag{2.46}$$

By rearranging equation (2.46) we get

$$P(s) = 1 \Big/ \Big(s + \sum_{i=1}^{n} \lambda_i \Big) \tag{2.47}$$

The inverse Laplace transform of Eq. (2.47), by utilizing the corresponding inverse Laplace transform of (2.38), is:

$$P(t) = e^{-\sum_{i=1}^{n} \lambda_i t} \tag{2.48}$$

2.7 USEFUL THEOREMS AND DEFINITIONS

This section presents the commonly used initial and final value theorems, mean value, μ, and variance, σ^2, definitions, respectively, as follows:

$$\lim_{t \to 0} f(t) = \lim_{s \to \infty} s F(s) \tag{2.49}$$

$$\lim_{t \to \infty} f(t) = \lim_{s \to 0} s F(s) \tag{2.50}$$

$$\mu = \int_0^\infty t f(t) \, dt \tag{2.51}$$

and

$$\sigma^2 = \int_0^\infty (t - \mu)^2 f(t) \, dt \tag{2.52}$$

2.8 SUMMARY

In this chapter the basic mathematical concepts necessary to understand the subsequent chapters of the book are presented. The following concepts are briefly discussed:

1. Basic probability.
2. Probability distributions.
3. Optimization mathematics.
4. Integrals of special functions.
5. Laplace transforms method.
6. Useful theorems and definitions.

A number of references are listed at the end of the chapter for further reading.

2.9 REFERENCES

1. H. S. Bean, *Differential Equations,* Addison-Wesley, Reading, MA, 1962.
2. Z. W. Birnbaum and S. C. Saunders, "A New Family of Life Distributions," *J. Appl. Probability,* 1969, pp. 319–327.
3. D. J. Davis, "An Analysis of Some Failure Data," *J. Amer. Stat. Assoc.,* 1952, pp. 113–150.
4. B. S. Dhillon, "A Hazard Rate Model," *IEEE Trans. on Reliability,* Vol. 28, June 1979, pp. 150.
5. B. S. Dhillon, "Life Distributions," *IEEE Trans. on Reliability,* Vol. 30, December 1981, pp. 457–460.
6. B. S. Dhillon, "Statistical Functions to Represent Various Types of Hazard Rates," *Microelectronics and Reliability,* Vol. 20, 1980, pp. 581–584.
7. G. Doetsch, *Theory and Application of the Laplace-Transform,* Chelsea, New York, 1965.
8. A. Drake, *Fundamentals of Applied Probability Theory,* McGraw-Hill, New York, 1967.
9. H. B. Dwight, *Tables of Integrals and Other Mathematical Data,* Macmillan, New York, 1957.
10. A. Erdelyi, *Tables of Integral Transforms, Vols. I and II,* McGraw-Hill, New York, 1954.
11. W. Feller, *An Introduction, To Probability Theory and its Applications, Vol. I,* Wiley, New York, 1957.
12. J. S. Gradshteyn and I. M. Ryzhik, *Table of Integrals, Series, and Products,* Academic, New York, 1980.
13. W. Grobner and N. Hofreiter, *Integraltafel, Vols. I and II,* Springer-Verlag OHG, Vienna, 1957.
14. *Handbook of Mathematical Functions,* National Bureau of Standards, U.S. Government Printing Office, Washington, D.C., 1964.
15. N. R. Mann, R. E. Shafer, and N. D. Singpurwalla, *Methods for Statistical Analysis of Reliability and Life Data,* Wiley, New York, 1974.
16. M. L. Shooman, *Probablistic Reliability: An Engineering Approach,* McGraw-Hill, New York, 1968.
17. M. R. Spiegel, *Laplace Transforms,* Schaum's Series, McGraw-Hill, New York, 1965.
18. M. R. Spiegel, *Advanced Calculus,* Schaum's Series, McGraw-Hill, New York, 1963.
19. W. Weibull, "A Statistical Distribution Function of Wide Applicability," *J. Appld. Mech.,* Vol. 18, 1951, pp. 293–297.
20. D. V. Widder, *The Laplace Transform,* Princeton University Press, Princeton, NJ, 1941.

3
Introductory Concepts in Engineering Reliability

3.1 INTRODUCTION

This chapter presents the basic concepts of engineering reliability. These concepts are considered to be useful for an understanding of the chapters that follow. The areas discussed briefly in the chapter are the bathtub failure rate curve, two-state device reliability networks, three-state device reliability networks, and reliability evaluation techniques. For further reading a list of books and appropriate literature associated with each topic of the chapter is presented at the end of the chapter.

3.2 CONCEPT OF BATHTUB HAZARD RATE CURVE

A bathtub hazard rate curve is shown in Fig. 3.1, at $b = 0.5$. It is called the bathtub curve because of its resemblance to a bathtub.

This curve is used to represent the failure rate pattern of components. Electronic components failure rate representation is a prime example, in which case only the middle portion (useful life period or the constant failure rate region) of the curve is used. The bathtub curve concept may or may not be used to represent the hazard rate pattern of mechanical components.

As it can be seen in Fig. 3.1 the hazard curve may be divided into three regions or parts (i.e., decreasing, constant, and increasing hazard rate). The decreasing hazard rate part of the curve, is known as the "burn-in period" or "debugging period" or "infant mortality period" or "break-in period." The "burn-in period" failures are the result of design or manufacturing defects in a new product. As the "burn-in period" increases, the product failures decrease, until the beginning of the constant failure rate region of the curve.

The middle portion (the constant hazard rate region) of the curve is known

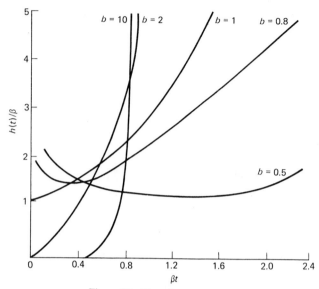

Figure 3.1. Hazard rate curves.

as the "useful life period" of the product. This period starts from the end of the "burn-in period" and finishes at the beginning of the "wear-out" region. The failures occurring during the "useful life period" of an item are known as the "random failures" because they occur unpredictably.

The last portion (the increasing hazard rate region) is known as the "wear-out region" of the product. It starts when a product has bypassed its useful operating life and begins to wear out. During this phase the number of failures begins to increase, as depicted in Fig. 3.1.

The following hazard rate function, $\lambda(t)$ [104] taken from Chapter 2 can be used to represent the bathtub hazard rate curve:

$$\lambda(t) = b\beta(\beta t)^{b-1}\, e^{(\beta t)^{b}} \tag{3.1}$$

for b, $\beta > 0$; $t \geq 0$
where b is the shape parameter
β is the scale parameter
t is time

Special cases:

$b = 1$: extreme value distribution
$b = 0.5$: bathtub hazard rate curve

The hazard rate curves are shown in Fig. 3.1 for several values of b.

3.3 RELIABILITY EVALUATION OF TWO-STATE DEVICE NETWORKS

A device is said to have two states if it either operates or fails. This section presents the reliability evaluation of the following two-state device networks:

3.3.1 Series Network

This network denotes a system whose components/subsystems are connected in a series. In a network, if any one of the system components malfunctions, it will cause system failure. A 'k' component series system is shown in Fig. 3.2.

For the k independent and nonidentical-units series-system time-t-dependent reliability, $R_s(t)$, is

$$R_s(t) = \{1 - F_1(t)\} \cdot \{1 - F_2(t)\}\{1 - F_3(t)\} \ldots \{1 - F_k(t)\} \tag{3.2}$$

and

$$\{1 - F_i(t)\} \equiv R_i(t) \tag{3.3}$$

where $F_i(t)$ is the ith unit/component failure probability for $i = 1, 2, 3 \ldots k$

$R_i(t)$ is the ith unit/component reliability, for $i = 1, 2, 3, \ldots k$

The ith component cumulative distribution function (failure probability) is defined by

$$F_i(t) = \int_0^t f_i(t) \cdot dt \tag{3.4}$$

where $f_i(t)$ is the ith component failure density function

By definition [89]

$$f_i(t) = \lim_{\Delta t \to 0} \frac{\alpha_s(t) - \alpha_s(t + \Delta t)}{\alpha_0 \cdot \Delta t} = \frac{d F_i(t)}{dt} \tag{3.5}$$

Figure 3.2. A block diagram of the series network.

where Δt is the time interval

α_0 is the total number of items put on test at $t = 0$

α_s is the number of items surviving at time t or $t + \Delta t$

Also the ith component failure density function is given by the relationship

$$f_i(t) = -\frac{d\,R_i(t)}{dt} \tag{3.6}$$

Substituting Eq. (3.4) in Eq. (3.3) leads to

$$R_i(t) = 1 - \int_0^t f_i(t) \cdot dt \tag{3.7}$$

However, an alternative commonly used expression for $R_i(t)$ is

$$R_i(t) = e^{-\int_0^t \lambda_i(t)\,dt} \tag{3.8}$$

where $\lambda_i(t)$ is the ith component hazard rate or instantaneous failure rate. In this case a component failure time can follow any statistical distribution function whose hazard rate is known.

By definition

$$\lambda_i(t) = -\lim_{\Delta t \to 0} \cdot \frac{\alpha_s(t) - \alpha_s(t + \Delta t)}{\alpha_s(t) \cdot \Delta t} \tag{3.9}$$

The ith component hazard rate, $\lambda_i(t)$, is also given by the following relationship

$$\lambda_i(t) = \frac{f_i(t)}{R_i(t)} \tag{3.10}$$

Furthermore, by substituting Eq. (3.6) in Eq. (3.10) we get the alternative hazard rate expression

$$\lambda_i(t) = -\frac{1}{R_i(t)} \cdot \frac{d\,R_i(t)}{dt} \tag{3.11}$$

For ith component constant failure rate

$$\lambda_i(t) = \lambda_i \tag{3.12}$$

Substituting Eq. (3.12) in Eq. (3.8) and then substituting the resulting expression in Eq. (3.3) leads to

$$R_i(t) = 1 - F_i(t) = e^{-\lambda_i t} \tag{3.13}$$

Utilizing Eq. (3.13) in Eq. (3.2) leads to the series system reliability expression

$$R_s(t) = e^{-\lambda_1 t} \cdot e^{-\lambda_2 t} \cdot e^{-\lambda_3 t} \ldots e^{-\lambda_k t} \tag{3.14}$$

A redundant configuration or a single component mean time to failure (MTTF) is defined by

$$\text{MTTF} = \int_0^\infty R(t) \cdot dt \tag{3.15}$$

Thus, substituting Eq. (3.14) in Eq. (3.15) and integrating results in the series system

$$\text{MTTF} = \left[\sum_{i=1}^n \lambda_i \right]^{-1} \tag{3.16}$$

EXAMPLE 1

Ten independent and identical subsystems form a series system. The failure times of each subsystem are exponentially distributed with mean time to failure (MTTF) of 2000 hours. Calculate the series system reliability for a 50-hour mission, if the system starts operating at time $t = 0$.

For known data $k = 10$, $t = 50$ hours and $\lambda = (2000)^{-1}$ failure/hour. Thus substituting the above data in Eq. (3.14) leads to

$$R_s(50) = \{e^{-(2000)^{-1} \cdot (50)}\}^{10}$$

$$= 0.7788$$

The series system reliability is 0.7788.

3.3.2 Parallel Network

This type of redundancy can be used to improve system reliability. The redundant system will only fail if all of its components fail.

In order to develop the reliability equation for the model it is assumed that (1) all units of the system are active and load sharing; (2) units are statistically independent. The parallel structure with nonidentical components unreliability, $F_p(t)$, at time t is

$$F_p(t) = \prod_{i=1}^{k} F_i(t) \tag{3.17}$$

where $F_i(t)$ is the ith component unreliability (failure probability)
  k is the number of components in the system

Since $R_p(t) + F_p(t) = 1$, the parallel structure reliability, $R_p(t)$, utilizing Eq. (3.17) is

$$R_p(t) = 1 - \prod_{i=1}^{k} F_i(t) \tag{3.18}$$

Similarly as for the series network components with constant failure rates, substituting [from Eq. (3.13)] for $F_i(t)$ in Eq. (3.18) we get:

$$R_p(t) = 1 - \prod_{i=1}^{k} (1 - e^{-\lambda_i t}) \tag{3.19}$$

In order to obtain the system mean time to failure (MTTF) substitute Eq. (3.19) for identical components in Eq. (3.15) and integrate as follows:

$$\mathrm{MTTF} = \int_{0}^{\infty} [1 - \sum_{j=0}^{k} \binom{n}{j} (-1)^j e^{-j\lambda t}] \, dt$$
$$= \frac{1}{\lambda} + \frac{1}{2\lambda} + \frac{1}{3\lambda} + \ldots + \frac{1}{k\lambda} \tag{3.20}$$

The expression for the MTTF of a parallel system with nonidentical components is given in reference 29. For example at $k = 2$, the Eq. (3.19) reduces to:

$$R_p(t) = e^{-\lambda_1 t} + e^{-\lambda_2 t} - e^{-(\lambda_1 + \lambda_2) t} \tag{3.21}$$

Utilizing Eq. (3.21) in Eq. (3.15), the two nonidentical unit system MTTF is given by

$$\mathrm{MTTF} = \int_{0}^{\infty} \{e^{-\lambda_1 t} + e^{-\lambda_2 t} - e^{-(\lambda_1 + \lambda_2) t}\} \, dt$$
$$= \frac{(\lambda_2 + \lambda_1)}{\lambda_1 \lambda_2} - \frac{1}{\lambda_1 + \lambda_2} \tag{3.22}$$

EXAMPLE 2

An aircraft has three identical and independent engines forming an active parallel redundant configuration. If any one of the engine is functioning the aircraft can accomplish its mission successfully. Each engine constant failure rate, λ, is 0.001 failure/hour. Calculate the parallel system mean time to failure.

By substituting $\lambda = 0.001$ failure/hour and $k = 3$ in Eq. (3.20) we get

$$\text{MTTF} = \frac{1}{\lambda} \sum_{j=1}^{k} \frac{1}{j} = \frac{1}{0.001} [1 + \frac{1}{2} + \frac{1}{3}] = 1833.3 \text{ hours}$$

3.3.3 k-out-of-m Unit Network

This type of redundancy is used when a certain number k of units in an active parallel redundant system must work for the system success. By utilizing the binomial distribution, the independent and identical unit system reliability, $R_{k/m}(t)$, at time t is

$$R_{k/m}(t) = \sum_{i=k}^{m} \binom{m}{i} [R(t)]^i [1 - R(t)]^{k-i} \qquad (3.23)$$

where $R(t)$ is the unit reliability
$\quad m \quad$ is the total number of system units
$\quad k \quad$ is the number of units required for the system success

Special cases of the k-out-of-m unit system are at

$k = 1$: parallel network
$k = m$: series network

For exponentially distributed failure times (constant failure rate) of a unit, substituting Eq. (3.13) in Eq. (3.23) for $k = 2$ and $m = 4$ the resulting equation becomes

$$R_{2/4}(t) = 3 e^{-4\lambda t} - 8 e^{-3\lambda t} + 6 e^{-2\lambda t} \qquad (3.24)$$

EXAMPLE 3

Determine a 2-out-of-4 identical and independent unit system reliability for a 100-hour mission. Each unit constant failure rate, λ, is 0.005 failure/hour.

Substituting the given data for $\lambda = 0.005$ failure/hour and $t = 100$ hours in Eq. (3.24) results in the following:

$$R_{2/4}(100) = 3 e^{-4(0.005)(100)} - 8 e^{-3(0.005)(100)} + 6 e^{-2(0.005)(100)}$$

$$= 0.8282$$

3.3.4 Standby Redundant System

In this case one unit is functioning and k units are on a standby mission. In other words, the k number of units are not active.

To develop the following system reliability expression the system units identical and independent; perfect switching, standby units remains as good as new and the general unit hazard rate were assumed:

$$R_s(t) = \sum_{i=0}^{k} \left\{ \int_0^t \lambda(t) \cdot dt \right\}^i e^{-\int_0^t \lambda(t)\, dt} (i!)^{-1} \qquad (3.25)$$

where k is the number of standbys

EXAMPLE 4

For known $k = 2$ and $\lambda(t) = \lambda = 0.003$ failures/hour in Eq. (3.25), calculate the system mean-time-to-failure. Therefore, using the above data in Eq. (3.25) results in the following:

$$R_s(t) = \sum_{i=0}^{2} \frac{(\lambda t)^i e^{-\lambda t}}{i!} \qquad (3.26)$$

Integrating Eq. (3.26) over the time interval from 0 to ∞, we get

$$\text{MTTF} = \int_0^\infty \sum_{i=0}^{2} \frac{(\lambda t)^i e^{-\lambda t}}{i!} dt = \frac{3}{\lambda}$$

Thus, for $\lambda = 0.003$ failure/hour the

$$\text{MTTF} = \frac{3}{0.003} = 1000 \text{ hours.}$$

3.4 RELIABILITY EVALUATION OF THREE-STATE DEVICE NETWORKS

A three-state device has one operational and two failure states. Devices such as a fluid flow valve, an electronic diode, and so on are examples of a three-state device. These devices have failure modes which can be described as closed (or shorted, for electrical and electronic devices) and opened. In this section we present reliability evaluation of networks composed of such devices.

Supplementary material on three-state device networks may be found in reference 105.

3.4.1 Parallel Network

A parallel network composed of active independent three-state devices will only fail, if all the devices fail in the open mode or at least one of the devices must fail in the short mode. The network (with nonidentical devices) time dependent reliability, $R_p(t)$, is

$$R_p(t) = \prod_{i=1}^{k} (1 - F_{si}(t)) - \prod_{i=1}^{k} F_{oi}(t) \tag{3.27}$$

where t is time
    k is the number of three state devices in parallel
    $F_{si}(t)$ is the short mode probability of the ith device at time t
    $F_{oi}(t)$ is the open mode probability of the ith device at time t

The derivation of Eq. (3.27) is given in references 29 and 106.
The network short mode failure probability, $F_{sp}(t)$, is given by

$$F_{sp}(t) = 1 - \prod_{i=1}^{k} (1 - F_{si}(t)) \tag{3.28}$$

Similarly, the network open mode probability, F_{op}, is given by

$$F_{op}(t) = \prod_{i=1}^{k} F_{oi}(t) \tag{3.29}$$

One should note that the Eqs. (3.28) and (3.29) are for nonidentical devices. Furthermore by adding Eqs. (3.28) and (3.29) and then subtracting the result from unity will yield Eq. (3.27).

For the ith device constant open and short (close) mode failure rates, λ_{oi} and λ_{si}, respectively, the time dependent open and short mode failure probability Eqs. from references [29, 106] are:

$$F_{ci} = \frac{\lambda_{ci}}{A_i} (1 - e^{-A_i t}) \qquad \text{for } c = 0, r \ s \tag{3.30}$$

where $A_i \equiv \lambda_{oi} + \lambda_{si}$ (3.31)
Thus, by substituting Eq. (3.30) in Eq. (3.27) we get:

$$R_p(t) = \prod_{i=1}^{k} \frac{1}{A_i} \{\lambda_{oi} + \lambda_{si}e^{-A_i t}\} - \prod_{i=1}^{k} \frac{\lambda_{oi}}{A_i} \{1 - e^{-A_i t}\} \tag{3.32}$$

EXAMPLE 5

For identical devices given $\lambda_0 = 0.001$ failures/hour, $\lambda_s = 0.002$ failure/hour and $k = 2$ use Eq. (3.32) to calculate network reliability for a 100-hour mission.

For known data utilizing Eq. (3.32) we get

$$R_p(100) = 0.68428 - 0.00746 = 0.6768$$

3.4.2 Series Network

A series network is the reverse of the parallel network. A series system will only fail if all of its independent elements fail in a short mode or any one of the component fail in open mode. Because of duality the series network with nonidentical and independent devices time dependent reliability, $R_s(t)$, is

$$R_s(t) = \prod_{i=1}^{k} (1 - F_{0i}(t)) - \prod_{i=1}^{k} F_{si}(t) \tag{3.33}$$

where k is the number of devices in the series configuration

EXAMPLE 6

Three independent and identical electronic diodes are connected in series. The diode open and short mode failure probabilities are 0.1 and 0.2, respectively. Find the network reliability using Eq. (3.33).

For identical components and components time independent failure probabilities, Eq. (3.33) reduces to

$$R_s = (1 - F_0)^k - F_s^k \tag{3.34}$$

Since $k = 3$, $F_0 = 0.1$ and $F_s = 0.2$, the network reliability from the above Eq. is

$$R_s = (1 - 0.1)^3 - 0.2^3 = 0.721$$

3.5 RELIABILITY DETERMINATION METHODS

This section presents a number of reliability evaluation techniques. The methods are as follows:

3.5.1 Network Reduction Technique

This is a simple and straightforward method to evaluate the reliability of systems composed of independent series and parallel subsystems.

Other subsystems forming bridge configurations can also be handled by applying the delta-star transformations to convert the bridge configuration to series and parallel equivalents [106]. One should note that some approximation is involved when using these conversions.

The network reduction technique consists of sequentially reducing the series and parallel structures to equivalent hypothetical components until the whole system network becomes a single hypothetical component. This method is demonstrated by solving the following example:

EXAMPLE 7

An independent two-state network is shown in Fig. 3.3 (a). By applying the network reduction technique calculate the system reliability.

As shown in Fig. 3.3 this example is solved in two stages. Step one consists of reducing the subsystems A and B to equivalent single units as follows:

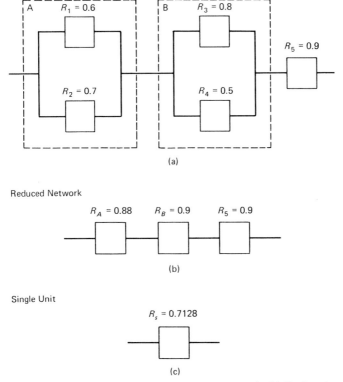

Figure 3.3. (a) Original network; (b) Reduced network; (c) Single unit.

Subsystem A. For components time independent reliability, utilizing Eqs. (3.3) and (3.18) for $k = 2$ we get the subsystem A reliability

$$R_A = 1 - (1 - R_1)(1 - R_2) = 1 - (1 - 0.6)(1 - 0.7) = 0.88$$

Subsystem B. Similarly by utilizing Eqs. (3.3) and (3.18) for $k = 2$ we get the subsystem B reliability

$$R_B = 1 - (1 - R_3)(1 - R_4) = 1 - (1 - 0.8)(1 - 0.5) = 0.9$$

Thus the reduced network is shown in Fig. 3.3 (b). For components constant reliabilities utilizing Eqs. (3.3) and (3.2), the series network shown in Fig. 3.3 (b), reliability, R_s is

$$R_s = R_A \cdot R_B \cdot R_5 = 0.88 \times 0.9 \times 0.9 = 0.7128$$

3.5.2 Path-Tracing Technique

This method [89] is concerned with reliability network successful paths identification.

The reliability expression is given by the probability of union of all events containing the successful paths.

EXAMPLE 8

Obtain a reliability expression for the independent two-state component network shown in Fig. 3.4.

In Fig. 3.4 the success paths of the network are

AC and BC

Thus, the network reliability R_s is obtained by taking the probability of the union of paths AC and BC:

$$R_s = P(AC + BC)$$

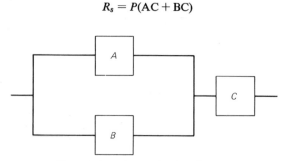

Figure 3.4. A three unit network.

After expanding the above expression and eliminating redundancy we get:

$$R_s = P(A) \cdot P(C) + P(B) \cdot P(C) - P(A) \cdot P(B) \cdot P(C) \qquad (3.36)$$

where $P(A)$, $P(B)$, $P(C)$ are the probabilities of success of components A, B, and C.

3.5.3 Decomposition Technique

This is another method used to evaluate the reliability of complex systems. It decomposes a complex system into simpler subsystems by the application of conditional probability theory [90]. The system reliability is obtained by combining the reliability measures of subsystems. The efficiency of the technique depends on selection of the key element used to decompose a complex network. The poor choice of the key element will lead to tedious computation of the system reliability.

First, this technique begins by assuming that the key element, say y, is replaced by another element that never fails, and second, it assumes that the key element is totally removed from the complex system. The system reliability, R_d, is given by

$$R_d = P(y)P(\text{system good}/y \text{ good}) + P(\bar{y})P(\text{system good}/y \text{ fails}) \qquad (3.37)$$

where $P(\cdot)$ denotes probability

$\quad\quad P(y)$ denotes success probability of event y

$\quad\quad P(\bar{y})$ denotes failure probability of event y

Similarly, the equation for system unreliability, F_d, is given by

$$F_d = P(y)P(\text{system fails}/y \text{ good}) + P(\bar{y})P(\text{system fails}/y \text{ fails}) \qquad (3.38)$$

EXAMPLE 9

Develop a reliability expression for a two-state device bridge network shown in Fig. 3.5 (a) by applying the decomposition method. From past experience we choose the element C as our key element called y. The next step is to replace the element C with a perfect element which never fails. Thus the Fig. 3.5 (a) network is reduced to the one shown in Fig. 3.5 (b).

Similarly by replacing the element C with a bad element, Fig. 3.5 (a) reduces to Fig. 3.5 (c).

By utilizing the network reduction technique and Eqs. (3.3), (3.18), and (3.2) for components' constant reliabilities, in the Fig. 3.5 (b) network with independent elements, the reliability, R_{sp}, is

$$R_{sp} = [1 - (1 - R_A)(1 - R_B)][1 - (1 - R_D)(1 - R_E)] \qquad (3.39)$$

where R_A, R_B, R_D, and R_E are the reliabilities of elements A, B, D, and E.

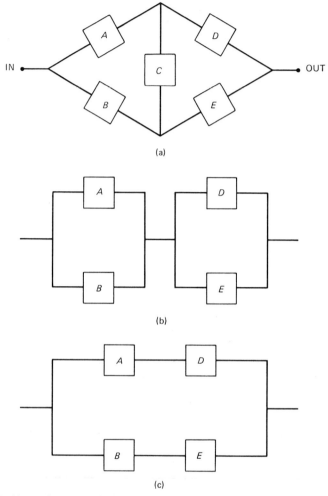

Figure 3.5. (a) A bridge network; (b) Reduced network because of a perfect key element; (c) Reduced network because of a failed (bad) key element.

For identical elements the above equation reduces to

$$R_{sp} = [1 - (1 - R)^2]^2 = (2R - R^2)^2 \qquad (3.40)$$

Similarly, by applying the network reduction technique and Eqs. (3.3), (3.2), and (3.18) for components' constant reliabilities, in the Fig. 3.5 (c) network with independent elements, the reliability, R_{ps}, is

$$R_{ps} = 1 - (1 - R_A R_D)(1 - R_B R_E) \qquad (3.41)$$

For identical elements the above expression reduces to

$$R_{ps} = 1 - (1 - R^2)^2 = 2R^2 - R^4 \qquad (3.42)$$

If the key element is identical to the other elements of the bridge then

$$P(y) = P(C) = R_C = R \qquad (3.43)$$

where R_C is the reliability of the element C

Thus

$$P(\bar{y}) = P(\bar{C}) = 1 - R_C = 1 - R \qquad (3.44)$$

By substituting Eqs. (3.40), and (3.42) to (3.44) in Eq. (3.37) we get

$$R_d = R\ (2R - R^2)^2 + (1 - R)\ (2R^2 - R^4)$$
$$= 2R^2 + 2R^3 - 5R^4 + 2R^5 \qquad (3.45)$$

3.5.4 Minimal Cut Set Techinque

This technique [89] is quite useful for computer application. However, the method is not suitable for evaluating the reliability of a system incorporating dependent failures. A reliability network's minimal cut sets may be obtained by using special algorithms.

A cut set is a collection of basic events whose presence will cause system failure. A distinct group of cut sets consisting of a minimum number of terms are called minimum cut sets. In other words, a cut set is minimal if it cannot be further minimized but still insures system failure. The system reliability equation is given by

$$R_s = 1 - P(\bar{C}_1 + \bar{C}_2 + \bar{C}_3 + \dots \bar{C}_k) \qquad (3.46)$$

where $P(\cdot)$ denotes failure probability
$C_1, C_2, C_3 \dots C_k$ denotes the k minimal cut sets
$\bar{C}_k$ denotes the kth cut set (failure)

EXAMPLE 10

Use the minimal cut set approach to obtain the reliability expression for the independent component network shown in Fig. 3.4.

As can be seen from the network diagram the following cut sets will cause system failure:

$$C_1 = AC \qquad C_2 = BC \qquad C_3 = C \qquad C_4 = AB \qquad C_5 = ABC$$

The cut sets C_3 and C_4 are minimal because of cut sets C_5. C_1 and C_2 are present in cut set C_3.

Therefore, for the minimal cut sets, utilizing Eq. (3.46) we get

$$R_s = 1 - P(\overline{C}_3 + \overline{C}_4) = 1 - P(\overline{C} + \overline{AB}) \tag{3.47}$$

By expanding the right side of the above Eq. the following expression results:

$$R_s = 1 - [P(\overline{C}) + P(\overline{A}) \cdot P(\overline{B}) - P(\overline{A}) \cdot P(\overline{B}) \cdot P(\overline{C})] \tag{3.48}$$

EXAMPLE 11

If the failure probabilities of components A, B, and C in Fig. 3.4 are 0.2, 0.2, and 0.4, respectively, then find the reliability of the system using Eq. (3.48)

$$R_s = 1 - [0.4 + 0.2 \times 0.2 - 0.2 \times 0.2 \times 0.4] = 0.576 \tag{3.49}$$

3.5.5 Delta-Star Method

This technique is used to evaluate the reliability of complex networks such as bridges. It is one of the simplest techniques to evaluate the reliability of bridge networks. A small error however, is introduced due to transformation. But for practical purposes it should be neglected. With the application of this technique a bridge structure is transformed to its equivalent series and parallel form. Then the reliability of the series and parallel subsystem network can be evaluated by applying network reduction or any other technique.

In the following section we have derived the open and short mode delta-to-star transformation formulas which can be used to evaluate reliability of *three-state device bridge* networks. However, in the case of *two-state device* networks one has to use only the open mode failure transformation formulas.

The open and short mode transformation formulas are developed as follows [106]:

Open Mode Failure Probability Transforms. The relationship for the transformation of open failure mode probabilities is shown in Fig. 3.6.

The equivalent delta-to-star leg diagrams for Fig. 3.6 are shown in Fig. 3.7.

In Figures 3.6–3.7 the $F_{0(\cdot)}$ denotes the respective open mode failure probability.

In the case of a series independent unit network [106], the system open mode failure probability, F_{0s}, is given by

$$F_{0s} = 1 - \prod_{i=1}^{k} (1 - F_{0i}) \tag{3.50}$$

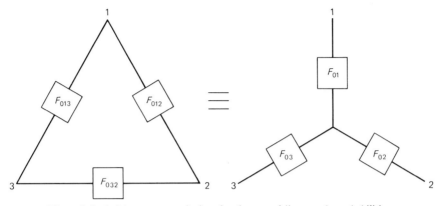

Figure 3.6. A delta to star equivalent for the open failure mode probabilities.

where F_{oi} is the ith element open mode failure probability
  k is the number of non-identical elements

Likewise, the parallel network open mode failure probability [106] (for components constant failure probabilities) from Eq. (3.29) is

$$F_{op} = \prod_{i=1}^{k} F_{oi} \tag{3.51}$$

By utilizing Eqs. (3.50) and (3.51), the following equations are associated with each leg and its equivalent in Fig. 3.7:

$$1 - (1 - F_{01})(1 - F_{02}) = [1 - (1 - F_{013})(1 - F_{032})] F_{012} \tag{3.52}$$

$$1 - (1 - F_{03})(1 - F_{01}) = [1 - (1 - F_{032})(1 - F_{012})] F_{013} \tag{3.53}$$

$$1 - (1 - F_{03})(1 - F_{02}) = [1 - (1 - F_{013})(1 - F_{012})] F_{032} \tag{3.54}$$

Solving Eqs. (3.52) to (3.54) we get

$$F_{01} = 1 - [(1 - A\ F_{013})(1 - B\ F_{012})/(1 - C\ F_{032})]^{1/2} \tag{3.55}$$

where $A \equiv 1 - (1 - F_{032})(1 - F_{012})$
  $B \equiv 1 - (1 - F_{013})(1 - F_{032})$
  $C \equiv 1 - (1 - F_{013})(1 - F_{012})$

$$F_{02} = 1 - [(1 - B\ F_{012})(1 - C\ F_{032})/(1 - A\ F_{013})]^{1/2} \tag{3.56}$$

$$F_{03} = 1 - [(1 - C\ F_{032})(1 - A\ F_{013})/(1 - B\ F_{012})]^{1/2} \tag{3.57}$$

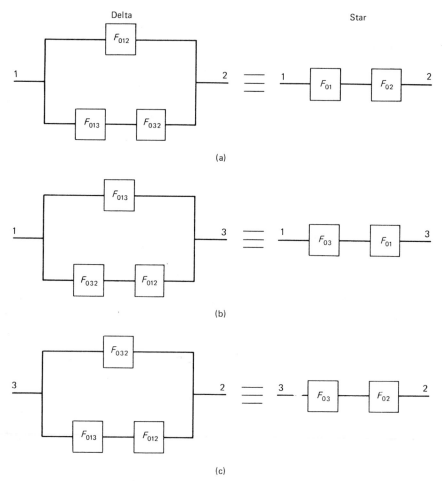

Figure 3.7. Delta to star equivalent legs.

Short Mode Failure Probability Transforms. Similarly, as for the open mode failure case, the relationship for the transformation of short failure mode probabilities is shown in Fig. 3.8.

In Fig. 3.8, the $F_{s(\cdot)}$ represents the respective short mode failure probability.

In the case of a series independent unit network [106], the configuration short mode failure probability, F_{ss} is given by

$$F_{ss} = \prod_{i=1}^{k} F_{si} \qquad (3.58)$$

where F_{si} is the ith element short mode failure probability

  k is the number of nonidentical elements

Similarly, the parallel network short mode failure probability [106] from Eq. (3.28) is

$$F_{sp} = 1 - \prod_{i=1}^{k} (1 - F_{si}) \tag{3.59}$$

Equation (3.59) is for the components constant short mode failure probabilities. By using Eqs. (3.58) to (3.59), the following equations are obtained for each arm and its equivalent in Fig. 3.8:

$$F_{s1} \cdot F_{s2} = 1 - (1 - F_{s13}F_{s32})(1 - F_{s12}) = A \tag{3.60}$$

$$F_{s3} \cdot F_{s1} = 1 - (1 - F_{s32}F_{s12})(1 - F_{s13}) = B \tag{3.61}$$

$$F_{s3} \cdot F_{s2} = 1 - (1 - F_{s13} \cdot F_{s12})(1 - F_{s32}) = C \tag{3.62}$$

Solving Eqs. (3.60) to (3.62), we get

$$F_{s1} = (A \cdot B/C)^{1/2} \tag{3.63}$$

$$F_{s2} = (A \cdot C/B)^{1/2} \tag{3.64}$$

$$F_{s3} = (B \cdot C/A)^{1/2} \tag{3.65}$$

A three-state device bridge network can be transformed to its equivalent series and parallel network by using Eqs. (3.55) to (3.57) and (3.63) to (3.65).

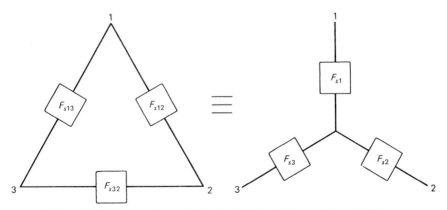

Figure 3.8. A delta to star equivalent for the short failure mode probabilities.

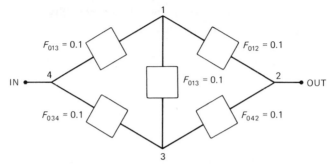

Figure 3.9. A bridge network.

Furthermore, Eqs. (3.55) to (3.57) can be utilized to transform a two-state device bridge network to its equivalent series and parallel network.

EXAMPLE 12

For the bridge network shown in Fig. 3.9, each independent bridge element failure probability (open mode) is 0.1. Calculate the bridge system reliability by using the delta-to-star Eqs. (3.55) to (3.57).

In order to transform the delta portion labelled as points 1, 2, and 3 in Fig. 3.9 we used the Eqs. (3.55) to (3.57) to get:

$$F_{01} = F_{02} = F_{03} = 0.0095456 \tag{3.66}$$

Thus, the network shown in Fig. 3.9 is transformed to the one shown in Fig. 3.10.

For components constant failure probabilities using Eqs. (3.2) and (3.18) to calculate the Figure 3.10 reliability, R, we get

$$R = [1 - \{1 - (1 - F_{013})(1 - F_{01})\}^2](1 - F_{02}) = 0.9788 \tag{3.67}$$

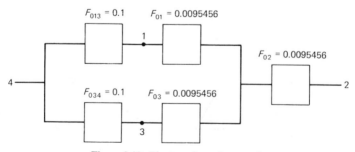

Figure 3.10. The transformed network.

3.5.6 Markov Modeling (Continuous-Time and Discrete States)

This method can be used in more cases than any other technique [89]. For example, Markov modeling is quite useful when modeling systems with dependent failure and repair modes as well as when components are independent. Furthermore, it can handle the modeling of multistate devices and common-cause failures without any conceptual difficulty.

The method is quite appealing when the system failure and repair rates are constant. Nevertheless, a problem may arise when solving a set of linear algebraic equations for large systems. In addition, with the exception of a few special situations, the method breaks down for a system with nonconstant failure and repair rates.

In order to formulate a set of Markov state equations, the rules associated with transition probabilities are:

1. The probability of more than one transition in time interval Δt from one state to the next state is negligible.
2. The transitional probability from one state to the next state in the time interval Δt is given by $\lambda \Delta t$, where the λ is the constant failure rate associated with the Markov states.
3. The occurrences are independent.

EXAMPLE 13

For the system state space diagram shown in Fig. 3.11, develop an expression for the system reliability.

The equations associated with Fig. 3.11 are

$$P_0(t + \Delta t) = P_0(t) \ \{1 - \lambda \Delta t\} \tag{3.68}$$

where $P_0(t)$ is the probability that the system is in operating state 0 at time t
$\quad\quad \lambda$ is the constant failure rate of the system
$\quad\quad (1 - \lambda \Delta t)$ is the probability of no failure in time interval Δt when the system is in state 0.
$\quad\quad P_0(t + \Delta t)$ is the probability of the system being in operating state 0 at time $t + \Delta t$

Figure 3.11. System transition diagram.

Similarly,

$$P_1(t + \Delta t) = P_0(t)\,(\lambda\Delta t) + P_1(t) \tag{3.69}$$

where $\lambda\Delta t$ denotes the probability of failure in time Δt

In the limiting case the Eqs. (3.68) and (3.69) become

$$\underset{\Delta t \to 0}{\text{limit}}\; \frac{P_0(t + \Delta t) - P_0(t)}{\Delta t} = \frac{dP_0(t)}{dt} = -\lambda P_0(t) \tag{3.70}$$

and

$$\underset{\Delta t \to 0}{\text{limit}}\; \frac{P_1(t + \Delta t) - P_1(t)}{\Delta t} = \frac{dP_1(t)}{dt} = \lambda P_0(t) \tag{3.71}$$

The initial condition is that when $t = 0$, $P_0(0) = 1$ and $P_1(0) = 0$.
Solving the Eqs. (3.70) and (3.71) by using Laplace transforms, we get

$$P_0(s) = \frac{1}{s + \lambda} \tag{3.72}$$

and

$$P_1(s) = \frac{\lambda}{s(s + \lambda)} \tag{3.73}$$

By using the inverse transforms, Eqs. (3.72) and (3.73) become

$$P_0(t) = e^{-\lambda t} \tag{3.74}$$

$$P_1(t) = 1 - e^{-\lambda t} \tag{3.75}$$

3.5.7 Binomial Method

This technique is used to evaluate the reliability of relatively simple systems such as series and parallel systems. For such systems' reliability evaluation, this is one of the simplest techniques. In the case of complex systems the method becomes a trying task. The binomial technique can be applied to systems with independent identical or nonidentical elements. The basis for the technique is the following formula:

$$\prod_{i=1}^{k} (R_i + F_i) \tag{3.76}$$

where k is the number of nonidentical components
$\quad\quad R_i$ is the ith component reliability
$\quad\quad F_i$ is the ith component unreliability

The technique is better understood by solving the following example.

EXAMPLE 14

Develop reliability expressions for series and parallel networks with two nonidentical and independent components each.
  Since $k = 2$ from Eq. (3.76) we get

$$(R_1 + F_1)(R_2 + F_2) = R_1R_2 + R_1F_2 + R_2F_1 + F_1F_2 \tag{3.77}$$

Series Network.   Two unit series network reliability, R_s, from Eq. (3.78) is

$$R_s = R_1R_2 \tag{3.78}$$

One should note here that Eq. (3.78) simply represents the first right hand term of (3.77).

Parallel Network.   Similarly for a two component parallel network reliability, R_p, (the first three terms of Eq. [3.33]) is:

$$R_p = R_1R_2 + R_1F_2 + R_2F_1 \tag{3.79}$$

Since $(R_1 + F_1) = 1$ and $(R_2 + F_2) = 1$, the above equation becomes

$$R_p = R_1R_2 + R_1(1 - R_2) + R_2(1 - R_1) \tag{3.80}$$

By rearranging Eq. (3.80) we get:

$$R_p = R_1R_2 + R_1 - R_1R_2 + R_2 - R_1R_2 \tag{3.81}$$
$$= R_1 + R_2 - R_1R_2 = 1 - (1 - R_1)(1 - R_2)$$

Similarly this procedure can be extended to k component system.

3.6 SUMMARY

The following main areas are briefly presented in the chapter:

1. Bathtub hazard rate curve
2. Mean-time-to-failure, Failure density, and hazard rate functions

3. Reliability evaluation of two-state device networks
4. Reliability evaluation of three-state device networks
5. Reliability evaluation techniques.

In this chapter the author has emphasized the structure of reliability evaluation concepts rather than the minute details. The references at the end of the chapter are provided for further reading.

3.7 REFERENCES

Reliability Engineering

Books

1. *Aerospace Reliability and Maintainability Conference Proceedings.* Available from the Society of Automotive Engineers, New York.
2. W. H. Von Alven, (ed.), *Reliability Engineering,* Prentice-Hall, Englewood Cliffs, NJ, 1964.
3. B. L. Amstadter, *Reliability Mathematics,* McGraw-Hill, New York, 1971.
4. R. T. Anderson, *Reliability Design Handbook,* (Cat. RDH-376), IIT Research Institute (RAC) RADC, Griffiths Air Force Base, Rome, NY, 1976.
5. F. L. Ankenbrandt, *Electronic Maintainability, Vol. 3* Engineering Publishers, Dist. by Reinhold, New York, 1960.
6. J. E. Arsenault and J. A. Roberts, *Reliability and Maintainability of Electronic Systems,* Computer Science Press, Potomac, MD, 1980.
7. L. J. Bain, *Statistical Analysis of Reliability and Life Testing Models:* Theory and Methods, Marcel Dekker, New York, 1978.
8. R. E. Barlow and F. Proschan, *Mathematical Theory of Reliability,* Wiley, New York, 1965.
9. R. E. Barlow and F. Proschan, *Statistical Theory and Life Testing,* Holt, Rinehart and Winston, New York, 1975.
10. W. R. Barlow, L. Hunter, and F. Proschan, *Probablistic Models in Reliability Theory,* Wiley, New York, 1962.
11. R. E. Barlow, J. B. Fussell, and N. D. Singpurwalla, *Reliability and Fault Tree Analysis: Theoretical and Applied Aspects of System Reliability and Safety Assessment,* SIAM, Philadelphia (1975). Available from the Society for Industrial and Applied Mathematics (SIAM).
12. I. Bazovsky, *Reliability: Theory and Practice,* Prentice-Hall, Englewood Cliffs, NJ, 1961.
13. B. S. Blanchard, *Design to Manage Life Cycle Cost,* M/A Press, Portland, OR, 1978.
14. B. S. Blanchard and E. E. Lawery, *Maintainability,* McGraw-Hill, New York, 1969.
15. R. Billinton, *Power-System Reliability Evaluation,* Gordon and Breach, New York, 1970.
16. R. Billinton, R. J. Ringlee, and A. J. Woods, *Power-System Reliability Calculations,* MIT Press, Cambridge, MA, 1973.
17. J. H. Bompas-Smith, *Mechanical Survival: The Use of Reliability Data,* McGraw-Hill, London, 1973.
18. R. H. W. Brook, *Reliability Concepts in Engineering Manufacture,* Wiley, New York, N.Y., 1972.
19. W. R. Buckland, *Statistical Assessment of the Life Characteristic, A Bibliographic Guide,* Charles Griffin Publishing, London, 1964.
20. J. J. Burns, *Advances in Reliability and Stress Analysis,* 1979. Available from the ASME.

21. R. H. Caflen, *A Practical Approach to Reliability,* Business Books Limited, London, 1972.
22. S. R. Calabro, *Reliability Principles and Practice,* McGraw-Hill, New York, 1962.
23. A. D. S. Carter, *Mechanical Reliability,* Wiley, New York, 1972.
24. P. Chapouille, *La fiabilite,* Presses Universitaires de France, Paris (1972).
25. D. N. Chorafas, *Statistical Processes and Reliability Engineering,* Van Nostrand, New York, 1960.
26. J. C. Cluley, *Electronic Equipment Reliability,* Macmillan, London, 1974.
27. D. R. Cox, *Renewal Theory,* Methuen, London, 1962.
28. C. E. Cunningham and W. Cox, *Applied Maintainability Engineering,* Wiley, New York, 1972.
29. B. S. Dhillon and C. Singh, *Engineering Reliability: New Techniques and Applications,* Wiley, New York, 1981.
30. G. W. A. Dummer and R. L. Winton, *A Elementary Guide to Reliability,* Pergamon Press, Oxford, 1968.
31. G. W. A. Dummer and N. B. Griffin, *Electronics Reliability-Calculations and Design,* Pergamon Press, Long Island City, N.Y., 1966.
32. J. Endrenyi, *Reliability Modeling in Electric Power Systems,* Wiley, New York, 1979.
33. J. M. English, *Cost Effectiveness: The Economic Evaluation of Engineered Systems,* Wiley, New York, 1968.
34. N. L. Enrich, *Quality Control and Reliability, Practice-Tested Methods and Procedures,* 6th ed., Industrial Press, New York, 1972.
35. B. Epstein, *Mathematical Models for System Reliability,* Student Association, Technion— Israel Institute of Technology, Halifa, Israel, 1969.
36. J. B. Fussell and G. R. Burdick (eds.) *Nuclear Systems, Reliability Engineering and Risk Assessment,* SIAM, 1977. Available from the same as ref. 82.
37. G. R. Gedye, *A Manager's Guide to Quality and Reliability,* Wiley, London, 1968.
38. I. B. Gertsbakh and K. H. Kordonskiy, *Models of Failure,* Translation from the Russian, Springer Verlag, Berlin, 1969.
39. I. B. Gertsbakh, *Models of Preventive Maintenance,* North-Holland, Amsterdam, 1977.
40. H. L. Gilmore and H. C. Schwartz, *Integrated Product Testing and Evaluation, A Systems Approach to Improve Reliability and Quality,* Wiley-Interscience, New York, 1969.
41. B. V. Gnedenko, Y. K. Belyaev, and A. D. Solovyew, *Mathematical Methods of Reliability Theory,* Academic, New York, 1969.
42. A. S. Goldman and T. B. Slattery, *Maintainability,* Wiley, New York, 1964.
43. D. Grouchko (ed.), *Operations Research and Reliability,* Gordon and Breach, New York, 1969.
44. D. Grouchko (ed.), *Operations Research and Reliability,* Gordon and Breach, New York, 1971.
45. F. M. Gryna, N. J. Ryerson, and S. Zwerling (eds.), "Reliability Training Text," 2nd Edition, Institute of Radio Engineers, New York, 1960.
46. G. J. Hahn and S. S. Shapiro, *Statistical Models in Engineering,* Wiley, New York, N.Y., 1967.
47. E. B. Haugen, *Probablistic Approaches to Design,* Wiley, New York, 1962.
48. R. T. Haviland, *Engineering Reliability and Long-Life Design,* Van Nostrand, Princeton, NJ, 1964.
49. E. J. Hensley and J. W. Lynn (eds.), *Generic Techniques in Systems Reliability Assessment,* Noordhoff Publishing, Leyden, The Netherlands, (1976).
50. K. Henney (ed.), *Reliability Factors for Ground Equipment,* McGraw-Hill, 1956.
51. R. A. Howard, *Dynamic Probablistic Systems, Vols. I and II,* Wiley, New York, 1971.
52. W. G. Ireson, (ed.), *Reliability Handbook,* McGraw-Hill, New York, 1966.

53. A. K. S. Jardine, *Maintenance, Replacement and Reliability,* Wiley, New York, 1973.
54. A. K. S. Jardine (ed.), *Operational Research in Maintenance,* Manchester University Press, Manchester and Barnes and Nobles, New York, 1970.
55. F. C. Jelen, *Cost and Optimization Engineering,* McGraw-Hill, New York, 1970.
56. D. W. Jorgenson, J. J. McCall, and R. Radner, *Optimal Replacement Policy,* Rand-McNally, Chicago, 1967.
57. C. E. Jowett, *Reliability of Electronic Components,* Iliffe Book Ltd., London, 1966.
58. K. C. Kapur and L. R. Lamberson, *Reliability in Engineering Design,* Wiley, New York, 1977.
59. P. Kenney, *Application of Reliability Techniques,* Argyle Publishing, New York, 1966. (A self-instructional, programmed training course book).
60. G. Kivenson, *Durability and Reliability in Engineering Design,* Hayden, Rochelle Park, NJ, 1971.
61. J. Kogan, *Crane Design: Theory and Calculations of Reliability, Halsted-Wiley,* New York, 1975.
62. A. Kaufmann, D. Grouchko, and R. Cruon, *Mathematical Models for the Study of the Reliability Systems,* Academic Press, New York, 1977.
63. A. Kaufmann, *La Confiance technique, theorie mathematique de la fiabilite,* Dunod, Paris, 1969.
64. B. A. Kozlov and I. A. Ushakov, *Reliability Handbook,* Holt, Rinehart & Winston Inc., New York, 1970.
65. R. R. Landers, *Reliability and Products Assurance: A Manual for Engineering and Management,* Prentice-Hall, Englewood Cliffs, NJ, (1963).
66. C. E. Leake, *Understanding Reliability,* United Testing Lab, Pasadena, CA, 1960.
67. A. Little, *Reliability of Shell Buckling Predictions,* MIT Press, Cambridge, MA, 1964.
68. D. K. Lloyd and M. Lipow, *Reliability: Management, Methods and Mathematics,* Prentice-Hall, Englewood Cliffs, NJ, 1962.
69. O. Locks, *Reliability, Maintainability, and Availability Assessment,* Hayden, Rochelle Park, NJ, 1973.
70. N. R. Mann, R. E. Shafer, and N. D. Singpurwalla, *Methods for Statistical Analysis of Reliability and Life Data,* Wiley, New York, 1974.
71. *Mechanical Reliability Concepts,* ASME, 1965. Available from the American Society of Mechanical Engineers, New York.
72. P. M. Morse, *Queues, Inventories and Maintenance,* Wiley, New York, 1958.
73. R. H. Myers, K. L. Wong, and H. M. Gordy (eds.), *Reliability Engineering for Electronic Systems,* Wiley, New York, 1964.
74. G. J. Myers, *Reliable Software Through Composite Design,* Petrocelli Books, New York, 1975.
75. G. J. Myers, *Software Reliability, Principles and Practices,* Wiley, New York, 1976.
76. F. Nixon, *Managing to Achieve Quality and Reliability,* McGraw-Hill, London (1971).
77. E. Pieruschaka, *Principles of Reliability,* Prentice-Hall, Englewood Cliffs, NJ, 1963.
78. W. H. Pierce, *Failure-Tolerant Computer Design,* Academic, New York, 1965.
79. A. M. Polvoko, *Fundamentals of Reliability Theory,* Academic, New York, 1968.
80. *Proceedings of the Annual Reliability and Maintainability Symposia* and its predecessors (available from the Annual Reliability and Maintainability Symposium, 6411 Chillum Place NW, Washington, D.C. 20012, or from the IEEE).
81. A. S. Pronikov, *Dependability and Durability of Engineering Products, Halsted-Wiley,* New York, 1973.
82. F. Proschan and R. Serfling, *Reliability and Biometry: Statistical Analysis of Life Length,* SIAM, 1974. Available from the Society for Industrial and Applied Mathematics (SIAM), Philadelphia, PA.

83. J. G. Rau, *Optimization and Probability in Systems Engineering,* McGraw-Hill, New York, 1970.
84. *Reliability Control in Aerospace Equipment Development,* SAE Technical Progress Series, Vol. 4, Society of Automotive Engineers, New York, 1963.
85. *Reliability of Military Electronic Equipment,* AGREE Report, Office of the Assistant Secretary of Defense (Research and Engineering), US Government Printing Office, Washington, D.C., 1957.
86. N. H. Roberts, *Mathematical Methods in Reliability Engineering,* McGraw-Hill, New York, 1965.
87. G. H. Sandler, *System Reliability Engineering,* Prentice-Hall, Englewood Cliffs, NJ, 1963.
88. J. Shwop and H. Sullivan, *Semiconductor Reliability,* Engineering Publishers, Elizabeth, New Jersey, 1961.
89. M. L. Shooman, *Probablistic Reliability: An Engineering Approach,* McGraw-Hill, New York, 1968.
90. C. Singh and R. Billinton, *System Reliability Modeling and Evaluation,* Hutchinson, London, 1977.
91. C. O. Smith, *Introduction to Reliability in Design,* McGraw-Hill, New York, 1976.
92. C. S. Smith, *Quality and Reliability: An Integrated Approach,* Pitman, New York, 1969.
93. C. J. Smith, *Reliability Engineering,* Barnes and Noble, New York, 1972.
94. D. J. Smith and A. H. Babb, *Maintainability Engineering,* Pitman, New York, 1973.
95. D. J. Smith, *Reliability Engineering,* Pitman, New York, 1973.
96. S. K. Srinivasan, R. Subramanian, *Probablistic Analysis of Redundant Systems,* Springer-Verlag, Berlin, 1980.
97. R. Thomason, *An Introduction to Reliability and Quality,* Machinery Publishing, Brighton, England, 1969.
98. C. P. Tsokos and I. N. Shimi (eds.), *The Theory and Applications of Reliability: With Emphasis on Bayesian and Nonparametric Methods,* Academic, New York, 1977.
99. W. G. Vogt and M. H. Mickle (eds.), *Modeling and Simulation Conference Proceedings,* Instrument Society of America, 1969–1981.
100. S. Weinberg, *Profit Through Quality: Management Control of Q and R Activities,* Gower Publishing, London, 1969.
101. *West Coast Annual Reliability Proceedings.* Available from the Western Periodical Company, 13000 Rayner Street, Northwood, CA, 91605.
102. R. H. Wilcox and W. C. Mann (eds.), *Redundancy Techniques for Computing Systems,* Spartan Press, Washington, D.C., 1962.
103. M. Zelen (ed.), *Statistical Theory of Reliability,* University of Wisconsin Press, Madison, WI, 1963.

Bathtub Hazard Rate Curve

104. B. S. Dhillon, "Life Distributions," *IEEE Trans. on Reliability,* Vol. 30, December 1981, pp. 457–460.

Multi-State Device Networks

105. B. S. Dhillon, "Literature Survey on Three-State Device Reliability Systems," *Microelectronics and Reliability,* Vol. 16, 1977, pp. 601–602.
106. B. S. Dhillon, *The Analysis of the Reliability of Multistate Device Networks,* Ph.D. Dissertation. Available from the National Library of Canada, 1975.
107. J. P. Lipp, "Topology of Switching Elements vs Reliability," *Trans. IRE Reliability Qual. Control,* Vol. 7, 1957, pp. 21–34.

4
Advanced Reliability
Evaluation Concepts

4.1 INTRODUCTION

The field of reliability engineering is wide in scope. It has applications across various areas such as aerospace, transportation, electric power generation, medical care, and so on. Researchers in these areas have been working to advance the field of reliability. As a result of the effort of these workers, many advances have been made in recent years.

This chapter presents some of the modern concepts used in reliability evaluation. The areas covered in this chapter are

1. Supplementary variables technique
2. Interference theory
3. Human reliability
4. Common-cause failures
5. Fault trees
6. Software reliability prediction
7. Renewal theory

At the end of the chapter several references important to each of the above areas are presented.

4.2 SUPPLEMENTARY VARIABLES TECHNIQUE

Markovian modeling is a widely used method to predict the reliability of systems, especially when the system failure and repair rates are constant. For many systems the assumption of constant failure rate may be acceptable, although the assumption of a constant repair rate may not be valid in as many cases. Repair times are often nonexponentially distributed.

The method of supplementary variables [1–7] can be used to model systems with constant failure rates and nonexponential repair times. These systems are sometimes known as the non-Markovian systems since the stochastic process is non-Markovian. The inclusion of sufficient supplementary variables [1] in the specification of the state of the system, can make a process Markovian. The method is demonstrated by solving an example.

4.2.1 An Application of the Technique

A system transition diagram is shown in Fig. 4.1. This diagram represents a system [4] with operational, partially operational, and failed states. The model represents a situation where when some of the system components fail, the system operates partially, and if a catastrophic failure occurs the system fails. When the system is operating partially a repair process is initiated to put back the system to its fully operational state. However, the system may have a catastrophic failure from the partially operating state. Once the system completely fails, it is repaired back to its normal operating state.

The following assumptions are associated with the model:

1. System failures are statistically independent.
2. A partially or fully failed system is restored to a like or as good as new state.
3. System failure rates are constant.
4. The partially failed system repair rate is constant. Failed system repair times are arbitrarily distributed.

The following notation is used in the development of this model:

i denotes the ith state of the system: $i = 0$ (system operating normally), $i = 1$ (system operating partially), $i = 2$ (system failed)

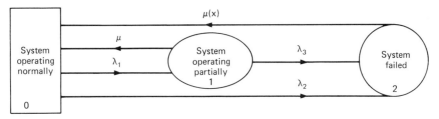

Figure 4.1. System transition diagram.

$P_i(t)$ denotes the probability that system is in state i at time t

$p_2(x, t) \cdot \Delta x$ denotes the probability that at time t, the system is in state 2 and the elapsed repair time lies in the interval $[x, x + \Delta x]$

λ_j represents the jth constant failure rate of the system: $j = 1$ (normal to partial state), $j = 2$ (normal to failed), $j = 3$ (partial to failed)

μ represents the system constant repair rate from partial to normal operating state

$\mu(x), q(x)$ represents the repair rate and probability density of repair times, respectively, when system is in state 2 and has an elapsed repair time of x.

s is the laplace transform variable

From Fig. 4.1, the forward equations for the process are

$$P_0(t + \Delta t) = P_0(t)(1 - \lambda_1\Delta t)(1 - \lambda_2\Delta t) + P_1(t)\mu\Delta t$$
$$+ \left[\int_0^\infty p_2(x, t)\mu(x)dx \right] \Delta t \tag{4.1}$$

$$P_1(t + \Delta t) = P_1(t)(1 - \lambda_3\Delta t)(1 - \mu\Delta t) + P_0(t)\lambda_1\Delta t \tag{4.2}$$

$$p_2(x + \Delta t; t + \Delta t) = p_2(x, t)[1 - \mu(x)\Delta t] \tag{4.3}$$

The boundary condition is

$$p_2(0, t) = \lambda_3 P_1(t) + \lambda_2 P_0(t) \tag{4.4}$$

At $t = 0$, $P_0(0) = 1$, $P_2(0) = 0$, $p_2(x, 0) = 0$.

The differential equations from Eqs. (4.1) to (4.3) as $\Delta t \to 0$ are

$$\frac{dP_0(t)}{dt} + (\lambda_1 + \lambda_2)P_0(t) - P_1(t)\mu = \int_0^\infty p_2(x, t)\mu(x) \, dx \tag{4.5}$$

$$\frac{dP_1(t)}{dt} + (\lambda_3 + \mu)P_1(t) - P_0(t)\lambda_1 = 0 \tag{4.6}$$

$$\frac{\partial p_2(x, t)}{\partial x} + \frac{\partial p_2(x, t)}{\partial t} + \mu(x)p_2(x, t) = 0 \tag{4.7}$$

The laplace transform of a function is

$$F(s) = \int_0^\infty e^{-st}f(t) \, dt \tag{4.8}$$

where $f(t)$ is a function of t.

By using laplace transforms and the initial condition $P_0(0) = 1$, Eqs. (4.4) to (4.7) become

$$sP_0(s) - 1 + (\lambda_1 + \lambda_2)P_0(s) - P_1(s)\mu = \int_0^\infty p_2(x, s)\mu(x)\, dx \qquad (4.9)$$

$$sP_1(s) + (\lambda_3 + \mu)P_1(s) - P_0(s)\lambda_1 = 0 \qquad (4.10)$$

$$\frac{\partial p_2(x, s)}{\partial x} + [s + \mu(x)]p_2(x, s) = 0 \qquad (4.11)$$

$$p_2(0, s) = \lambda_3 P_1(s) + \lambda_2 P_0(s) \qquad (4.12)$$

Solving the differential equation [Eq. (4.11)] by using the integrating factor method, the resulting expression is

$$p_2(x, s) = p_2(0, s)e^{-sx}e^{-\int_0^x \mu(x)\, dx} \qquad (4.13)$$

Substituting Eqs. (4.13) and (4.12) in equation (4.9)

$$[s + (\lambda_1 + \lambda_2)]P_0(s) - P_1(s)\mu - 1$$

$$= [\lambda_3 P_1(s) + \lambda_2 P_0(s)] \int_0^\infty e^{-sx}\mu(x)e^{-\int_0^x \mu(x)\, dx}\, dx \qquad (4.14)$$

Rewriting Eq. (4.14)

$$[s + (\lambda_1 + \lambda_2)]P_0(s) - P_1(s)\mu - 1 = [\lambda_3 P_1(s) + \lambda_2 P_0(s)]N(s) \qquad (4.15)$$

$$N(s) \equiv \int_0^\infty e^{-sx}q(x)\, dx$$

where $q(x) = \mu(x)e^{-\int_0^x \mu(x)\, dx}$

From Eq. (4.10)

$$P_1(s) = \frac{P_0(s)\lambda_1}{(s + \lambda_3 + \mu)} \qquad (4.16)$$

By substituting Eq. (4.16) in Eq. (4.15) we get

$$[s + (\lambda_1 + \lambda_2)]P_0(s) - \left(\frac{P_0(s)\lambda_1\mu}{s + \lambda_3 + \mu}\right) - 1$$

$$= \left[\frac{\lambda_3\lambda_1 P_0(s)}{(s + \lambda_3 + \mu)} + \lambda_2 P_0(s)\right]N(s) \qquad (4.17)$$

Rearranging Eq. (4.17)

$$P_0(s) = \left[s + (\lambda_1 + \lambda_2) - \frac{\lambda_1\mu}{(s + \lambda_3 + \mu)} - \left(\frac{\lambda_3\lambda_1}{s + \lambda_3 + \mu} + \lambda_2\right)N(s)\right]^{-1} \qquad (4.18)$$

The laplace transform of the probability, $P_2(t)$, that due to a failure the system is under repair, is

$$P_2(s) = \int_0^\infty p_2(x, s) \, dx \qquad (4.19)$$

By substituting Eqs. (4.13) and (4.12) in Eq. (4.19), the resulting expression is

$$P_2(s) = (\lambda_3 P_1(s) + \lambda_2 P_0(s))\int_0^\infty e^{-sx} e^{-\int_0^x \mu(x)\,dx} \qquad (4.20)$$

The integral of

$$\int_0^\infty e^{-sx} e^{-\int_0^x \mu(x)\,dx} = \frac{1 - N(s)}{s} \qquad (4.21)$$

since

$$e^{-\int_0^x \mu(x)\,dx} = 1 - \int_0^x \mu(x)e^{-\int_0^x \mu(x)\,dx} \qquad (4.22)$$

Utilizing Eqs. (4.16) and (4.21) in Eq. (4.20)

$$P_2(s) = \left[\frac{\lambda_1\lambda_3}{(s + \lambda_3 + \mu)} + \lambda_2\right]\left(\frac{1 - N(s)}{s}\right)P_0(s) \qquad (4.23)$$

For known repair times probability density function $q(x)$ [$N(s) = \int_0^\infty e^{-sx}q(x)\ dx$] one can invert Eqs. (4.18), (4.16), and (4.23) to get the corresponding expressions for probabilities $P_0(t)$, $P_1(t)$, and $P_2(t)$.

4.3 INTERFERENCE THEORY

Some of the item failures are not necessarily dependent upon the usage time, especially in the case of mechanical components, one-shot devices, and so on. In such cases, generally a failure occurs when the stress exceeds the strength. Therefore, to predict reliability of such items the nature of the stress and strength random variables must be known. In other words, this method assumes that the probability density functions of stress and strength are known and the variables are statistically independent. Before presenting the basis of this approach [8–17] we will first define the stress and strength.

Stress. Stress is defined as the load which will produce a failure of a component, device, or material. The term load may be defined as mechanical load, environment, temperature, electric current, etc.

Strength. Strength is defined as the ability of a component, device, or material to accomplish its required mission satisfactorily without a failure when subject to the external loading and environment. A stress-strength interference diagram is shown in Fig. 4.2. The darkened area in the diagram represents the interference area.

The reliability is defined as the probability that the failure governing stress will not exceed the failure governing strength. In mathematical form this can be stated as

$$R_c = P(s < S) = P(S > s) \tag{4.24}$$

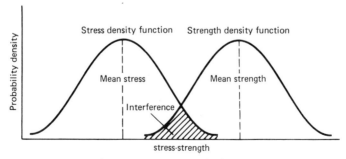

Figure 4.2. Stress-strength diagram.

where R_c is the reliability of a component or a device, p is the probability, S is the strength and, s is the stress.

Equation (4.24) can be rewritten in the following form

$$R_c = \int_{-\infty}^{\infty} f_2(s) \left[\int_{s}^{\infty} f_1(S) \, dS \right] ds \tag{4.25}$$

where $f_2(s)$ is the probability density function of the stress, s and $f_1(S)$ is the probability density function of the strength, S.

The stress-strength mathematical models are presented in the following sections.

4.3.1 Single Item Stress-Strength Reliability Modeling

The following stress-strength mathematical models are presented to determine single component reliability.

Model I: Exponentially Distributed Stress and Strength.   The strength and stress probability density functions are

$$f_1(S) = \alpha e^{-\alpha S} \qquad 0 \leq S < \infty \tag{4.26}$$

and

$$f_2(s) = \lambda e^{-\lambda s} \qquad 0 \leq s < \infty \tag{4.27}$$

where $\alpha = 1/S_m$, where S_m is the mean strength
$\qquad \lambda = 1/s$, where s_m is the mean stress

By substituting Eqs. (4.26) and (4.27) in Eq. (4.25) the resulting expression is

$$R_c = \int_{0}^{\infty} \lambda e^{-\lambda s} \left[\int_{s}^{\infty} \alpha e^{-\alpha S} \, dS \right] ds \tag{4.28}$$

Then integrating Eq. (4.28) we get

$$R_c = \frac{\lambda}{\lambda + \alpha} \tag{4.29}$$

Since $\lambda = 1/s_m$, $\alpha = 1/S_m$ the component reliability, R_c, from Eq. (4.29), becomes

$$R_c = \frac{\text{mean strength}}{\text{mean strength} + \text{mean stress}} = \frac{S_m}{S_m + s_m} \tag{4.30}$$

Model II: Rayleigh Distributed Stress and Strength. The stress and strength probability density functions are

$$f_2(s) = 2\lambda s e^{-\lambda s^2} \qquad 0 \le s < \infty \tag{4.31}$$

and

$$f_1(S) = 2\alpha S e^{-\alpha S^2} \qquad 0 \le S < \infty \tag{4.32}$$

where λ is the stress parameter
   α is the strength parameter
   Similarly, as for model I, substituting the above probability density functions in Eq. (4.25) and integrating

$$R_c = \frac{2\sqrt{\pi}}{1 + \rho} \tag{4.33}$$

where $\rho = \alpha/\lambda$.

Model III: Component Reliability Determination with Power Series Hazard Rate (n = 1) Distributed Strength and Maxwellian Distributed Stress. Both stress and strength probability density functions [9] are

$$f_{st}(s) = \frac{4}{\sqrt{\pi}} \frac{1}{\alpha_s^3} s^2 e^{(-s^2/\alpha_s^2)} \qquad \alpha_s > 0, s \ge 0 \tag{4.34}$$

and

$$f_{Sth}(S) = (k_0 + k_1 S) e^{-(k_0 S + k_1 S^2/2)} \qquad S \ge 0 \tag{4.35}$$

where k_0 and k_1 are the positive or negative constants.
   By substituting the probability density expressions of Eqs. (4.34) and (4.35) in Eq. (4.25) we get the reliability expression

$$R_c = \frac{4}{\sqrt{\pi}} \frac{1}{\alpha_s^3} \int_0^\infty s^2 e^{-(s^2/\alpha_s^2)} \, ds \int_s^\infty (k_0 + k_1 S) e^{-[k_0 S + (k_1 S^2/2)]} \, dS \tag{4.36}$$

By evaluating Eq. (4.36) and simplifying its result, the expression for the component reliability is obtained

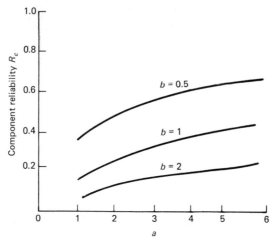

Figure 4.3. Component reliability plot for various values of a and b.

$$R_c = 2e^{(b^2/a)}\left(\frac{1}{2}+\frac{b^2}{a}\right)\left[1-\text{erf}\left(\frac{b}{\sqrt{a}}\right)\right]-\frac{2}{\sqrt{\pi a}} \qquad (4.37)$$

where $a=\left(\dfrac{k_1}{2}+\dfrac{1}{\alpha_s^2}\right)$ $b=\dfrac{k_0}{2}$

The reliability plots of the above expression are shown in Fig. 4.3 for the various values of a and b. This plot shows that as b takes increasing values, the component reliability decreases. In contrast, when a increases, the component reliability increases irrespective of values of parameter b.

Model IV: Component Reliability Determination When Stress and Strength Follows Power Series Hazard Rate Distribution. The stress and strength probability density functions [9] are given by

$$f_{st}(s)=\left(\sum_{i=0}^{n}k_{si}s^i\right)e^{-\left(\sum_{i=0}^{n}k_{si}\frac{s^{(1+i)}}{1+i}\right)} \qquad s\geq 0 \qquad (4.38)$$

and

$$f_{sth}(S)=\left(\sum_{i=0}^{n}k_{Si}S^i\right)e^{-\left\{\sum_{i=0}^{n}k_{Si}\frac{S^{(1+i)}}{1+i}\right\}} \qquad s\geq 0 \qquad (4.39)$$

where the k's are the positive or negative constants.

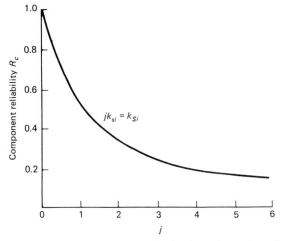

Figure 4.4. Component reliability plot for the various values of j.

To obtain the component reliability we substitute the probability density functions of Eqs. (4.38) and (4.39) in Eq. (4.25)

$$R_c = \int_0^\infty \left\{ \left(\sum_{i=0}^n k_{si} s^i \right) \right\} e^{-\left[\sum_{i=0}^n k_{si} (s^{i+1}/i+1) \right]}$$

$$\times \left[\int_s^\infty \left(\sum_{i=0}^n k_{Si}\, S^i \right) e^{-\sum_{i=0}^n (k_{Si} S^{1+i}/i+1)} dS \right] ds \qquad (4.40)$$

By solving the first right-hand integral of Eq. (4.40) we get

$$R_c = \int_0^\infty \left(\sum_{i=0}^n k_{si} s^i \right) e^{-\sum_{i=0}^n (k_{Si} + k_{si})(s^{i+1}/i+1)} ds \qquad (4.41)$$

The reliability of the above expression can be approximated numerically. If we let $k_{Si} = jk_{si}$ in Eq. (4.41), then the following reliability equation results

$$R_c = 1/(1 + j) \qquad (4.42)$$

The reliability plot of the above equation is shown in Fig. 4.4 for various values of j. As can be seen from Fig. 4.4, as the value of j increases the component reliability decreases.

If we set $n = 1$ in Eqs. (4.38) and (4.39), then the following stress and strength probability expressions result

$$f_{st}(s) = (k_{s0} + k_{s1}s)e^{-(k_{s0}s + \frac{s^2}{2}k_{s1})} \tag{4.43}$$

and

$$f_{Sth}(S) = (k_{S0} + k_{S1})e^{-(k_{S0}S + \frac{S^2}{2}k_{S1})} \tag{4.44}$$

Again by substituting Eqs. (4.43) and (4.44) in Eq. (4.25) the component reliability expression becomes

$$R_c = \int_0^\infty (k_{s0} + k_{s1}s)e^{-[(k_{s0} + k_{S0})s + (k_{s1} + k_{S1})\frac{s^2}{2}]} \, dS \tag{4.45}$$

The final result of the Eq. (4.45) is

$$R_c = \frac{k_{s1}}{2a} + \left(\frac{k_{s0}a - k_{s1}b}{a^{3/2}}\right)e^{(b^2/a)}\int_{b/\sqrt{a}}^\infty e^{-y^2} dy \tag{4.46}$$

where $a = (k_{s1} + k_{S1})/2$ and $b = (k_{s0} + k_{S0})/2$

The second term of the above equation can be evaluated by using the standard error function tables. The reliability plots of the above equation are shown in Fig. 4.5 (at $k_{s0} = k_{s1} = 1$) for various values of k_{S0} and k_{S1}.

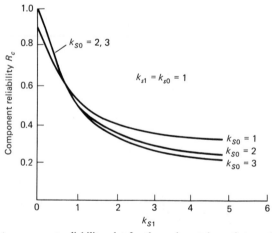

Figure 4.5. A component reliability plot for the various values of strength parameter.

These plots show that as k_{S1} and k_{S0} increase the component reliability decreases quite rapidly.

4.3.2 Redundant Items Stress-Strength Reliability Modeling

This section deals with determining the reliability of identical and independent unit redundant systems. The following two types of configurations are discussed:

1. Parallel
2. k-out-of-n units

Parallel Configurations.   Here we are concerned with measuring an identical unit, parallel redundant system reliability which is subject to some form of stress. The second right-hand part of Eq. (4.25) represents the single item probability that strength is greater than imposed stress. In other words, it is a single item reliability which may be expressed as

$$R(s) = P(S > s) = \int_s^\infty f_{Sth}(S)\, dS \qquad (4.47)$$

To measure a parallel redundant system reliability one should obtain a general expression for the system reliability in terms of its item reliabilities. This can be done if we substitute the result of Eq. (4.47) into Eq. (4.48)

$$R_p(s) = 1 - \{1 - R(s)\}^n \qquad (4.48)$$

where n is the number of identical mechanical devices or components and R_p is the parallel system reliability.

To compute the overall reliability, R_0, of a parallel redundant system under stress [19], the following equation can be utilized

$$R_0 = \int_0^\infty (R_p(s) f_{st}(s))\, ds \qquad (4.49)$$

where $R_p(s)$ is the parallel system reliability given by Eq. (4.48). The above formulas are applied to the following cases.

Model I: Reliability Determination When (System) Stress and Unit Strength Follow Exponential Distribution.   The stress and strength probability density functions [12] are defined by

$$f_{st}(s) = \lambda_s e^{-\lambda_s s} \qquad s \geq 0 \tag{4.50}$$

and

$$f_{sth}(S) = \lambda_S e^{-\lambda_S S} \qquad S \geq 0 \tag{4.51}$$

where λ_s is the reciprocal of the mean stress, $\bar{s}$
$\qquad \lambda_S$ is the reciprocal of the mean strength, $\bar{S}$

By substituting the probability function of Eq. (4.51) into Eq. (4.47) and evaluating its integral, the resulting expression is

$$R(s) = e^{-\lambda_S s} \tag{4.52}$$

The parallel system reliability is obtained by substituting Eq. (4.52) into Eq. (4.48)

$$R_p(s) = 1 - (1 - e^{-\lambda_S s})^n \tag{4.53}$$

To obtain the redundant system reliability subject to exponential stress, we substitute Eqs. (4.50) and (4.53) into Eq. (4.49) to get

$$R_0 = \int_0^{\infty} \lambda_s e^{-\lambda_s s}[1 - (1 - e^{-\lambda_S s})^n] \, ds \tag{4.54}$$

By performing a few steps in between and evaluating the above integral, the end result simplifies to

$$R_0 = 1 - \lambda_s \sum_{j=0}^{n} \binom{n}{j} (-1)^j / (\lambda_s + \lambda_S j) \tag{4.55}$$

Figure 4.6 shows reliability plots of Eq. (4.55) for the various values of n and K ($K = \lambda_S/\lambda_s$). These plots show that as K increases the system reliability decreases. However, when a number of redundant units are added, the system reliability is enhanced.

Model II: System Reliability Determination When Both (System) Stress and Unit Strength Follow Power Series Hazard Rate Distribution. Both stress and strength distribution probability density functions [12] are defined

$$f_{st}(s) = (K_{s0} + K_{s1}s)e^{-(K_{s0}s + (s^2/2)K_{s1})} \tag{4.56}$$

and

$$f_{Sth}(S) = (K_{S0} + K_{S1}S)e^{-(K_{S0}S + (S^2/2)K_{S1})} \tag{4.57}$$

where K_{s0}, K_{s1}, K_{S0}, and K_{S1} are the respective distribution parameters.

Similarly, as for model I and by carrying out a few steps in between, the reliability expression [13] is obtained

$$R_0 = 1 - \sum_{j=0}^{n} \binom{n}{j} (-1)^j \left[\frac{K_{s1}}{2a} + \frac{1}{\sqrt{a}} e^{b^2/a} \right.$$
$$\left. \left(K_{s0} - \frac{K_{s1}b}{a} \right) \frac{\sqrt{\pi}}{2} \operatorname{erf} c \left(\frac{b}{\sqrt{a}} \right) \right] \tag{4.58}$$

where $s = j(K_{s1} + K_{S1})$ and $b = [j(K_{s0} + K_{S0})]/2$

k-out-of-n Units System.   To measure a k-out-of-n unit system reliability we should obtain a general expression for the system reliability in terms of its item reliability. This can be done if we substitute the result of Eq. (4.47) into Eq. (4.59)

$$R_s(s) = \sum_{m=k}^{n} \binom{n}{m} \sum_{j=0}^{n-m} \binom{n+m}{j} (-1)^j (R(s))^{m+j} \tag{4.59}$$

where k is the number of identical mechanical devices or components required for mission success

n is the number of identical components

R_s is the k-out-of-n unit system reliability

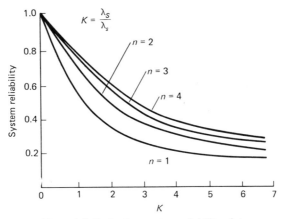

Figure 4.6. Redundant system reliability plot.

Two k-out-of-n unit system models are as follows:

Model I: k-out-of-n Unit System with Exponentially Distributed Unit Strength and System Stress. From Eqs. (4.51) and (4.50), both unit strength and system stress are defined by the following probability density functions [10]

$$f_{Sth}(S) = \lambda_S e^{-\lambda_S S} \qquad S \geq 0 \tag{4.60}$$

and

$$f_{st}(s) = \lambda_s e^{-\lambda_s s} \qquad s \geq 0 \tag{4.61}$$

Substituting Eq. (4.60) in Eq. (4.47) and evaluating its integral we get

$$R(s) = e^{-\lambda_S s} \tag{4.62}$$

The following expression results by substituting Eq. (4.62) in Eq. (4.59).

$$R_s(s) = \sum_{m=k}^{n} \binom{n}{m} \sum_{j=0}^{n-m} \binom{n-m}{j} (-1)^j e^{-(\lambda_S m + \lambda_S j)s} \tag{4.63}$$

By substituting Eqs. (4.61) and (4.63) in Eq. (4.49) the resulting expression becomes

$$R_0 = \sum_{m=k}^{n} \binom{n}{m} \sum_{j=0}^{n-m} \binom{n-m}{j} (-1)^j \frac{1}{\rho m + \rho j + 1} \tag{4.64}$$

where $\rho = \lambda_S/\lambda_s$

The plots of reliability versus ρ of the above expression are shown in Fig. 4.7 for the various values k and n. These plots show that at a very large value of the mean strength, the system reliability approaches unity. Figure 4.7 clearly shows that 2-out-of-2 unit system reliability is lower than the 2-out-of-3 unit system. However, the 1-out-of-2 unit system reliability is greater than that of the 2-out-of-3 unit system.

Model II: k-out-of-n Unit System With Rayleigh Distributed Unit Strength and System Stress. The stress and strength probability density functions [10] are

$$f_{st}(s) = \frac{2s}{\theta} e^{-s^2/\theta} \qquad \theta > 0, \ s \geq 0 \tag{4.65}$$

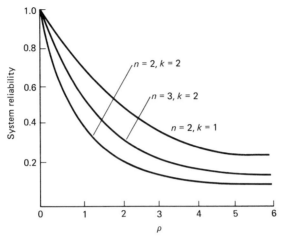

Figure 4.7. Reliability plots of a k-out-of-n unit system.

and

$$f_{Sth}(S) = 2\lambda_{Sth} S e^{-\lambda_{Sth} S^2} \qquad S \geq 0 \qquad (4.66)$$

where θ, and λ_{Sth} are the distribution parameters.

By following the same procedure as for model I, the resulting system reliability equation is

$$R_0 = \sum_{m=k}^{n} \binom{n}{m} \sum_{j=0}^{n-m} \binom{n-m}{j} (-1)^j \, \Gamma(1)/\theta \left(\lambda_{Sth} m + \lambda_{Sth} j + \frac{1}{\theta} \right) \qquad (4.67)$$

4.3.3 Graphical Method to Determine Item Reliability

This technique is used to determine the reliability of an item. The graphical method may be applied to any distribution. It is based on Mellin transforms. The transforms of Eq. (4.25) are

$$\begin{aligned}
P &= \int_{s}^{\infty} f_1(S) \, dS \\
&= 1 - \int_{0}^{s} f_1(S) \, dS \\
&= 1 - F_1(s)
\end{aligned} \qquad (4.68)$$

and

$$Q = \int_0^s f_2(s)\, ds = F_2(s) \tag{4.69}$$

Equation (4.69) is rewritten to the form

$$dQ = f_2(s)\, ds \tag{4.70}$$

By substituting Eqs. (4.68) and (4.70) in Eq. (4.25) we get

$$R_c = \int_0^1 P\, dQ \tag{4.71}$$

A hypothetical graph of Eq. (4.71) is shown in Fig. 4.8, which is a plot of P against Q. It is obvious from Eq. (4.71) that Q takes values from 0 to 1. The area under the curve represents the component reliability, R_c. Simpson's rule may be applied to compute the area under the curve.

EXAMPLE 1

An item strength is exponentially distributed with a scale parameter value of 20,000 psi. The stress of the item follows the Rayleigh distribution with a scale parameter value of 10,000 psi. Calculate the item reliability.

The stress and strength probability density functions of the item are

$$f_2(s) = \frac{2}{10,000}\left(\frac{s}{10,000}\right) e^{-(s/10,000)^2} \tag{4.72}$$

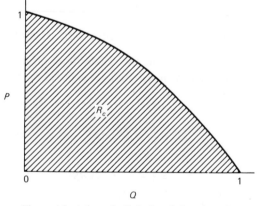

Figure 4.8. A hypothetical plot of P against Q.

and

$$f_1(S) = \frac{1}{20,000} e^{-(S/20,000)} \tag{4.73}$$

Utilizing Eqs. (4.72) and (4.73) in Eqs. (4.69) and (4.68), respectively, the resulting equations are

$$Q = F_2(s) = 1 - e^{-(s/10,000)^2} \tag{4.74}$$

and

$$P = 1 - F_1(s) = e^{-(s/20,000)} \tag{4.75}$$

For the various values of stress, s, a tabulation of values for P and Q is presented in Table 4.1.

Table 4.1. Values for P and Q.

s	P	Q
0	1	0
2,000	0.9048	0.0392
4,000	0.8187	0.1479
6,000	0.7408	0.3023
8,000	0.6703	0.4727
10,000	0.6065	0.6321
12,000	0.5488	0.7631
14,000	0.4966	0.8591
16,000	0.4493	0.9227
18,000	0.4066	0.9608
20,000	0.3679	0.9817
22,000	0.3329	0.992
∞	0	1

A plot of data given for P and Q in Table 4.1 is shown in Fig. 4.9. Following Simpson's formula to calculate the component reliability, R_c, (area under the curve from Fig. 4.9) we obtain

$$R_c = \frac{h}{3}\{P_0 + 4P_1 + 2P_2 + 4P_3 + P_4\} \tag{4.76}$$

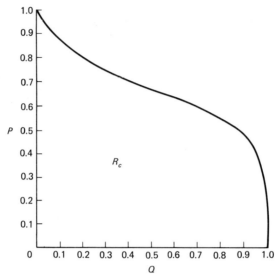

Figure 4.9. A plot of P verses Q.

and

$$h = \frac{b-a}{n} = \frac{1-0}{4} = 0.25 \qquad (4.77)$$

b and a are the upper and lower integral limits.

n is number of divisions of the curve. For example Q axis was divided into four parts ($n = 4$). It gives us Q_0, Q_1, Q_2, Q_3, and Q_4 corresponding to 0, 0.25, 0.5, 0.75, and 1, respectively

By substituting the result of Eq. (4.77) and estimated values of P_0, P_1, P_2, P_3, and P_4 corresponding to Q_0, Q_1, Q_2, Q_3, and Q_4 in Eq. (4.76), we get

$$R_c = \frac{0.25}{3} (1 + 4 \times 0.77 + 2 \times 0.66 + 4 \times 0.55 + 0)$$

$$R_c = 0.6333$$

For further reading on mechanical reliability, consult reference [9] which presents an extensive list of selected literature.

4.4 HUMAN RELIABILITY

Human links interconnect numerous systems. Human element reliability was neglected in the early days of reliability analysis, and attention was given to the equipment hardware reliability only. In the late fifties this shortcoming was recognized by Williams [27]. He pointed out that in order to predict a realistic reliability of an equipment, the human reliability must be included.

About 10 to 15 percent of the total equipment failures are directly due to human errors [21]. Occurrence of such failures are mainly due to misinterpretation of instruments, wrong actions, maintenance errors, and so on.

Since the late fifties several papers on human reliability have been published [20]. This research is concerned with various aspects of human reliability such as reliability modeling in continous time, development of human error data banks, human error allocation, and so on.

Human Reliability Definition.   This is the probability [49] of completing a task successfully by personnel within a minimum time at any required stage in system/equipment operation.

4.4.1 Human Error Categories

The following are the classifications [23] of human errors:

1. Maintenance errors: These errors occur in the field due to wrong repair or installation of the item.
2. Fabrication errors: This category of errors occurs at the fabrication stage of an item, and occur, for example, due to poor workmanship, wrong material usage, and so on.
3. Design errors: Occurrence of these errors result from the design inadequacy.
4. Operator errors: These are due to the failure of equipment operators to follow correct procedures.
5. Handling errors: Occurrence of these errors are due to poor transport or storage facilities.
6. Inspection errors: These are due to rejecting in-tolerance items and accepting out-of-tolerance items.
7. Contributing errors: Those errors which are difficult to differentiate either to human or hardware errors.

4.4.2 Task Reliability Prediction Procedure

The main objective of this procedure is to obtain subtask reliability estimates for which no previous reliability data is available [25]. To obtain a total task estimate subtask estimates may be combined. This method is outlined in the following six steps:

1. Task identification: These are the tasks which are to be performed. A complete operation is represented by each task. For example, "Prepare for motor check." A series of subtasks may be performed sequentially in order to complete a task or an operation. In other words, a task is composed of series of subtasks.
2. Subtask identification: Once the tasks to be performed are identified then the next logical step is to identify the subtasks of each task.
3. Obtain empirical data: This type of subtask data may be available from a number of sources such as in-house operations, experimental literature, laboratory, and so on.
4. Establish subtask rate: This is concerned with rating each subtask in accordance with its potential for error or level of difficulty. A scale from 1 to 10 points corresponding from least error to most error can be used to rate a subtask. One should note that this type of rating is purely based on individual (human) judgment.
5. Develop the equation for the straight line: To obtain a subtask reliability estimate, express the empirical data and the judged rating of that data in the form of a straight line. Test the line for the goodness of fit. This line can then be used to estimate the subtask reliability.
6. Determine each task reliability: The total task reliability is given by the product of subtask reliabilities obtained from the straight line equation.

Finally one should note that this procedure is applicable only to evaluate one person's performance when he or she is acting alone.

4.4.3 Human Reliability Measures

A basic equation [24] to measure human reliability, R_h, is

$$R_h = 1 - e/n \qquad (4.78)$$

where n denotes the number of times the certain task was accomplished
e denotes the number of job noncompletion occurrences (errors)

EXAMPLE 2

A task is performed by a person 500 times out of which seven occurrences lead to job noncompletion due to the human error. Find the human reliability.
By using Eq. (4.78)

$$R_h = 1 - 7/500 = 0.986$$

Human Reliability Measures: Continuous Time. Tasks such as aircraft maneuvering, scope monitoring, automobile operating, missile countdown, and so on are known as the time continuous tasks. In this section we are concerned with evaluating the reliability of humans performing such continuous time tasks. The reliability modeling of such tasks are analogous to the classical time continuous reliability modeling. The human reliability, R_h, [26] is given by

$$R_h(t) = e^{-\int_0^t e(t)\, dt} \qquad (4.79)$$

where $e(t)$ denotes the human error rate at time t; in the classical reliability theory this is analogous to hazard rate, $\lambda(t)$ or $z(t)$.

Equation (4.79) can be used to estimate human reliability of a time continous task for a known human error rate. An extensive list of selected references on human reliability is presented in reference 20.

EXAMPLE 3

A person performing a time continous task has an constant error rate $e(t) = 0.0001$ error/hour. Calculate the human reliability for a 500-hour mission. By utilizing Eq. (4.79) we get

$$R_h(500) = e^{-\int_0^t e(t)\, dt} = e^{-(0.0001)(500)} = 0.9512$$

4.5 COMMON-CAUSE FAILURES

In recent years common-cause failures which were overlooked some years ago, have been receiving wide attention. This may be due to the fact that the assumption of statistically-independent failure of redundant items is easily violated in real life [34]. A definition of a common-cause failure is any instance where multiple items malfunction due to a single cause. The following are some causes of common-cause failures:

1. External abnormal environments: dust/dirt, temperature, humidity/moisture, vibrations.
2. Equipment failure resulting from some unforseen external event: fires, floods, earthquakes, tornados.
3. Design deficiencies: During the design phase of the system some failures may have been overlooked. For example, the interdependence between electrical and mechanical items of a redundant system may have been overlooked during the design phase of a system.

4. Operation and maintenance errors: Occurrence of these errors may be due to improper maintenance, carelessness, improper calibration, same person performing maintenance on all redundant units may repeat the same mistake on all of them, etc.
5. Multiple items purchased from the same manufacturer: All these items may have same manufacturing defects.
6. Common external power source to redundant units.
7. Functional deficiency: misunderstanding of process variable behavior, inadequacy of designed protective action, inappropriate instrumentation, etc.

Several examples of hardware failure due to common-cause failures are given in reference 30. For example, spring-loaded relays forming a parallel network failed simultaneously due to a common-cause.

Some of the methods and models used to incorporate common-cause failures are as follows.

4.5.1 Common-Cause Failure Analysis of Redundant Systems

This section presents a method to incorporate common-cause failures in redundant network analysis [30, 28]. Here we assume that network units are identical and independent; in addition, the same portion of common-cause failures are associated with one or more of the redundant network units. Furthermore, in this model we assume that the unit and common-cause failure rates are constant.

In order to develop a model for common-cause failure analysis it is assumed that

$$\gamma \equiv \text{fraction of unit or system failures that are common-cause}$$

In addition, a unit failure rate is composed of two mutually components

$$\lambda_u = \lambda_i + \lambda_c \tag{4.80}$$

where λ_u is the unit constant failure rate
λ_i is the unit independent constant failure rate
λ_c is the unit or system constant common-cause failure rate

Since

$$\gamma = \frac{\lambda_c}{\lambda_u} \tag{4.81}$$

$$\therefore \quad \lambda_c = \gamma \lambda_u \tag{4.82}$$

By substituting Eq. (4.82) in Eq. (4.80) and rearranging we get

$$\lambda_i = (1 - \gamma)\lambda_u \tag{4.83}$$

In the following mathematical model Eqs. (4.81) to (4.83) are used.

Parallel Network. A series-parallel network is shown in Fig. 4.10. This is actually a modified parallel network to incorporate common-cause failures. The parallel portion of the network represents n independent failure units and the single unit in series is a hypothetical unit representing system common-cause failures. The failure of the hypothetical common-cause failure unit will cause the system failure.

The reliability, R_s, of the network shown in Fig. 4.10 is

$$R_s = [1 - (1 - R_i)^n]R_c \tag{4.84}$$

where R_i is the independent failure mode reliability of a unit
$\quad R_c$ is the common-cause failure mode system reliability
$\quad n$ is the number of identical units

The time dependent reliability of the ith independent unit with constant failure rate is

$$R_i(t) = e^{-\lambda_i t} \tag{4.85}$$

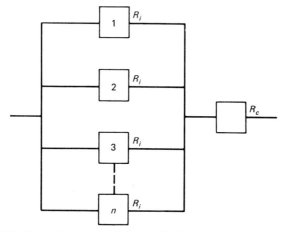

Figure 4.10. A identical unit parallel network with a common-cause failure unit.

Similarly, the hypothetical common-cause failure unit reliability is

$$R_c = e^{-\lambda_c t} \tag{4.86}$$

By substituting Eqs. (4.85) and (4.86) and (4.82) and (4.83) in Eq. (4.84)

$$R_s(t) = [1 - (1 - e^{-(1-\gamma)\lambda_u t})^n] e^{-\gamma \lambda_u t} \tag{4.87}$$

To obtain the mean time to failure (MTTF), integrate $R_s(t)$ over the time interval $[0, \infty]$

$$\text{MTTF} = \int_0^\infty R_s(t) = \sum_{i=1}^n (-1)^{i+1} \binom{n}{i} / [\lambda_u \gamma + n\lambda_u(i - \gamma_j)] \tag{4.88}$$

EXAMPLE 4

For the following data, calculate the system reliability using Eq. (4.87).

$$\gamma = 0.8$$
$$\lambda_u = 0.0001 \text{ failures/hour}$$
$$n = 2$$
$$t = 100 \text{ hours}$$

Utilizing Eq. (4.87)

$$R_s(100) = [1 - (1 - e^{-(1-0.8)0.0001(100)})^2] e^{-0.8(0.0001)100}$$
$$= 0.9920$$

This technique can be extended to other networks and probability density functions [29, 31]. In the following section two stochastic models are developed utilizing the supplementary variable method.

4.5.2 Multistate Device Redundant Systems with Common-Cause Failures and One Standby Unit

This section presents two mathematical models [30]. Model I represents a two identical unit, active redundant system whose units may fail in either of the two mutually exclusive failure modes. Similarly, model II represents a two nonidentical unit, active parallel system whose units either fail or survive. In addition, models I and II consist of one standby unit and take into consideration the occurrence of common-cause failures. Systems are only repaired when all the system units fail (including the standby unit). Since the failed system repair times are arbitrarily distributed, the supplementary variables method is utilized (Section 4.2).

The assumptions used to develop models I and II were

1. Common-cause and other failures are statistically independent.
2. Common-cause failures can only occur with more than one unit.
3. Units are repaired only when the system fails. A failed system is restored to like-new or good as new, condition. This includes the stand-by unit.
4. The standby remains as good as new until its first operational use.
5. All unit and system common-cause failure rates are constant.
6. System repair times are arbitrarily distributed.
7. In Model I, all the three system units are identical. However, in Model II, the active parallel redundant units are nonidentical. The standby unit can replace either one of them.

Model I. Model I deals with a two identical unit, active redundant system with one standby unit. Each unit of the system may fail in either of the two mutually exclusive failure modes. These may be open or closed (short) failure modes. A typical example of such a device would be a fluid flow valve or an electrical switch. Literature concerning such devices may be found in reference [3]. In addition, the two active identical unit system may also fail due to common-cause failures. When both units of the active redundant system fail (due to a common-cause failure, a closed mode failure, or the last active unit failure in the open mode), then one of the failed units of the system is immediately replaced with the identical standby unit. When the operating standby unit (the last system unit) fails in either open or closed (short) mode, then all the units (three) of the system are repaired. The state-space diagram of the system is shown in Fig. 4.11.

i State of the system; $i = 0$ (both units operating), $i = 1$ (one unit failed in open mode, other operating), $i = 2$ (both units failed due to a closed (short) mode failure), $i = 3$ (both system unit failed in open mode), $i = 4$ (the stand-by unit is operating), $i = 5$ (the stand-by unit failed in open mode), $i = 6$ (the stand-by unit failed in closed (short) mode), $i = c$ (both active units failed due to a common-cause). At $i = 0$, 1, 4 (system operating) and for $i = 2$, 3, 5, 6 (system failed).

j At $j = 0$, means stand-by unit failure in open mode; $j = c$ means stand-by unit failure in closed (short) mode.

$P_i(t)$ Probability that system is in state i at time t.

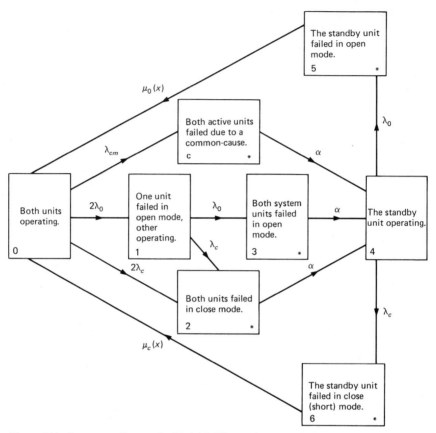

Figure 4.11. State-space diagram for Model I. The numbers and the letter c denote the system states. Stars represent failed states of the system.

$p_j(x,t)$	Probability density (with respect to repair time) that the failed system is in state j and has an elapsed repair time of x.
$\mu_j(x), G_j(x)$	Repair rate and probability density function of repair time when system is in state j and has an elapsed repair time of x.
λ_j	Unit constant failure rate.
λ_{cm}	Redundant system constant common-cause failure rate.
α	Unit constant replacement rate to replace a failed unit of the failed system with the standby unit.
s	Laplace transform variable.

The supplementary variables technique was utilized to develop the equations for the model [1].

The system of differential equations associated with Figure 4.11 is

$$\left(\frac{d}{dt} + \lambda_{cm} + 2\lambda_0 + 2\lambda_0 + 2\lambda_c\right) P_0(t)$$

$$= \int_0^\infty p_5(x, t)\mu_0(x)\, dx + \int_0^\infty p_6(x, t)\mu_c(x)\, dx \tag{4.89}$$

$$\left(\frac{d}{dt} + \lambda_0 + \lambda_c\right) P_1(t) = P_0(t)2\lambda_0 \tag{4.90}$$

$$\left(\frac{d}{dt} + \alpha\right) P_2(t) = P_1(t)\lambda_c + P_0(t)2\lambda_c \tag{4.91}$$

$$\left(\frac{d}{dt} + \alpha\right) P_3(t) = P_1(t)\lambda_0 \tag{4.92}$$

$$\left(\frac{d}{dt} + \alpha\right) P_c(t) = P_0(t)\lambda_{cm} \tag{4.93}$$

$$\left(\frac{d}{dt} + \lambda_0 + \lambda_c\right) P_4(t) = \alpha(P_c(t) + P_3(t) + P_2(t)) \tag{4.94}$$

$$\left(\frac{\partial}{\partial x} + \frac{\partial}{\partial t} + \mu_0(x)\right) p_5(x, t) = 0 \tag{4.95}$$

$$\left(\frac{\partial}{\partial x} + \frac{\partial}{\partial t} + \mu_c(x)\right) p_6(x, t) = 0 \tag{4.96}$$

$$p_5(0, t) = \lambda_0 P_4(t) \tag{4.97}$$

$$P_6(0, t) = \lambda_c P_4(t) \tag{4.98}$$

At $P_0(0) = 1$, $P_c(0) = P_1(0) = P_2(0) = P_3(0) = P_4(0) = p_5(x, 0) = p_6(x, 0) = 0$.

The laplace transforms of the solution are shown in Eqs. (4.99) to (4.106)

$$P_0(s) = \left\{ s + \lambda_{cm} + 2\lambda_0 + 2\lambda_c - (\lambda_0 G_0(s) - \lambda_c G_c(s)) \right.$$

$$\left. \left[\frac{\lambda_{cm}}{A_1} + \frac{2\lambda_0^2}{A_1 A_2} + \left(\frac{\lambda_0}{A_2} + 1\right)\frac{2\lambda_c}{A_1}\right]\frac{\alpha}{A_2}\right\}^{-1} \tag{4.99}$$

$$G_j(s) \equiv \int_0^\infty e^{-sx} G_j(x)\, dx \qquad \text{for } j = 0 \text{ or } c$$

$$G_j(x) \equiv \mu_j(x) \, e^{-\int_0^x \mu_j(x)\,dx} \qquad \text{for } j = 0 \text{ or } c$$

$$A_1 \equiv (s + \alpha)$$

$$A_2 \equiv (s + \lambda_0 + \lambda_c)$$

$$P_1(s) = P_0(s) \, 2\lambda_0 / A_2 \tag{4.100}$$

$$P_2(s) = \left(\frac{\lambda_0}{A_2} + 1\right) 2\lambda_c P_0(s) / A_1 \tag{4.101}$$

$$P_3(s) = 12\lambda_0^2 \, P_0(s) / A_1 A_2 \tag{4.102}$$

$$P_4(s) = \left[\frac{\lambda_{cm}}{A_1} + \frac{2\lambda_0^2}{A_1 A_2} + \left(\frac{\lambda_0}{A_2} + 1\right)\frac{2\lambda_c}{A_1}\right]\frac{\alpha P_0(s)}{A_2} \tag{4.013}$$

$$P_c(s) = P_0(s)\lambda_{cm} / A_1 \tag{4.104}$$

$$P_5(s) = \lambda_0 P_4(s)(1 - G_0(s))/s \tag{4.105}$$

$$P_6(s) = \lambda_c P_4(s)(1 - G_c(s))/s \tag{4.106}$$

For a given repair times probability density function $G_j(x)$, one can invert Eqs. (4.99) to (4.106) to obtain the corresponding time dependent probability equations.

Model II. In the case of model II, it is assumed that two nonidentical units form an active parallel redundant system. All units of the system can only be in either the operational or the failed state. This system may also fail due to common-cause failures. When both of the redundant system units fail due to common-cause failures or otherwise, then one unit of the system is immediately replaced by the standby unit. The standby unit can replace either of the two system units. Whenever the standby unit fails, then all three units of the system are repaired to their as good as new state. The repair times of the system are arbitrarily distributed. The transition diagram of the system is shown in Fig. 4.12.

The following notation is associated with Fig. 4.12:

i State of the system; $i = 0$ (both nonidentical units of the system are operating), $i = 1$ (means unit 1 failed and unit 2 operating), $i = 2$ (means unit 2 failed and unit 1 operating), $i = 3$ (means both units 1 and 2 failed), $i = r$ (means either unit 1 or 2 is replaced with the stand-by unit), $i = f$ (means the standby unit has failed), $i = c$ (means both the redundant system units 1 and 2 have failed due to a common-cause)

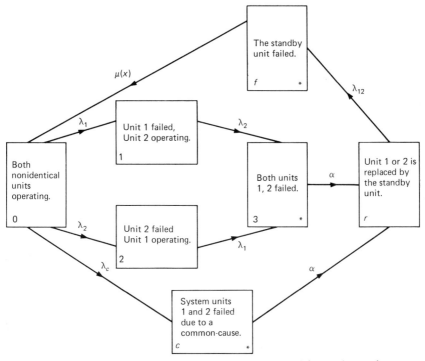

Figure 4.12. State-space diagram for Model II. The numbers and letters denote the system states. Asterisks represent failed states of the system.

$P_i(t)$	Probability that system is in state i at time t; $i = 0, 1, 2, 3, c, r$
$p_f(x, t)$	Probability density (with respect to repair time) that the failed system is in state f and has an elapsed repair time of x. f means the standby unit has failed
$\mu(x), G(x)$	Repair rate and probability density function of repair time when system is in state f and has an elapsed repair time of x
λ_i	Constant failure rate of the active redundant system unit i for $i = 1, 2$
λ_c	Redundant system constant common-cause failure rate
α	Constant unit replacement rate to replace any one of the failed units with the standby unit when system is in state $i = 3$ or c
λ_{12}	Constant failure rate of the standby unit
s	Laplace transform variable

The supplementary variables method [1] was used to develop equations for the model. The system of differential equations associated with Fig. 4.12 is

$$\left(\frac{d}{dt} + \lambda_c + \lambda_1 + \lambda_2\right) P_0(t) = \int_0^\infty p_f(x, t)\mu(x)\, dx \qquad (4.107)$$

$$\left(\frac{d}{dt} + \lambda_2\right) P_1(t) = P_0(t)\lambda_1 \qquad (4.108)$$

$$\left(\frac{d}{dt} + \lambda_1\right) P_2(t) = P_0(t)\lambda_2 \qquad (4.109)$$

$$\left(\frac{d}{dt} + a\right) P_3(t) = P_1(t)\lambda_2 + P_2(t)\lambda_1 \qquad (4.110)$$

$$\left(\frac{d}{dt} + \lambda_{12}\right) P_r(t) = P_3(t)a + P_c(t)a \qquad (4.111)$$

$$\left(\frac{d}{dt} + a\right) P_c(t) = P_0(t)\lambda_c \qquad (4.112)$$

$$\left(\frac{\partial}{\partial x} + \frac{\partial}{\partial t} + \mu(x)\right) p_f(x, t) = 0 \qquad (4.113)$$

$$p_f(0, t) = p_r(t)\lambda_{12} \qquad (4.114)$$

At $P_0(0) = 1$, other initial condition probabilities are equal to zero and $p_f(x, 0) = 0$.

The Laplace transforms of the solution are

$$P_0(s) = \{(s + \lambda_c + \lambda_1 + \lambda_2) - G(s)\lambda_{12}a[(\lambda_2 A_1 + A_2\lambda_1 + \lambda_c)/A_3 A_4]\}^{-1}$$

$$G(s) \equiv \int_0^\infty e^{-sx} G(x)\, dx \qquad (4.115)$$

$$G(x) \equiv \mu(x) e^{-\int_0^x \mu(x)\, dx}$$

$$A_1 \equiv \lambda_1/(s + \lambda_2) \qquad A_2 \equiv \lambda_2/(s + \lambda_1),$$
$$A_3 \equiv (s + a) \qquad A_4 \equiv s + \lambda_{12}$$

$$P_1(s) = P_0(s)A_1 \qquad (4.116)$$

$$P_2(s) = P_0(s)A_2 \qquad (4.117)$$

$$P_3(s) = (A_1\lambda_2 + A_2\lambda_1)P_0(s)/A_3 \qquad (4.118)$$

$$P_r(s) = (A_1\lambda_2 + A_2\lambda_1 + \lambda_1 + \lambda_c)\alpha \, P_0(s)/A_3 A_4 \qquad (4.119)$$

$$P_c(s) = \lambda_c P_0(s)/A_3 \qquad (4.120)$$

$$P_f(s) = P_r(s)\lambda_{12}(1 - G(s))/s \qquad (4.121)$$

For a known repair times probability density function, $G(x)$, one can invert Eqs. (4.115) to (4.121) to get the corresponding probability equations.

4.6 FAULT TREES

This is a widely used technique in industry to evaluate the reliability of complex systems. This method was originated in the early sixties by the personnel of Bell Laboratories to evaluate the reliability and safety of the Minuteman Launch Control System. Since then several experts have contributed to the technique. A literature list is presented in references 35 and 36. The commonly used symbols and definitions associated with the technique are as shown below.

4.6.1 Symbols and Definitions

This section presents some selected fault tree symbols and definitions. These symbols are taken from references 37, and 40.

OR Gate. This symbol denotes that an output event occurs if any one or more of the n input events occur.

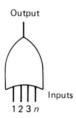

Output

Inputs

1 2 3 n

AND Gate. This is the dual of the OR gate. An AND gate denotes that an output event only occurs if all of the n input events occur.

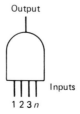

Output

Inputs

1 2 3 n

Priority AND Gate. This is basically an AND gate with the exception that the input events must occur in a certain specified order.

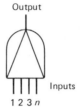

Output

Inputs

1 2 3 n

Delay Gate. The output of the gate only occurs after a specified delay time has elapsed.

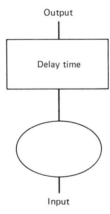

Output

Delay time

Input

Resultant Event. This event is represented by a rectangle. The rectangle represents an event which is a result of the combination of fault events that precede it.

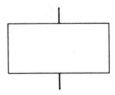

Basic Fault Event. This event is denoted by a circle. It represents the failure of an elementary component or a basic fault event. The event parameters such as the probability of occurrence, the failure rate, and the repair rate are obtained from the field failure data or other reliable sources.

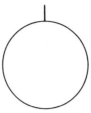

Incomplete Event. This event is denoted by a diamond. It denotes a fault event whose cause has not been fully determined either due to lack of interest or due to lack of information or data.

Conditional Event. This event is represented by an ellipse. This symbol indicates any condition or restriction to a logic gate.

Trigger Event. This event is denoted by a house-shaped symbol. This represents a fault event which is expected to occur.

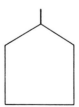

Transfer-in and transfer-out. These situations are denoted by a triangle. A line from the top of the triangle denotes a transfer-in, whereas a line from the side represents the transfer-out.

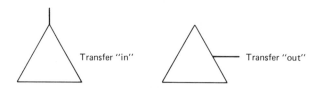

Transfer "in"   Transfer "out"

4.6.2 Fault Tree Development Procedure

To develop fault trees [35], the four steps that should be followed are

1. System definition: This defines the system, its undesired event (the top event), and boundaries.
2. Fault tree construction: This is the representation of system conditions symbolically.
3. Qualitative evaluation: Once a system fault tree has been constructed, the next step is to evaluate the tree qualitatively; for example, obtaining the minimal cut sets of the fault tree. Several system weaknesses may be sighted during this phase.
4. Quantitative evaluation: This phase obtains the numerical information relating to the fault tree top event; for example, the failure rate, the repair rate, the unavailability, the failure probability, and so on. There are basically two procedures to evaluate fault trees:

    a. The analytical approach
    b. Monte Carlo simulation

4.6.3 Gate Analytical Developments

AND Gate. An n-input AND gate is shown in Fig. 4.13. In Fig. 4.13 the resultant event and the basic events are represented by a rectangle and circles, respectively. In Boolean algebra the output event expression, X_0, is

$$X_0 = X_1 \cdot X_2 \cdot X_3 \cdots X_n \qquad (4.122)$$

where the symbol $\cdot$ (dot), represents the intersection of events. $X_1, X_2, \cdots, X_n$ are the input events.

OR Gate. Figure 4.14 shows an n-input OR gate. In Boolean algebra the output event expression, Y_0, is

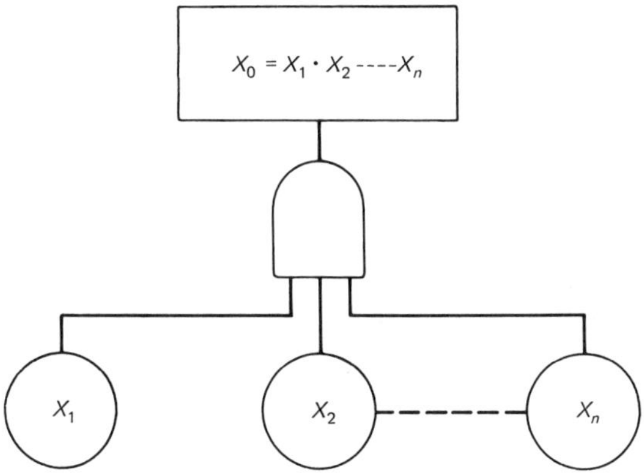

$$X_0 = X_1 \cdot X_2 \text{----} X_n$$

X_1 X_2 ------ X_n

Figure 4.13. A n-input AND gate.

$$Y_0 = Y_1 + Y_2 + \cdots + Y_n \tag{4.123}$$

where the symbol $+$ denotes the union of events. Y_1, Y_2, $\cdots$, Y_n are the input events.

4.6.4 Boolean Algebra Properties

The commonly used Boolean algebra properties are

1. Identities

$$X_1 + X_1 = X_1 \tag{4.124}$$

$$X_1 \cdot X_1 = X_1 \tag{4.125}$$

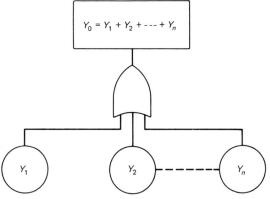

$$Y_0 = Y_1 + Y_2 + \text{---} + Y_n$$

Figure 4.14. A n-input OR gate.

2. Absorption Laws

$$X_1 \cdot (X_1 \cdot X_2) = X_1 \cdot X_2 \qquad (4.126)$$

$$X_1 + (X_1 \cdot X_2) = X_1 \qquad (4.127)$$

3. Distributive Laws

$$X_1 + X_2 \cdot X_3 = (X_1 + X_2) \cdot (X_1 + X_3) \qquad (4.128)$$

These properties can be used to eliminate repetitive events in a fault tree. An example of such a case is given below.

EXAMPLE 5

A fault tree with a repeated event is shown in Fig. 4.15. As shown in Fig. 4.15, Y_1 to Y_6 denote the basic events. Intermediate events are denoted by X_0 to X_4. The top event is represented X_5.

The fault tree of Fig. 4.15 is represented by Boolean expressions as

$$X_5 = X_3 \cdot X_4 \cdot Y_4 \qquad (4.129)$$

$$X_4 = (Y_5 + Y_6) \qquad (4.130)$$

$$X_3 = (Y_1 + Y_2) \cdot (Y_1 + Y_3) \qquad (4.131)$$

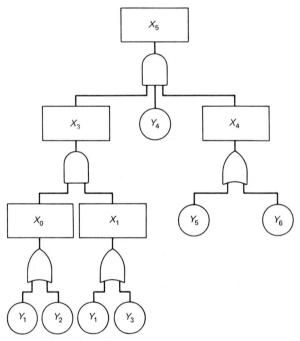

Figure 4.15. A fault tree with a repeated event.

As can be seen from Fig. 4.15 and Eq. (4.131), event Y_1 is the basic repeated event. To eliminate the repetitiveness of Y_1, the Boolean property of Eq. (4.128) was used in Eq. (4.131)

$$Y_1 + Y_2 \cdot Y_3 = (Y_1 + Y_2) \cdot (Y_1 + Y_3) \tag{4.132}$$

By substituting Eqs. (4.131) and (4.132) and Eq. (4.130) in Eq. (4.129) we get

$$X_5 = (Y_1 + Y_2 \cdot Y_3) \cdot (Y_5 + Y_6) \cdot Y_4 \tag{4.133}$$

The repeated-event-free fault tree of Eq. (4.133) is given in Fig. 4.16.

References 34, 38, 49, and 41 present algorithms to obtain the repeated-event-free fault tree. One should note that before calculating quantitative results, a fault tree must be repeated-event free, otherwise these results will be incorrect.

4.6.5 Probability Evaluation of Fault Trees

This section is concerned with calculating the occurrence probability of the repeated-event-free fault tree top and intermediate events. To obtain the AND and OR gate top event probability, the following formulas are used along with the laws of probability.

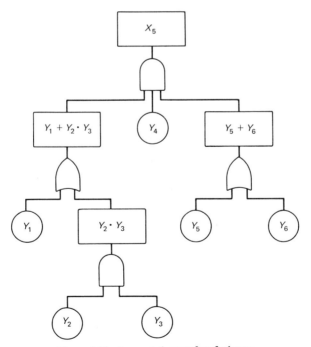

Figure 4.16. A repeated-event-free fault tree.

AND Gate. For the n statistically independent input event AND gate shown in Fig. 4.13, the top event occurrence probability expression is

$$P(X_1 \cdot X_2 \cdot X_3 \cdot \cdots \cdot X_n) = P(X_1) \cdot P(X_2) \cdot P(X_3) \cdot \cdots \cdot P(X_n) \qquad (4.134)$$

Where $P(X_1)$, $P(X_2)$, $\cdots$, $P(X_n)$ are the occurrence probabilities of events X_1, X_2, $\cdots$, X_n, respectively.

OR Gate. A n-input event fault tree is presented in Fig. 4.14. In order to obtain an approximation of the top event occurrence probability expression, consider the two statistically independent input OR gate probability expression

$$P(Y_1 + Y_2) = P(Y_1) + P(Y_2) - P(Y_1) \cdot P(Y_2) \qquad (4.135)$$

where $P(Y_1)$ and $P(Y_2)$ are the occurrence probabilities of events Y_1 and Y_2, respectively.

For very small $P(Y_1)$ and $P(Y_2)$, Eq. (4.135) may be approximated as

$$P(Y_1 + Y_2) \simeq P(Y_1) + P(Y_2) \qquad (4.136)$$

In the case of the n-input event OR gate, Eq. (4.136) is generalized

$$P(Y_1 + Y_2 + Y_3 + \cdots + Y_n)$$
$$\simeq P(Y_1) + P(Y_2) + P(Y_3) + \cdots + P(Y_n) \qquad (4.137)$$

EXAMPLE 6

A fault tree is shown in Fig. 4.17. Calculate the probability of occurrence for the top fault event, T, if the each basic independent fault event probability of occurrence is 0.02.

For statistically independent events, the intermediate and top event occurrence probability expressions and quantitative results are obtained by utilizing Eqs. (4.134) to (4.136) and are

$$P(I_1) = P(A) + P(B) - P(A) \cdot P(B)$$
$$= 0.02 + 0.02 - 0.02 \times 0.02 = 0.0396 \qquad (4.138)$$

or

$$P(I_1) \simeq P(A) + P(B) = 0.02 + 0.02 = 0.04 \qquad (4.139)$$

$$P(I_2) = P(I_1) \cdot P(C) \cdot P(D)$$
$$= 0.0396 \times 0.02 \times 0.02 = 0.0000158 \qquad (4.140)$$

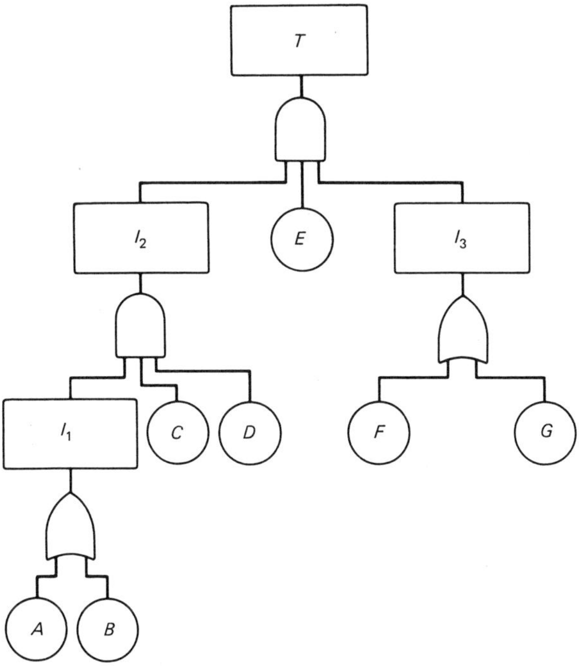

Figure 4.17. A fault tree.

$$P(I_3) = P(F) + P(G) - P(F) \cdot P(G)$$
$$= 0.02 + 0.02 - 0.02 \times 0.02 = 0.0396 \qquad (4.141)$$

$$P(T) = P(I_2) \cdot P(I_3) \cdot P(E)$$
$$= 0.0000158 \times 0.0396 \times 0.02 = 1.25136 \times 10^{-8} \qquad (4.142)$$

The top event probability of occurrence is 1.25136×10^{-8}. For further reading on fault trees consult references 35 and 36.

4.7 SOFTWARE RELIABILITY PREDICTION

Today increasing numbers of computers are being used in various applications. Since the seventies large and complex computer programs have been introduced to perform sophisticated real-time tasks. In the future the software cost is expected to increase many times over as compared to the computer hardware cost [57]. The magnitude of attention required to develop software of computers is quite large. Since the last decade, an emphasis has been placed on the software reliability in order to develop reliable computer operations. Errors in the software may cause the catastrophic failure of computer systems. For example, a failure of NORAD air and missile defense computer system due to an software error could have started World War III.

Software Reliability Definition. This is the probability [207] that a software program functions without an external error for some time period on the system it is to be used under the actual working conditions.

This section briefly discusses hardware and software reliabilities and software reliability models. An extensive literature list [42–234] on software reliability is presented at the end of this chapter.

4.7.1 Hardware and Software Reliability

The following are some of the main points to be noted:

<table>
<tr><td>SOFTWARE RELIABILITY</td><td>HARDWARE RELIABILITY</td></tr>
<tr><td>1. No bathtub hazard rate curve [131]</td><td>1. Has the bathtub hazard rate curve [131]</td></tr>
<tr><td>2. The software will not wear out [131]</td><td>2. The hardware will wear out</td></tr>
<tr><td>3. The software field is relatively new</td><td>3. The hardware field is well established (especially in the area of electronic components)</td></tr>
<tr><td>4. Useful data collection is a problem</td><td>4. Same, as for software</td></tr>
<tr><td>5. Basically the software reliability is design oriented</td><td>5. The hardware reliability is affected by design, production, and operation</td></tr>
<tr><td>6. Has the potential for monetary savings</td><td>6. Same as for software</td></tr>
<tr><td>7. Redundancy in the software may not be effective</td><td>7. Generally, item redundancy is effective</td></tr>
</table>

For further discussion on software and hardware reliability one should consult reference 131.

4.7.2 Software Reliability Modeling

Once the software is put into operational use mathematical models can be developed to predict its reliability. These models utilize information such as the number of errors debugged when developing a software program.

In the past number of years many software models have been developed [219]. Each has its advantages, disadvantages, and particular area of use. This section presents some of the models available in the literature. A list of selected literature is also presented at the end of this chapter which contains most of the available software reliability models.

Shooman Model. This model was developed by Shooman [207]. The hazard function, $z(t)$, of the model is

$$z(t) = C_p E_r(y) \qquad (4.143)$$

where $E_r(y) = (e/i) - E_c(y)$.

y denotes the debugging time since the beginning of machine or system integration

C_p represents the proportionality constant

$E_r(y)$ denotes the total remaining number of errors in the program at time y

e represents the total number of initial errors, at $y = 0$

i denotes the total number of machine language instructions

$E_c(y)$ denotes the cumulative number of errors corrected in interval y

t is the system operating time

The following basic assumptions are associated with Eq. (4.143):

1. In the software program, the total number of machine language instructions remains constant.
2. At the start of integration testing, the number of errors in a program is constant. Furthermore, their number decreases directly as errors are corrected. The process of testing does not introduce any new errors.
3. The number of residual errors is the difference between the errors initially present and the total sum of errors corrected.
4. The hazard rate is proportional to the number of residual errors.

The reliability, $R(t)$, at time t of an item is [207]

$$R(t) = e^{-\int_0^t z(x)\,dx} \qquad (4.144)$$

where $z(t)$ is the item hazard rate.

By substituting Eq. (4.143) in Eq. (4.144) we get

$$R(t;y) = e^{-C_p[(e/i) - E_c(y)]t} \qquad (4.145)$$

The above expression is obtained by assuming that the hazard rate is independent of time t.

The mean time to failure (MTTF) is

$$\text{MTTF} = \int_0^\infty R(t;y)\,dt = \int_0^\infty e^{-C_p[(e/i) - E_c(y)]t}\,dt \qquad (4.146)$$

$$= \frac{1}{C_p[(e/i) - E_c(y)]}$$

Suppose we know i and $E_c(y)$ from the size of the program and the collection of error data, respectively. Then we can estimate the constants C_p and e by using the moment matching method [207–209].

Jelinski-Moranda (JM) Model. This mathematical model assumes that the software errors are exponentially distributed [124, 125], as does the Shooman model. The following assumptions were used to develop the model [219]:

1. The occurrence rate between errors is constant.
2. Each error detected is corrected, therefore the total number of errors is reduced by one each time.
3. The hazard rate function of debugging time between error occurrences is proportional to the number of errors remaining.

The hazard rate function $z_{jm}(y_i)$, of the model is given by

$$z_{jm}(y_i) = \theta[I - (i - 1)] \tag{4.147}$$

where I denotes the total number of initial errors
y_i represents the time between the ith and $(i - 1)$st errors discovered
θ denotes the constant of proportionality

By substituting Eq. (4.147) in Eq. (4.144), the resulting reliability expression, $R(t_i)$, is

$$R(t_i) = e^{-\theta(I - i + 1)t_i} \tag{4.148}$$

By substituting Eq. (4.148) in the left-hand side of Eq. (4.146), that is, integrating Eq. (4.148) over the time interval [0, 00], we get the mean time to failure

$$\text{MTTF} = 1/\theta(I - m) \tag{4.149}$$

where $m \equiv i - 1$. In other words, m denotes the number of errors discovered to date.

Schick-Wolverton (SW) Model. This model [232, 199] is similar to Jelinski-Moranda (JM) model. The following assumptions are associated with this model:

1. Each error detected is corrected, therefore the total number of errors is reduced by one each time.

2. The debugging times between error occurrences follow the Rayleigh distribution.
3. The error rate is proportional to the debugging time spent and the total number of errors remaining.

The model hazard rate function, $z_{sw}(t_i)$, is

$$z_{sw}(t_i) = \theta[I - (i - 1)]t_i \qquad (4.150)$$

where θ is the constant of proportionality
 I denotes the total number of initial errors
 t_i is the time interval between the $(i - 1)$st and the ith errors

As for JM model, the reliability, $R(t_i)$, and MTTF expressions associated with the SW model are

$$R(t_i) = e^{-\theta[I - m]t_i^2/2} \qquad (4.151)$$

where $m = i - 1$, and

$$\text{MTTF} = \pi^{1/2}[2\theta(I - m)]^{-1/2} \qquad (4.152)$$

Geometric Model. This model also was developed by Jelinski-Moranda [219] and is similar to the JM model. One of the assumptions of the model is that a geometric progression is formed by the hazard rate between successive errors. Other assumptions of the model are similar to the JM model. The hazard rate, $z(t)$, is given by

$$z(t) = LC^{i-1} \qquad (4.153)$$

where L and C are the constants
 t is time

As for the JM model, the reliability and MTTF functions are

$$R(t) = e^{-LC^m t} \qquad (4.154)$$

and

$$\text{MTTF} = C^{-m}/L \qquad (4.155)$$

where $m = i - 1$, the total number of time intervals.
 Other software reliability models may be found in reference [219].

4.8 RENEWAL THEORY

Some of the earlier published work on the renewal theory may be found in references 238, 239, and 242. A review paper entitled "Renewal Theory and Its Ramifications" was published by Smith [246] in 1958. Since then several contributions on the subject have been published.

Renewal theory is utilized in the field of reliability engineering to determine, for example, how many equipment failure-repair cycles have occurred [3, 11]. This knowledge is vital in order to schedule the work load of maintenance personnel and determine a proper supply of spares. The basic renewal theory is discussed in reference 237. In this section we present one example to present the concept of renewal theory.

Suppose that we have one equipment labeled 1 and it is operating. Equipment 1 failed for the first time after x_1 hours of operation. It was replaced immediately by an identical equipment 2. Due to very small equipment down time during replacement, it was neglected. Therefore, equipment 2 starts operating at time x_1 and fails after time x_2'. Similarly equipment 3 (identical) fails after operating for x_3 hours. The equipment 2 operating time is $x_2 = x_2' - x_1$ hours. For $(k-1)$ renewals or replacements, the total system operating time X_k, is

$$X_k = \sum_{i=1}^{k} x_i \qquad (4.156)$$

In other words, X_k is the kth failure time.

Suppose each operating time x_1, x_2, x_3, $\cdots$, x_k has a corresponding probability density function $f(x_1)$, $f(x_2)$, $f(x_3)$, $f(x_4)$, $\cdots$, $f(x_k)$. Furthermore, for k replacements (renewals), the system operating time probability density function is $f(x_k)$. This is obtained by the multiple convolution integral from $f(x_1)$, $f(x_2)$, $f(x_3)$, $\cdots$, $f(x_k)$. Using the Laplace transform convolution theorem, we get

$$F_{X_k}(s) = \prod_{i=1}^{k} F_i(s) \qquad (4.157)$$

where s is the laplace transform variable.

For identical equipment or units, the Eq. (4.157) reduces to

$$F_{X_k}(s) = [F(s)]^k \qquad (4.158)$$

EXAMPLE 7

Assume constant failure rates of equipment and use Eq. (4.158) to obtain a probability density function expression in the time domain for k renewals.

For the case of a constant failure rate, λ, the single unit probability density is given by

$$f(x) = \lambda e^{-\lambda x} \qquad (4.159)$$

where x is time.

The Laplace transform of Eq. (4.159) becomes

$$F(s) = \frac{\lambda}{s + \lambda} \qquad (4.160)$$

By substituting Eq. (4.160) in Eq. (4.158) and taking the inverse Laplace transform of the resulting expression we get the k-stage special Erlangian distribution

$$f_{X_k} = \lambda(\lambda x)^{k-1} e^{-\lambda x}/(k-1)! \qquad (4.161)$$

For further reading on the subject consult references 235 to 246.

4.9 SUMMARY

This chapter has been concerned with the following areas:

1. Supplementary variables technique
2. Interference theory
3. Human reliability
4. Common-cause failures
5. Fault trees
6. Software reliability prediction
7. Renewal theory

References to useful literature related to all of the above areas are presented at the end of this chapter. A comprehensive list of selective references on interference theory, human reliability, common-cause failures, and fault trees may be found in reference 3. Furthermore, a comprehensive list of selected literature on software reliability [42–234] is also included at the end of this chapter. Readers will find all of these references quite useful. Finally, the reader should note that the author has emphasized the structure of the advanced concepts rather than minute detail in this chapter.

4.10 REFERENCES

Supplementary Variables Method

1. D. R. Cox, "The Analysis of Non-Markovian Stochastic Processes, by the Inclusion of Supplementary Variables," *Proc. Camb. Phil. Soc.*, Vol. 51, 1955, pp. 433–441.

2. D. R. Cox, "A Use of Complex Probabilities in the Theory of Stochastic Processes," *Proc. Camb. Phil. Soc.,* Vol. 51, 1955, pp. 313–319.
3. B. S. Dhillon and C. Singh, *Engineering Reliability: New Techniques and Applications,* Wiley, New York, 1981.
4. R. C. Garg, "Dependability of a Complex System Having Two Types of Components," *IEEE Transactions on Reliability,* Vol. R-12, 1963, pp. 11–15.
5. D. P. Gaver, "Time to Failure and Availability of Paralleled Systems with Repair," *IEEE Transactions on Reliability,* Vol. R-12, 1963, pp. 30–38.
6. J. Keilson and A. Kooharian, "On Time Dependent Queuing Processes," *Annals of Mathematical Statistics,* Vol. 31, 1960, pp. 104–112.
7. C. Singh and R. Billinton, *System Reliability Modeling and Evaluation,* Hutchinson, London, 1977.

Interference Theory

8. B. S. Dhillon and C. Singh, *Engineering Reliability: New Techniques and Applications,* Wiley, New York, 1981.
9. B. S. Dhillon, "Mechanical Reliability: Interference Theory Models," *Proceedings Annual Reliability and Maintainability Symposium,* pp. 462–467, 1980. Available from the IEEE.
11. B. S. Dhillon, "Mechanical Component Reliability Under Environmental Stress," *Microelectronics and Reliability,* Vol. 20, 1980, pp. 153–162.
12. B. S. Dhillon, "Stress-Strength Reliability Models," *Microelectronics and Reliability,* Vol. 20, pp. 513–516, 1980.
13. I. S. Gradshteyn and I. M. Ryzhik, *Table of Integrals, Series, and Products,* Academic, New York, 1965.
14. K. C. Kapur and L. R. Lamberson, *Reliability in Engineering Design,* Wiley, New York, 1977.
15. D. Kececioglu, "Why Design By Reliability?," *1968 Annals of Reliability And Maintainability Conference,* IEEE, New York, 1968, p. 491.
16. D. Kececioglu, "Reliability Analysis of Mechanical Components and Systems," *Nucl. Eng. Design,* 19, 1972, pp. 259–290.
17. D. Kececioglu, "Fundamentals of Mechanical Reliability Theory and Applications to Vibroacoustic Failures," *Proceedings of Reliability Design for Vibroacoustic Environments,* ASME, New York, 1974, pp. 1–38.
18. D. Kececioglu, J. W. McKinley, and M. Saroni, "A Probablistic Method of Designing a Specified Reliability into Mechanical Components with Time Dependent Stress and Strength Distribution," The University of Arizona, Tucson, AZ, Jan. 1967. (NASA Report under Contract NGR 03-002-044).
19. *Practical Reliability, Vol. IV: Prediction.* N68.31979, August 1968. Available from NTIS, Springfield, VA 22161.

Human Reliability

20. B. S. Dhillon, "On Human Reliability—Bibliography," *Microelectronics and Reliability,* Vol. 20, 1980, pp. 371–374.
21. E. W. Hagen, (Ed.), "Human Reliability Analysis," *Nuclear Safety,* Vol. 17, 1976, pp. 315–326.
22. D. Meister, "Human Factors in Reliability," in *Reliability Handbook,* G. W. Ireson (ed.), McGraw-Hill, New York, 1966.

23. D. Meister, "The Problem of Human-Initiated Failures," *Eighth National Symposium on Reliability and Quality Control, 1962.*
24. D. Meister, "The Prediction and Measurement of Human Reliability," *Proceedings of the IAS Aerospace Systems Reliability Symposium,* 1962, pp. 205–212.
25. A. B. Pontecorvo, "A Method of Predicting Human Reliability," *Annals of Reliability and Maintainability,* 1965, pp. 337–342.
26. T. L. Regulinski and W. B. Askren, "Mathematical Modeling of Human Performance Reliability," *Proceedings of Annual Symposium on Reliability,* 1969.
27. H. L. Williams, "Reliability Evaluation of the Human Component in Man-Machine Systems," *Electrical Manufacturing,* April 1958.

Common-Cause Failures

28. B. S. Dhillon, "On Common-Cause Failures — Bibliography," *Microelectronics and Reliability,* Vol. 18, 1978, pp. 533–534.
29. B. S. Dhillon and C. L. Proctor, "Common-Cause Failure Analysis of Reliability Networks," *Proceedings of the Annual Reliability and Maintainability Symposium,* IEEE, New York, 1977, pp. 404–408.
30. B. S. Dhillon, "Multi-State Device Redundant Systems with Common-Cause Failures and One Standby Unit," *Microelectronics and Reliability,* Vol. 20, 1980, pp. 411–417.
31. K. N. Fleming, "A Redundant Model For Common-Mode Failures in Redundant Safety Systems," *Proceedings of the Sixth Pittsburgh Annual Modelling and Simulation Conference,* Instrument Society of America, Pittsburgh, 1975, pp. 579–581.
32. W. C. Gangloff, "Common-Mode Failure Analysis," *IEEE Transactions on Power Apparatus and Systems,* 94, 1975, pp. 27–30.
33. "Reactor Safety Study," WASH-1400 (NUREG-75/014), October 1975. Available from NTIS, Springfield, VA 22161.
34. J. R. Taylor, "A Study of Failure Causes Based on US Power Reactor Abnormal Occurrences Reports," *Reliable Nuclear Power Plants,* IAEA-SM-195/16, 1975.

Fault Trees

35. B. S. Dhillon and C. Singh, *Engineering Reliability: New Techniques and Applications,* Wiley, New York, 1981.
36. B. S. Dhillon and C. Singh, "Bibliography of Literature on Fault Trees," *Microelectronics and Reliability,* Vol. 17, 1978, pp. 501–503.
37. "Flow Research Report, Risk Analysis Using the Fault Tree Technique," Flow Research, Inc., 1973.
38. J. B. Fussell, "Fault Tree Analysis-Concepts and Techniques," *Proceedings of the NATO Advanced Study Institute on Generic Techniques of System Reliability Assessment,* Nordhoff, Leiden, Netherlands, 1975.
39. J. B. Fussell and W. E. Vesley, "A New Methodology for Obtaining Cut Sets for Fault Trees," *Trans. Am. Nucl. Sc.,* Vol. 15, 1972, pp. 262–263.
40. R. J. Schroder, "Fault Trees For Reliability Analysis," *Proceedings of the Annual Symposium on Reliability,* IEEE, New York, 1970.
41. S. N. Semanderes, "ELRAFT: A Computer Program for Efficient Logic Reduction Analysis of Fault Trees," *IEEE Trans. Nucl. Sci.,* Vol. 18, 1971.

Software Reliability

Books

42. *Computing Surveys:* Special Issues on Reliable Software I. "Software Validation," *Computing Surveys,* Vol. 8, No. 3, September 1976. II. "Fault-Tolerant Software," *Computing Surveys,* Vol. 8, No. 4, December 1976.
43. *Conference on Language Design for Reliable Software,* sponsored by The Association for Computing Machinery, Raleigh, NC, 1977, Available from the society.
44. J. D. Cooper and M. J. Fischer (eds.), *Software Quality Management,* Petrocelli Books, New York, 1979.
45. B. S. Dhillon and C. Singh, *Engineering Reliability: New Techniques and Applications,* Wiley, New York, 1981.
46. *International Conference on Reliable Software Proceedings,* Los Angeles, April 1975. Sponsored by IEEE. Available from the IEEE. New York.
47. *International Symposia on Fault-Tolerant Computing,* sponsored by the IEEE Computer Society, Fault Tolerant Technical Committee. Available from the IEEE, New York.
48. H. Kopetz, *Software Reliability,* Macmillan, New York, 1979.
49. D. K. Lloyd and M. Lipow, *Reliability: Management, Methods, and Mathematics,* 2nd ed. Prentice-Hall, Englewood Cliffs, NJ, 1979.
50. G. J. Myers, *Software Reliability: Principles and Practices,* Wiley-Interscience, New York, 1976.
51. G. J. Myers, *Reliable Software Through Composite Design,* Petrocelli/Charter, New York, 1975.
52. *Software Engineering Conference Proceedings,* sponsored by IEEE, September 1975, 1976. Available from the IEEE, New York.
53. "Software Reliability Vols. I & II," Infotech State of the Art Report, published by Infotech International Limited, Maidenhead, England, 1977.
54. *Symposium on Computer Software Reliability,* sponsored by IEEE, New York, 1973.
55. T. A. Thayer, M. Lipow, and E. C. Nelson, "Software Reliability: A Study of Large Project Reality," TRW Series of Software Technology, Vol. 2, North-Holland, New York, 1978.

Articles

56. Akiyama, "An Example of Software System Debugging," *Proceedings IFIP Congress, Vol. 1,* North-Holland, Amsterdam, August 1971, pp. 353–359.
57. S. J. Amster and M. L. Shooman, "Software Reliability: An Overview," in *Reliability and Fault-Tree Analysis,* Barlow, Fussel, and Singapurwallah, (eds.), SIAM, Philadelphia, 1975, pp. 655–685.
58. P. G. Anderson, "Redundancy Techniques for Software Quality," *Annual Reliability and Maintainability Symposium,* 1978, pp. 86–93.
59. T. Anderson, "Software Fault-Tolerance: A System Supporting Fault-Tolerant Software," Software Reliability, Infotech State-of-the-Art Report, Vol. 2, Infotech International Limited, Maidenhead, England, 1977, pp. 1–14.
60. A. Avizienis, "Computer System Reliability: An Overview," Computer Systems Reliability, Infotech State-of-the-Art Report, Maidenhead, 1974.
61. A Avizienis, "Fault-Tolerance and Fault-Intolerance: Complimentary Approaches to Reliable Computing," *Proceedings International Conference on Reliable Software,* 1975, pp. 458–464.
62. R. G. Bennetts, "A Comment on Reliability Evaluation of Software," *1973 NATO Generic Conference,* NATO Advanced Study Institute on Generic Techniques of System Reliability Assessment. North-Holland, New York, 1974.

63. W. R. Bezanson, "Reliable Software Through Requirements Definition Using Data Abstractions," *Microelectronics and Reliability,* Vol. 17, No. 1, 1976, pp. 85–92.

64. B. S. Bloom, M. J. McPheters, and S. H. Tsiang, "Software Quality Control," *IEEE Symposium on Computer Software Reliability,* 1973. Available from the IEEE, New York.

65. B. W. Boehm, "Software and Its Impact: A Quantitative Assessment," *Datamation,* vol. 19 May 1973. pp. 48–59.

66. B. W. Boehm, R. K. McClean, and D. B. Urfrig, "Some Experience with Automated Aids to the Design of Large-Scale Reliable Software," *Proceedings of the International Conference on Reliable Software,* 1975, pp. 105–113.

67. B. W. Boehm, J. R. Brown, and M. Lipow, "Quantitative Evaluation of Software Quality," *Proceedings of the Second International Conference on Software Engineering,* 1976.

68. B. W. Boehm, J. R. Brown, H. Kaspar, M. Lipow, G. J. McLeoad, and M. J. Merrit, "Characteristics of Software Quality," TRW, Redondo Beach, CA, TRW-SS-73-09, December 1973.

69. W. D. Brooks and P. W. Weiler, "Software Reliability Analysis," IBM Technical Report FSD-77-0009, International Business Machines Corporation, Federal Systems Division, Gaithersburg, MD.

70. J. R. Brown and M. Lipow, "Testing for Software Reliability," *Proceedings of the International Conference on Reliable Software,* 1975, pp. 518–527.

71. J. R. Brown and H. N. Buchanan, "The Quantitative Measurement of Software Safety and Reliability," TRW Report, SDP 1776, 2H Aug 1973. Redondo Beach, CA.

72. J. R. Brown, "A Case for Software Test Tools," *Proceedings of the TRW Symposium on Reliable Cost-Effective, Secure Software,* TRW Software Series SS-74-14, March 1974. Redondo Beach, CA.

73. F. J. Buckley, "Software Testing — A Report from the Field," *Record IEEE Symposium on Computer Software Reliability,* April 1973. Available from the IEEE, New York.

74. L. C. Carpenter and L. L. Tripp, "Software Design Validation Tool," *Proceedings International Conference on Reliable Software,* 1975, pp. 395–400.

75. H. Y. Chang, "Hardware Maintainability and Software Reliability of Electronic Switching Systems," Computer Systems Reliability, Infotech State-of-the-Art Report, C. J. Bunyan (ed.), 1974.

76. R. C. Cheung, "A User-Oriented Software Reliability Model," *IEEE Transactions on Software Engineering,* Vol. SE-6, No. 2, March 1980.

77. L. Chen and A. Avizienis, "*N*-Version Programming: A Fault-Tolerant Approach to Reliability of Software Operation," Digest of Papers, FTC-8, June 1978, pp. 3–9.

78. H. B. Chenoweth, "Modified Musa Theoretic Software Reliability," *Annual Reliability and Maintainability Symposium,* 1981, pp. 353–356.

79. J. R. Connet, E. J. Pasternak, and B. D. Wagner, "Software Defenses in Real-Time systems," *2nd IEEE Symposium on Fault-Tolerant Computing,* June 1972, pp. 94–99.

80. J. D. S. Coutinho, "Software Reliability Growth," *Record IEEE Symposium on Computer Software Reliability,* 1973, pp. 58–64.

81. G. R. Craig, W. L. Hetrick, M. Lipow, and T. A. Thayer, "Software Reliability Study," TRW Systems Group, Interim Technical Report, RADC-TR-74-250, October 1974. Redondo Beach, CA.

82. L. M. Culpepper, "A System for Reliable Engineering Software," *IEEE Transactions on Software Engineering,* Vol. SE-1, No. 2, June 1975, pp. 174–178.

83. P. J. Denning, "Fault-Tolerant Operating Systems," *Computing Surveys,* Vol. 8, No. 4, Dec 1976, pp. 359–389.

84. J. C. Dickson et al., "Quantitative Analysis of Software Reliability," *Annual Reliability and Maintainability Symposium,* 1972, pp. 148–157.

85. P. H. Dunn, R. S. Ullman, "A Workable Software Quality/Reliability Plan," *Annual Reliability and Maintainability Symposium,* 1978, pp. 210–217.
86. L. Duvall, J. Martens, D. Swearingen, J. and Donahoo, "Data Needs for Software Reliability Modeling," *Annual Reliability and Maintainability Symposium,* 1980, pp. 200–207.
87. W. R. Elmendorf, "Fault-Tolerant Programming," *International Symposium on Fault-Tolerant Computing,* 1972, pp. 79–83.
88. B. Elspas, K. N. Lewitt, and R. J. Waldinger, "An Assessment of Techniques for Proving Program Correctness," *Computing Surveys,* Vol. 4, No. 2, June 1972, pp. 99–147.
89. B. Elspas, M. W. Green, and K. N. Lewitt, "Software Reliability," *Computer,* Vol. 4, No. 1, 1971, pp. 21–27.
90. A. Endres, "An Analysis of Errors and Their Causes in System Programs," *IEEE Transactions on Software Engineering,* Vol. SE-1, No. 2, June 1975, pp. 140–149.
91. J. G. Estep, "A Software Reliability and Availability Model," *IEEE Symposium on Computer Software Reliability,* 1973, p. 101.
92. A. R. Ets and W. E. Thompson, "Specification and Control of Risk in Software Reliability Testing," *2nd International Parametrics Conference,* Cherry Hill, NJ, April 1980.
93. A. R. Ets and W. E. Thompson, "Combined Hardware and Software Reliability Specification and Test," *5th International Conference on Software Engineering,* San Diego, March 1981.
94. M. E. Fagan, "Design and Code Inspection to Reduce Errors in Program Development," *IBM Systems Journal,* No. 3, 1976, pp. 202–203.
95. K. F. Fischer and M. G. Walker, "Improved Software Reliability Through Requirements Verification," *IEEE Transactions on Reliability,* Vol. R-28, No. 3, August 1979, pp. 233–240.
96. I. Fischler, C. Firschem, and A. F. Drew," Distinct Software: An Approach to Reliable Computing," Lockheed Research Report 3251, Lockheed Space and Missile Labs., Hanover, Palo Alto, CA 94034.
97. D. Floh and T. Isutani, "FADEBUG-1, A New Tool for Program Debugging," *Record of the IEEE Symposium on Computer Software Reliability,* 1973.
98. E. Forman and N. Z. Singapurwalla, "Optimal Time Intervals for Testing Hypotheses on Computer Software Errors," *IEEE Transactions on Reliability,* Vol. R-28, No. 3, August 1979, pp. 250–253.
99. E. H. Forman, Statistical Models and Methods for Measuring Software Reliability, D. Sc. Dissertation, School of Engineering and Applied Science, George Washington University, Washington, D.C., 1974.
100. L. D. Fosdick and L. J. Osterwell, "Data Flow Analysis in Software Reliability," *Computing Surveys,* September 1976.
101. J. R. Fragola and J. F. Spahn, "Software Error Effect Analysis, a Quantitative Design Tool," *IEEE Symposium on Computer Software Reliability,* 1973, pp. 90–93.
102. J. D. Gannon and J. J. Horning, "The Impact of Language Design on Reliable Software," *International Conference on Reliable Software,* 1975, pp. 10–22.
103. S. L. Gerhart, "A Unified View of Current Program Testing and Proving—Theory and Practice," Software Reliability, Infotech State-of-the-Art Report, Infotech International Limited, Maidenhead, England, 1977, pp. 71–100.
104. S. L. Gerhart and L. Yelowitz, "Observations of Fallibility in Applications of Modern Programming Methodologies," *IEEE Transactions on Software Engineering* Vol. SE-2, No. 3, September 1976, pp. 195–207.
105. T. Gilb, "Software Metrics Technology: Some Unconventional Approaches to Reliable Software," Software Reliability, Infotech State-of-the-Art Report, Infotech International Limited, Maidenhead, England, 1977, pp. 101–115.

106. T. Gilb, "Distinct Software: A Redundancy Technology for Reliable Software," Software Reliability, Infotech State-of-the-Art Report, Infotech International Limited, Maidenhead, England, 1977, pp. 117–133.

107. A. L. Goel and K. Okumoto, "Time-Dependent Error-Detection Rate Model for Software Reliability and Other Performance Measures," *IEEE Transactions on Reliability*, Vol. R-28, No. 3, August 1979, pp. 206–211.

108. A. L. Goel and K. Okumoto, "Bayesian Software Prediction Models, Vol. 1: An Imperfect Debugging Model for Reliability and Other Quantitative Measures of Software Systems," Rome Air Development Center, N.Y., RADC-78-155, 1978.

109. D. I. Good, R. L. London, and W. W. Bledsoe, "An Interactive Program Verification System," *International Conference on Reliable Software*, 1975. Also published in *IEEE Transactions in Software Engineering*, Vol. SE-1, No. 1, March 1975, pp. 59–67.

110. J. B. Goodenough and S. L. Gerhart, "Toward a Theory of Test Data Selection," *International Conference on Reliable Software*, 1975. Available from the IEEE, New York.

111. "Guidelines for Software Quality Assurance," TRW Systems, Redondo Beach, CA, Group Report SDP-3055, September 1974.

112. P. A. Hamilton and J. D. Musa, "Measuring Reliability of Computer Center Software," *Proceedings of the Third International Conference on Software Engineering*, Atlanta, 1978, pp. 29–38.

113. S. L. Hantler and J. C. King, "An Introduction to Proving the Correctness of Programs," *Computing Surveys*, Vol. 8, No. 3, September 1976, pp. 331–353.

114. R. D. Haynes and W. E. Thompson, "Combined Hardware and Software Availability," *Annual Reliability and Maintainability Symposium*, 1981, pp. 365–370.

115. H. Hecht, "Measurement, Estimation, and Prediction of Software Reliability," Volume on Software Engineering Techniques, Infotech International Series, 1977.

116. H. Hecht, "Fault-Tolerant Software for Real-time Applications, *Computing Surveys*, Vol. 8, December 1976, pp. 391–407.

117. H. Hecht, W. A. Strum, and S. Trattner, "Reliability Measurement During Software Development," A Collection of Technical Papers; *AIAA/NASA/IEEE/ACM Computers in Aerospace Conference*, November 1977, pp. 404–412.

118. H. Hecht, "Fault-Tolerant Software," *IEEE Transactions on Reliability*, Vol. R-28, No. 3, August 1979, pp. 227–232.

119. W. C. Hetzel, An Experimental Analysis of Program Verification Methods, Thesis, University of North Carolina, Chapel Hill, NC, 1976.

120. J. Heutinck and D. Smith, "Software Reliability Report," PRC Systems Sciences Company, Los Angeles, Report PRC R-1470, September 1972.

121. J. J. Horning, H. C. Lauer, P. M. Melliar-Smith, and B. Randell, "A Program Structure for Error Detection and Recovery," Lecture Notes, *Computer Science*, Vol. 16, Springer-Verlag, pp. 171–187.

122. W. E. Howden, "Empirical Studies of Software Validation," *Microelectronics and Reliability*, Vol. 19, pp. 39–47, 1979.

123. P. P. Howley, "Software Quality Assurance for Reliable Software," *Annual Reliability and Maintainability Symposium*, 1968, pp. 73–78.

124. Z. Jelinski and P. B. Morando, "Status Report on Software Reliability Study for 1971," McDonnell-Douglas Astronautics Company Report No. G3004, April 1972.

125. Z. Jelinski and P. Moranda, "Software Reliability Research," in *Statistical Computer Performance Evaluation*, W. Freiberger (ed.), Academic, New York, 1972, pp. 465–484.

126. J. P. Johnson, "Software Reliability Measurement Study," GIDEP Report No. 195. 45.00.00-BA-01. Available from GIDEP Operations Center, Corona, CA, 91720, 1975.

127. S. Katz and Z. Manna, "Towards Automatic Debugging or Programs," *International Conference on Reliable Software*, 1975, pp. 143–155.

128. K. H. Kim and C. V. Ramamoorthy, "Fault-Tolerant Parallel Programming and Its Supporting system Architecture," *AFIP Conference Proceedings*, Vol. 45, NCC 1976, pp. 413–423.

129. J. King, "A New Approach to Program Testing," *International Conference on Reliable Software*, 1975.

130. J. C. King, "Proving Programs to Be Correct," *International Symposium on Fault-Tolerant Computing*, 1971, pp. 130–133.

131. M. B. Kline, "*Software and Hardware R & M: What Are the Differences?,*" *Annual Reliability and Maintainability Symposium*, 1980, pp. 179–185.

132. R. K. Klobert, "Quest for Reliable Software," *Proceedings of the Summer Computer Simulation Conference*, July 1977, pp. 700–704.

133. H. Kopetz, "System Reliability and Software Redundancy," Real-Time Software, Infotech State of the Art Report, J. C. Spencer, (ed.), 1976.

134. K. W. Krause, R. W. Smith, and M. A. Goodwin, "Optimal Software Test Planning Through Automated Network Analysis," *IEEE Symposium on Computer Software Reliability*, 1973.

135. C. Landrault and J. C. Laprie, "Reliability and Availability Modeling of Systems Featuring Hardware and Software Faults," *7th Annual International Conference on Fault-Tolerant Computing*, June 1977, pp. 10–15.

136. L. J. LaPadula, "Engineering of Quality Software Systems, Vol. III — Software Reliability Modelling and Measurement Techniques," MITRE Corporation, RADC-TR-74-325, Vol. VIII, Final Technical Report, January 1975. (Under RADC contract F19628-C-73-0001, "Software Reliability and Timeliness").

137. T. A. Linden, "A Summary of Progress Toward Proving Program Correctness," *1972 Fall Joint Comput. Conf., AFIPS Conf. Proc.*, Vol. 41, Washington, D.C., Spartan, 1972.

138. M. Lipow and T. A. Thayer, "Prediction of Software Failure," *Annual Reliability and Maintainability Symposium*, 1977, pp. 489–494.

139. B. H. Liskov, "A Design Methodology for Reliable Software Systems," presented at the *Fall Joint Computer Conference*, December 1972, *AFIPs Conference Proceedings*, Vol. 41, pp. 191–199.

140. B. Littlewood, "Software Reliability Model for Modular Program Structure," *IEEE Transactions on Reliability*, Vol. R-28, No. 3, August, 1979, pp. 241–246.

141. B. Littlewood, "A Critique of the Jelinski-Moranda Model for Software Reliability," *Annual Reliability and Maintainability Symposium*, 1981, pp. 357–364.

142. B. Littlewood, "How to Measure Software Reliability and How Not To," *IEEE Transactions on Reliability*, Vol. R-25, No. 2, June 1979, pp. 103–110.

143. B. Littlewood and J. L. Verrall, "A Bayesian Reliability Growth Model for Computer Software," *Record IEEE Symposium on Computer Software Reliability*, 1973, pp. 70–77.

144. B. Littlewood, "A Bayesian Differential Debugging Model for Software Reliability," *Proceedings Workshop on Quantitative Software Models*, Kiamesha, NY, October 1979, pp. 170–181.

145. B. Littlewood, "A Reliability Model for Markov Structured Software," *International Conference on Reliable Software*, 1975, pp. 204–207.

146. B. Littlewood, "A Semi-Markov Model for Software Reliability with Failure Costs," *Proceedings Symposium on Software Engineering*, New York, April 1976, pp. 281–300.

147. B. Littlewood, "Theories of Software Reliability: How Good Are They, and How Can They Be Improved?," *IEEE Transactions on Software Engineering*, Vol. SE-6, No. 5, September 1980.

148. R. L. London, "Bibliography on Proving the Correctness of Computer Programs," *Machine Intelligence*, 5, Edinburgh University Press, 1970, pp. 569–580.

149. R. L. London, "Proving Programs Correct: Some Techniques and Examples," *BIT*, Vol. 10, No. 2., May 1969, pp. 168–182.

150. R. L. London, "A View of Program Verification," *International Conference on Reliable Software*, 1975, pp. 534–545.

151. R. L. London, "Software Reliability Through Proving Programs Correct," *International Symposium on Fault-Tolerant Computing*, March 1971, pp. 125–129.

152. W. H. MacWilliams, "Reliability of Large Real Time Control Software Systems," *Record IEEE Symposium on Computer Software Reliability*, 1973, pp. 1–6.

153. Z. J. Manna, "The Correctness of Programs," *Journal of Computer and System Sciences*, Vol. 3, No. 2, May 1969, pp. 119–127.

154. J. A. McCall, "An Introduction to Software Quality Metrics," Chapter 8 in *Software Quality Management*, Cooper and Fischer (eds.), Petrocelli Books, New York, 1979.

155. J. A. McCall, P. K. Richards, and F. Walters, "Factors in Software Quality, Concept and Definitions of Software Quality," RADC-TR-77-369, 3 vols., November 1977, Rome Air Development Center, USA.

156. J. McKissick, and R. A. Price, "The Software Development Notebook — A Proven Technique," *Annual Reliability and Maintainability Symposium*, 1979, pp. 346–351.

157. M. J. Merritt et al., "Characteristics of Software Quality," Report 25201-6001-RU-00, TRW Systems, Redondo Beach, CA, December 1973.

158. E. F. Miller, "Program Testing Tools and Their Uses," Software Reliability, Infotech State-of-the-Art Report, Infotech International Limited, Maidenhead, England, 1977, pp. 183–216.

159. E. F. Miller, "Testing for Software Reliability," Software Reliability, Infotech-State-of-the-Art Report, Infotech International Limited, Maidenhead, England, 1977, pp. 217–241.

160. H. D. Mills, "On the Development of Large Reliable Programs," *Record IEEE Symposium on Computer Software Reliability*, 1973, pp. 155–159.

161. H. D. Mills, "How to Write Correct Programs and Know It," *International Conference on Reliable Software*, 1975, pp. 363–370.

162. I. Miyamato, "Software Reliability in On Line Real Time Environment," *International Conference on Reliable Software*, 1975, pp. 194–203.

163. P. B. Noranda and Z. Jelinski, "Software Reliability Predictions," *6th Triennial World Congress, IFAC*, in *Proceedings IFAC/75*, August 1975.

164. P. B. Moranda, "Prediction of Software Reliability During Debugging," *Annual Reliability and Maintainability Symposium Proceedings*, 1975, pp. 327–332.

165. R. B. Mulock, "Program Correctness, Software Reliability, Today's Capabilities," *Proceedings International Symposium on Fault-Tolerant Computing*, March 1971, pp. 137–139.

166. R. B. Mulock, "Software Reliability Engineering," *Proceedings of Annual Reliability and Maintainability Symposium*, 1972.

167. R. B. Mulock, "Software Reliability," *Annual Symposium on Reliability Proceedings*, 1969, p. 495.

168. J. D. Musa, "Software Reliability Measures Applied to Systems Engineering," *AFIPS Conference Proceedings; National Computer Conference*, June 1979, pp. 941–946.

169. J. D. Musa, "A Theory of Software Reliability and Its Application," *IEEE Transactions on Software Engineering*, Vol. SE-1, No. 3, Sept. 1975, pp. 312–327.

170. J. D. Musa, "Progress in Software Reliability Measurement," *2nd Software Life Cycle Management Workshop Proceedings*, Atlanta, August 1978.

171. J. D. Musa, "Measuring Software Reliability," *ORSA/TIMS Journal*, May 1977, pp. 1–25.

172. J. D. Musa, "Validation of Execution Time Theory of Software Reliability," *IEEE Transactions on Reliability*, Vol. R-28, No. 3, August 1979, pp. 181–191.

173. G. Myers, *Reliable Software Through Composite Design*, Petrocelli/Charter, New York, 1975.

174. P. Naur, "Software Reliability," Infotech State-of-the-Art Report, Infotech International Limited, Maidenhead, England, 1977, pp. 243–251.

175. E. C. Nelson, "A Statistical Basis for Software Reliability Assessment," TRW Software Series, TRW-SS-73-03, Redondo Beach, CA, March 1973.

176. E. C. Nelson, "Estimating Software Reliability from Test Data," *Microelectronics and Reliability*, Vol. 17, No. 1, January 1978, pp. 67–74.

177. J. L. Ogdin, "Designing Reliable Software," *Datamation*, 18, 1972, pp. 71–78.

178. M. R. Paige and E. F. Miller, "Methodology for Software Validation—A Survey of the Literature," General Research Co., RM 1549, March 1972.

179. D. L. Parnas, "The Influence of Software Structure on Reliability," *International Conference on Reliable Software*, 1975, pp. 353–362.

180. "Power Plant Computer Reliability Survey," *IEE Transactions on Power Apparatus and Systems*, Vol. PAS-97, No. 4, 1978, pp. 1115–1123.

181. C. V. Ramamoorthy, R. E. Meeker, and J. Turner, "Design and Construction of An Automated Software Evaluation System," *IEEE International Symposium on Computer Software Reliability*, 1973, pp. 28–37.

182. C. V. Ramamoorthy, R. C. Cheung, and K. H. Kim, "Reliability and Integrity of Large Computer Programs," Computer Systems Reliability, Infotech State-of-the-Art Report, C. J. Bunyan (ed.), 1974, pp. 617–710.

183. C. V. Ramamoorthy and S. F. Ho, "Testing Large Software With Automated Software Evaluation Systems," *IEEE Transactions on Software Engineering*, Vol. SE-1, No. 1, 1975, pp. 46–58.

184. C. V. Ramamoorthy, S. F. Ho, and W. T. Chen, "On the Automated Generation of Program Test Data," *IEEE Transactions on Software Engineering*, December 1976.

185. B. Randell, "System Structure for Software Fault-Tolerance," *IEEE Transactions on Software Engineering*, Vol. SE-1, June 1975, pp. 220–232.

186. J. C. Rault, "Extension of Hardware Fault Detection Models to the Verification of Software," *Program Test Methods*, Hetzel W. C. (ed.), Prentice-Hall, Englewood Cliffs, NJ, 1973, pp. 255–262.

187. D. J. Reifer, "Automated Aids for Reliable Software," *International Conference on Reliable Software*, 1975, pp. 131–142.

188. D. J. Reifer, "A New Assurance Technology for Computer Software," *Annual Reliability and Maintainability Symposium*, 1976, pp. 446–451.

189. L. Robinson, "On Attaining Reliable Software for a Secure Operating System," *International Conference on Reliable Software*, 1975, pp. 267–284.

190. H. Roggenbauer, "Software Reliability for Computerized Control and Safety Systems in Nuclear Power Plants," *Atomic Energy Review*, Vol. 15, No. 4, 1977, pp. 793–800.

191. J. A. Ronaback, "Crisis in Software Reliability," *16th Annual Quality Control Forum Proceedings*, April 1973, Montreal, pp. 36–46.

192. J. A. Ronaback, "Software Reliability — How it Affects System Reliability," *Microelectronics and Reliability*, Vol. 14, 1975, pp. 121–140.

193. J. P. Roth, "Diagnosis of Software Faults," *IEEE International Computer Society Conference*, 1971, p. 83.

194. R. J. Rubey, "Planning for Software Reliability," *Annual Reliability and Maintainability Symposium*, 1977, pp. 495–499.

195. R. J. Rubey, "Quantitative Aspects of Software Validation," *International Conference on Reliable Software*, 1975, pp. 246–251.

196. J. L. Sauter, "Reliability in Computer Programs," *Mech. Eng.,* Vol. 91, February 1969, pp. 24–27.
197. R. E. Schafer et al., "Validation of Software Reliability Models," Griffiths AFB, NY, Rome Air Development Center, RADC-TR-79-147, June 1979.
198. G. J. Schick and R. W. Wolverton, "Achieving Reliability in Large-Scale Software Systems," *Annual Reliability and Maintainability Symposium,* 1974, pp. 302–319.
199. G. J. Schick and R. W. Wolverton, "An Analysis of Competing Software Reliability Models," *IEEE Transactions on Software Engineering,* Vol. SE-4, No. 2, March 1978, pp. 104–120.
200. B. Schneiderman, "Human Factors Experiments for Developing Quality Software," Software Reliability, Infotech State-of-the-Art Report, Infotech International Ltd., Maidenhead, England, 1977, pp. 261–276.
201. N. F. Scheidewind, "Application of Program Graphs and Complexity Analysis to Software Development Testing," *IEEE Transactions on Reliability,* Vol. R-28, No. 3, August 1979, pp. 192–198.
202. N. F. Schneidewind, "The Application of Hardware Reliability Principles to Computer Software," Chapter II in *Software Quality Management,* Cooper and Fishcher (eds.), Petrocelli Books, New York, 1979.
203. N. F. Schneidewind, "An Approach to Software Reliability Prediction and Quality Control," *Fall Joint Computer Conference, AFIPS Conference Proceedings,* Vol. 41, 1972, pp. 837–847.
204. N. F. Schneidewind, "Methodology for Software Reliability Prediction and Quality Control," NTIS Report AD 754337, National Technical Information Service, Springfield, VA, 22161.
205. N. F. Scheidewind, "Analysis of Error Processes in Computer Software," *International Conference on Reliable Software,* 1975, pp. 337–346.
206. M. L. Shooman, "Structural Models for Software Reliability Prediction," *2nd International Conference on Software Engineering Proceedings,* San Francisco, 1976, pp. 268–275.
207. M. L. Shooman, "Software Reliability: Measurement and Models," *Annual Reliability and Maintainability Symposium Proceedings,* 1975, pp. 485–491.
208. M. L. Shooman and M. I. Bolsky, "Types, Distribution, and Test Correction Times for Programming Errors," *International Conference on Reliable Software,* 1975, pp. 347–362.
209. M. L. Shooman, "Operational Testing and Software Reliability During Program Development," *Record IEEE Symposium on Computer Software Reliability,* 1973, pp. 51–57.
210. M. L. Shooman, "Probabilistic Models for Software Reliability Prediction," in *Statistical Computer Performance Evaluation,* W. Freiberger (ed.), Academic, New York, 1972, pp. 485–502.
211. M. L. Shooman, "Software Reliability: Analysis and Prediction," *Proceedings of the NATO Advanced Study Institute on Systems Reliability,* July 17–28, 1973.
212. M. L. Shooman and A. K. Trivedi, "A Many State Markov Model for Computer Software Performance Parameters," *IEEE Transactions on Reliability,* Vol. R-25, No. 2, June 1976.
213. I. M. Soi and K. Gopal, "Some Aspects of Reliable Software Packages," *Microelectronics and Reliability,* Vol. 19, 1979, pp. 379–386.
214. I. M. Soi and K. Gopal, "Hardware vs. Software Reliability—A Comparative Study," *Microelectronics and Reliability,* Vol. 20, pp. 881–885.
215. L. J. Stucki and G. L. Foshee, "New Assertion Concepts for Self-Metric Software Validation," *International Conference on Reliable Software,* 1975, pp. 59–71.
216. N. Sugiura, M. Yamamoto, and T. Shiino, "On the Software Reliability," *Microelectronics and Reliability,* Vol. 13, 1973, pp. 529–533.
217. A. N. Sukert, "Empirical Validation of Three Software Error Prediction Models," *IEEE Transactions on Reliability,* Vol. R-28, No. 3, August 1979, pp. 199–205.

218. A. N. Sukert, "All Multi-Project Comparison of Software Reliability Models," *Proceedings of the AIAA Computers in Aerospace Conference,* October–November 1977, pp. 413–421.
219. A. N. Sukert, "An Investigation of Software Reliability Models," *Annual Reliability and Maintainability Symposium,* 1977, pp. 478–484.
220. S. G. Szabo, "A Schema for Producing Reliable Software," *International Symposium on Fault-Tolerant Computing,* 1975, pp. 151–155.
221. W. E. Thompson and P. O. Chelson, "Software Reliability Testing for Embedded Computer Systems," *IEEE Transactions on Reliability,* Vol. R-28, No. 3, September 1979, pp. 201–202.
222. A. K. Trivedi and M. L. Shooman, "A Many-State Markov Model for the Estimation and Prediction of Computer Software Performance Parameters," *International Conference on Reliable Software,* 1975, pp. 208–220.
223. C. R. Vick, "Specification for Reliable Software," *IEEE Electronics and Aerospace Systems Convention,* October 7–9, 1974.
224. F. W. Von Henke and D. C. Luckam, "A Methodology for Verifying Programs," *International Conference on Reliable Software,* 1975, pp. 156–164.
225. J. K. Wall and P. A. Ferguson, "Pragmatic Software Reliability Prediction," *Annual Reliability and Maintainability Symposium,* 1977, pp. 485–489.
226. G. F. Walters and J. A. McCall, "The Development of Metrics for Software R & M," *Annual Reliability and Maintainability Symposium,* 1978, pp. 79–85.
227. B. Wegbreit, "Some Remarks on Software Correctness and Program Verification," Software Reliability, Infotech State-of-the-Art Report, Infotech International Ltd., Maidenhead, England, 1977, pp. 397–401.
228. L. Weissman, "An Interface System for Improving Reliability of Software Systems," *IEEE International Symposium on Computer Software Reliability,* 1973, pp. 136–140.
229. H. E. Williams et al., "Software Systems Reliability: A Raytheon Project History," RADC-TR-77-188, Technical Report, Rome Air Development Center, Air Force Systems Command, Griffiths Air Force Base, NY, June 1977.
230. R. D. Williams, "Managing the Development of Reliable Software," *International Conference on Reliable Software,* 1975, pp. 3–8.
231. O. L. Williamson, G. G. Dorris, A. J. Rybert, and W. E. Straight, "A Software Reliability Program," *Reliability and Maintainability Symposium,* 1970, pp. 420–428.
232. R. W. Wolverton and G. J. Schick, *"Assessment of Software Reliability,"* TRW Software Series, TRW-SS-73-04, September 1972.
233. R. W. Wolverton, "Software Reliability Modeling, Prediction and Measurement Methodology," TRW Systems Group Rep. 6600.7–99/74, Redondo Beach, CA, July 1974.
234. E. Yourdon, "Reliability Measurement for Third Generation Computer Systems," *Proceedings Annual Reliability Symposium,* IEEE, 1972.

Renewal Theory

235. H. Ascher, "Hazard Functions, Renewal Rates and Peril Rates," *Proceedings Reliability and Maintainability Conference,* 1970, pp. 414–426.
236. R. E. Barlow and L. C. Hunter, "Reliability Analysis of A One-Unit System," *Operations Research,* Vol. 9, pp. 200–208.
237. D. R. Cox, *Renewal Theory,* Methuen, London, 1962.
238. W. Feller, "Fluctuation Theory of Recurrent Events," *Trans. Amer. Math. Soc.,* Vol. 67, 1949, pp. 98–119.
239. M. Frechet, *Statistical Self-Renewing Aggregates,* Fouad I University Press, Cairo, 1949.
240. J. H. K. Kao, "Discrete-Time Renewal Theory and Applications," *Proceedings Annual Symposium on Reliability,* 1969, pp. 522–534.

241. C. A. Krohn, "Hazard V. Renewal Rate of Electronic Items, *"IEEE Transactions on Reliability,* Vol. R-18, No. 2, May 1969, pp. 64–73.

242. A. Lotka, "A Contribution to the Theory of Self-Renewing Aggregates, with Special Reference to Industrial Replacement," *Ann. Math. Statist.,* Vol. 10, 1939, pp. 1–25.

243. M. Muntner, "Discrete Renewal Processes," *IEEE Transactions on Reliability,* Vol. R-20, May 1971, pp. 46–51.

244. R. Pyke, "Markov Renewal Processes: Definitions and Preliminary Properties," *Ann. Math. Stat.,* Vol. 32, 1961, pp. 1231–1242.

245. M. L. Shooman, *Probabilistic Reliability: An Engineering Approach,* McGraw-Hill, New York, 1968.

246. W. L. Smith, "Renewal Theory and Its Ramifications," *J. R. Statist. Soc.,* B, 20, 1958, pp. 243–302.

5
Reliability Optimization

5.1 INTRODUCTION

Reliability engineering has been an invaluable tool of the last few decades in solving many unreliability oriented system problems. One of the functions of a reliability analyst is to present recommendations to managers regarding the subsystem constraints such as weight, volume, cost, and so on, in order to achieve maximum system reliability.

Nowadays reliability is one of the key design factors when designing complex, critical, and expensive systems. To produce a highly reliable system there are basically two paths that can be followed. The first is to use redundancy, whereas the second is to manufacture high reliability components. In today's competitive economy to overdesign the system to assure high reliability is not as feasible as it once was. System reliability optimization, subject to various resource and other constraints, is now practiced in industry.

According to reference 96 system reliability optimization subject to cost constraints was perhaps first formulated in reference 112. Other authors [36, 147, 148, 102, 98, 96, 103, and 101] have studied the system reliability optimization problem subject to constraints. These researchers assume that the components are of high reliability.

A paper in reference 146 on optimization techniques for system reliability with redundancy lists most of the published literature on the subject. The following is the list of articles [146] on the system reliability optimization:

1. Series system [8, 14, 17, 19, 38, 43, 51, 55, 112, 131, 133, 147, 153, 154]
2. Parallel [7, 51, 55, 112, 131, 147, 153, 154]
3. Series-Parallel [51, 55, 101, 112, 131, 133, 153, 154]
4. Parallel-Series [3, 8, 16, 17, 19, 23, 35, 36, 38, 42, 43, 51, 55, 62, 69, 76, 82]

5. Stand-by [17, 153, 154, 55, 69, 108, 112, 127, 131, 133]
6. Non–Series-Parallel (including bridge networks) [1, 7, 23, 63, 148]

Other related literature is listed at the end of this chapter. This bibliography is reasonably complete. This chapter is divided into the following sections:

1. Optimization techniques
2. Redundancy optimization
3. Reliability optimization of redundant configurations with two failure mode components
4. Mechanical component reliability optimization
5. Optimum maintenance policy

5.2 OPTIMIZATION TECHNIQUES

This section briefly discusses the optimization techniques because the main intent of this chapter is reliability optimization. It is assumed here that most readers are familiar with the commonly used optimization methods. Many of these techniques are applied to optimize the system reliability. Optimization techniques such as dynamic programming, geometric programming, integer programming, linear programming, the maximum principle, the Lagrangian procedure, the generalized reduced gradient, the sequential unconstrained minimization technique, etc., are presented in references 156 to 170. Many of these methods are applied to the reliability problems in references 1 to 155.

5.3 REDUNDANCY OPTIMIZATION

This section presents parallel, series, series-parallel, non–series-parallel, etc., network reliability optimization. Statistical independence is always assumed among the redundant configuration components.

5.3.1 Parallel System

By introducing component redundancy, one can acquire the desired reliability of a system. To achieve the desired reliability of a system, the cost factor has to be considered.

We first ask, how much one should pay to increase reliability of a basic system from R_b to R_p when only parallel redundancy of the basic system can be used. Although, the basic system could have multiple elements in series form, in this text we consider only one element basic system.

The independent and identical n-units parallel system reliability R_p, is given by [112]

$$R_p = 1 - (1 - R_b)^n \qquad (5.1)$$

where R_b is the unit or basic system reliability.
Equation (5.1) can be rewritten

$$(1 - R_b)^n = 1 - R_p \qquad (5.2)$$

Taking the logarithm of Eq. (5.2) and rearranging

$$n = \frac{\log(1 - R_p)}{\log(1 - R_b)} \qquad (5.3)$$

Now suppose that

$$c_s = nc_b \qquad (5.4)$$

where c_s is the cost of the parallel or final (total) system
　　　　c_b is the unit cost or the cost of the original basic system

Substituting Eq. (5.4) in Eq. (5.3) we get

$$n = \left(\frac{c_s}{c_b}\right) = \frac{\log(1 - R_p)}{\log(1 - R_b)} \qquad (5.5)$$

The above relationship can be used to obtain the cost figure necessary to increase R_b to the desired final system reliability, R_p.

5.3.2 Series-Parallel Network (Homogeneous case)

A homogeneous case means that each element is identical with the same reliability and cost [112]. This type of network is shown in Fig. 5.1. The basic system has k series elements. Each element reliability is given by $R_b^{1/k}$. We assume that system elements are statistically independent.

Suppose that one would like to increase the basic system reliability from R_b to R_{sp} [the final system (series-parallel network) reliability], which can be obtained by paralleling each basic element with an additional $(n - 1)$ elements. This will achieve the group of parallel elements (subsystem) reliability, $R_{sp}^{1/k}$.

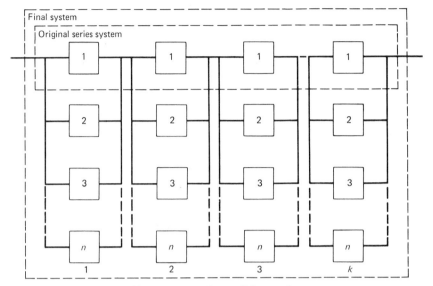

Figure 5.1. A series-parallel network.

Equation (5.6) is developed in the same way as Eq. (5.5)

$$n = \frac{c_s}{c_b} = \frac{\log(1 - R_{sp}^{1/k})}{\log(1 - R_b^{1/k})} \qquad (5.6)$$

The above equation can be used to obtain the cost estimate if one desires to increase R_b to R_{sp}.

5.3.3 Series-Parallel Network (General case)

This model is a generalized version of the model of Section 5.3.2. Here we assume that element costs and reliabilities are nonidentical. Our objective is to maximize reliability at a minimum cost for the optimum number of elements[112]. A basic series system with k nonidentical elements is shown in Fig. 5.2. In Fig. 5.2, R_i and c_i are the ith element reliability and cost, respectively, for $i = 1, 2, 3, \ldots, k$. Suppose the original cost and reliability of the Fig. 5.2 basic series system are c_b and R_b, respectively where the objective is to determine the minimum cost, c_s, for increasing the basic series system reliability from R_b to R_{sp}. To increase reliability, we add the identical elements in parallel to each series element of Fig. 5.2.

Assuming the system failures are statistically independent, the expressions

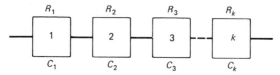

Figure 5.2. A basic series system.

for this standard variational problem are developed as follows. The Fig. 5.2 basic system reliability and cost are given by

$$R_b = \prod_{i=1}^{k} R_i \qquad (5.7)$$

and

$$c_b = \sum_{i=1}^{k} c_i \qquad (5.8)$$

If one assumes that the final system (series-parallel network) reliability is very large, then we could write

$$F_{sp} = 1 - R_{sp} \ll 1 \qquad (5.9)$$

where F_{sp} is the final system failure probability.

In order to obtain the specified system reliability, R_{sp}, parallel element 1, in Fig. 5.2, with n_1 identical elements; parallel element 2 with n_2 identical elements, and so on. Figure 5.3 shows the resulting overall configuration.

After paralleling the ith group (subsystem), the reliability is given by

$$R_{pi} = 1 - F_{pi}^{ni} \qquad (5.10)$$

where $F_{pi} = 1 - R_{pi}$
$\quad F_{pi} \quad$ is the ith parallel group or subsystem element unreliability
$\quad n_i \quad$ is the number of components or elements in the ith subsystem

Using Eq. (5.10), the final system reliability is

$$R_{sp} = \prod_{i=1}^{k} R_{pi} \qquad (5.11)$$

where k is the number of parallel subsystems in Fig. 5.3.

The ith parallel subsystem cost is $n_i c_i$, where c_i is the ith subsystem

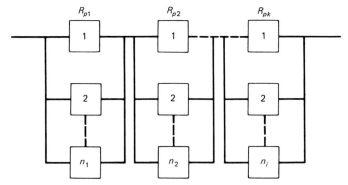

Figure 5.3. Improved configuration of Fig. 5.2; increases basic system reliability.

element cost. In other words, the $n_i c_i$ is the cost of ith group of parallel elements. The final system cost, c_s, is given by

$$c_s = \sum_{i=1}^{k} n_i c_i \qquad (5.12)$$

To find the minimum cost of a distribution of n_i's solve Eqs. (5.10) to (5.12) as a variational problem. This may not be as simple as it appears; therefore, it is desirable to introduce a new variable, β_i

$$R_{pi} = R_{sp}^{\beta_i} \qquad (5.13)$$

To satisfy Eq. (5.13), β_i has to be a real, positive number between zero and unity. Utilizing Eqs. (5.10) and (5.13), we get

$$n_i = \frac{\log(1 - R_{sp}^{\beta_i})}{\log F_{pi}} \qquad (5.14)$$

Substituting Eq. (5.14) in Eq. (5.12) results in

$$c_s = \sum_{i=1}^{k} \frac{c_i \log(1 - R_{sp}^{\beta_i})}{\log F_{pi}} \qquad (5.15)$$

To obtain the final system reliability expression, substitute Eq. (5.13) in Eq. (5.11) to get

$$R_{sp} = (R_{sp})^{\sum_{i=1}^{k} \beta_i} \qquad (5.16)$$

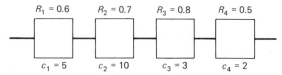

Figure 5.4. A basic series system.

Equation (5.16) is satisfied if the summation of β_i from $i = 1$ to $i = k$ is equal to unity. In other words, if

$$\sum_{i=1}^{k} \beta_i = 1 \tag{5.17}$$

The following expression for β_i is derived in reference 112

$$\beta_i = \frac{c_i / \log F_{pi}}{\sum_{j=1}^{k} (c_j / \log F_{pj})} \tag{5.18}$$

EXAMPLE 1

A basic series system is shown in Fig. 5.4. The cost and reliability data is given for each element in the diagram. By using the procedure presented in Section 5.3.3, calculate a minimum system cost if the desired final system reliability, R_{sp}, is to be 0.98. It is assumed that the system elements are statistically independent.

Using Eq. (5.7) and element reliability data given in Fig. 5.4, the basic system reliability

$$R_b = \prod_{i=1}^{4} R_i = 0.6 \times 0.7 \times 0.8 \times 0.5 = 0.168$$

To calculate values for β_i we use Eq. (5.18)

$$\beta_i = \frac{(c_i / \log F_{pi})}{\sum_{j=1}^{4} (c_j / \log F_{pj})} \qquad \text{for } i = 1, 2, 3, 4 \tag{5.19}$$

Using the element reliability data given in Fig. 5.4, the following results

$$F_{p_1} = 1 - 0.6 = 0.4 \qquad \therefore \log F_{p_1} = -0.398 \tag{5.20}$$

$$F_{p_2} = 1 - 0.7 = 0.3 \qquad \therefore \log F_{p_2} = -0.523 \tag{5.21}$$

$$F_{p_3} = 1 - 0.8 = 0.2 \qquad \therefore \log F_{p_3} = -0.699 \tag{5.22}$$

$$F_{p_4} = 1 - 0.5 = 0.5 \qquad \therefore \log F_{p_4} = -0.301 \tag{5.23}$$

Using the cost data given for each element in Fig. 5.4 and results of Eqs. (5.20) to (5.23), the following c_i' factor (i.e., $i = 1, 2, 3, 4$) values are calculated

$$c_1' = \frac{c_1}{\log F_{p_1}} = \frac{5}{-0.398} = -12.56 \qquad (5.24)$$

$$c_2' = \frac{c_2}{\log F_{p_2}} = \frac{10}{-0.523} = -19.12 \qquad (5.25)$$

$$c_3' = \frac{c_3}{\log F_{p_3}} = \frac{3}{-0.699} = -4.29 \qquad (5.26)$$

$$c_4' = \frac{c_4}{\log F_{p_4}} = \frac{2}{-0.301} = -6.64 \qquad (5.27)$$

The total value of Eqs. (5.24) to (5.27) is

$$\sum_{j=1}^{4} \frac{c_j}{\log F_{pj}} = -12.56 - 19.12 - 4.29 - 6.64 = -42.61 \qquad (5.28)$$

Substituting Eqs. (5.24) to (5.28) in Eq. (5.19) we get

$$\beta_1 = \frac{-12.56}{-42.61} = 0.295 \qquad (5.29)$$

$$\beta_2 = \frac{-19.12}{-42.61} = 0.449 \qquad (5.30)$$

$$\beta_3 = \frac{-4.29}{-42.61} = 0.1 \qquad (5.31)$$

$$\beta_4 = \frac{-6.64}{-42.61} = 0.156 \qquad (5.32)$$

To calculate number of elements for each group of parallel subsystems, we use Eq. (5.14)

$$n_i = \frac{\log(1 - R_{sp}^{\beta_i})}{\log F_{pi}} \qquad (5.33)$$

Since $R_{sp} = 0.98$

$$F_{sp} = 1 - 0.98 = 0.02 \qquad (5.34)$$

where

$$R_{sp} + F_{sp} = 1 \qquad (5.35)$$

Using Eq. (5.35), Eq. (5.33) can be rewritten

$$n_i = \frac{\log[1 - \{1 - F_{sp}\}^{\beta_i}]}{\log F_{pi}} \tag{5.36}$$

For small $F_{sp}(F_{sp} = 0.02)$, $(1 - F_{sp})^{\beta_i}$ approximates to

$$(1 - F_{sp})^{\beta_i} \approx 1 - F_{sp} \, \beta_i \tag{5.37}$$

Using the result of Eq. (5.37), Eq. (5.36) becomes

$$n_i \approx \frac{\log F_{sp} \, \beta_i}{\log F_{pi}} \tag{5.38}$$

Substituting the results of Eqs. (5.20) to (5.23), (5.34), and (5.29) to (5.32) in Eq. (5.38), the following results

$$n_1 \approx \frac{\log(F_{sp} \, \beta_1)}{\log F_{p_1}} = \frac{\log(0.02 \times 0.295)}{-0.398} = 5.6 \approx 6 \text{ elements} \tag{5.39}$$

$$n_2 \approx \frac{\log(F_{sp} \, \beta_2)}{\log F_{p_2}} = \frac{\log(0.02 \times 0.449)}{-0.523} = 3.9 \approx 4 \text{ elements} \tag{5.40}$$

$$n_3 \approx \frac{\log(F_{sp} \, \beta_3)}{\log F_{p_3}} = \frac{\log(0.02 \times 0.1)}{-0.699} = 3.86 \approx 4 \text{ elements} \tag{5.41}$$

$$n_4 \approx \frac{\log(F_{sp} \, \beta_4)}{\log F_{p_4}} = \frac{\log(0.02 \times 0.156)}{-0.301} = 8.3 \approx 8 \text{ elements} \tag{5.42}$$

The total minimum system cost, c_s, at the desired reliability level is calculated by using the data of Fig. 5.4 and Eqs. (5.39) to (5.42)

$$c_s = \sum_{i=1}^{4} c_i n_i = 5 \times 6 + 10 \times 4 + 3 \times 4 + 2 \times 8 = 98$$

The original cost, c_b of the four element basic series system of Fig. 5.3 is

$$c_b = c_1 + c_2 + c_3 + c_4 = 20$$

Therefore, the ratio of the final system minimum cost to the basic system cost is

$$\frac{c_s}{c_b} = \frac{98}{20} = 4.9$$

The above cost ratio indicates that to increase the four element basic system of Fig. 5.4 reliability from 0.168 to 0.98, the system cost increase will be fivefold.

For comparison, if we had simply paralleled the basic system of Figure 5.4 with n basic systems, then using Eq. (5.5)

$$n = \frac{c_s}{c_b} = \frac{\log(1 - R_{sp})}{\log(1 - R_b)} \tag{5.43}$$

where R_b is the basic system reliability
R_{sp} is the desired system reliability

Substituting the known data for R_{sp} (equal to 0.98) and R_b (0.168) in Eq. (5.43) we obtain

$$n = \frac{c_s}{c_b} = \frac{\log(1 - 0.98)}{\log(1 - 0.168)} = \frac{-1.7}{-0.0799} \approx 21$$

The above result indicates that to increase the basic system reliability from 0.168 to 0.98, the increase in the system cost could have been twenty-one fold.

5.3.4 Series Network Optimum Reliability Allocation

This problem is concerned with allocating the reliability to each component of an M-stage series system to achieve maximum overall reliability. Some of the related literature on the subject is presented in references 14, 101, and 146. It is assumed that the series system components are statistically independent. This problem is formulated [146] as follows:
Maximize the system reliability, R_{ss},

$$\text{Max } R_{ss} = \prod_{i=1}^{M} R_i \tag{5.44}$$

subject to the following resource constraints:

$$\sum_{i=1}^{M} p_{ki}(R_i) \leq a_k \qquad \text{for } k = 1, 2, 3, 4, \ldots, n \tag{5.45}$$

where a_k represents the total amount availability of the kth resource
R_{ss} is the reliability of the system under study
R_i denotes the ith stage component reliability
$p_{ki}(R_i)$ represents consumption of the kth resource at the ith stage
M represents the number of system stages

5.3.5 Network Cost Minimization Subject to the Desired Level of System Reliability

As its title states, this problem is concerned with the network cost minimization when the system reliability is specified. This concept is applied to parallel, series-parallel, and other configurations in references 8, 19, 63, 69, 112, 135, and 147.

The independent component problem [146] is formulated as follows: Minimize the total system cost K_M,

$$K_M = \sum_{i=1}^{M} K_i(Y_i) \tag{5.46}$$

subject to the system reliability constraint

$$R_{ss} = \prod_{i=1}^{M} R_i(Y_i) \geq R_{rd} \tag{5.47}$$

where R_{ss} is the system reliability
R_{rd} is the specified reliability of system
K_i is the ith stage cost ($K_i(Y_i)$ is function of Y_i, the number of components in each stage)
M is the number of system stages

5.3.6 System Reliability Maximization (with Optimal Redundancy Allocation) Under Cost Constraints

This section presents formulations to maximize system reliability with the optimum number of redundancies subject to cost constraints. Some of the literature dealing with this topic is listed in references 8, 16, 17, 23, 35, 38, 42, 51, 55, 62, 69, 86, 88, 105, and 97. We assume that the components' failures are statistically independent.

Maximize the [146] system reliability, R_{ss},

$$\text{Max } R_{ss} = \prod_{i=1}^{M} R_i(Y_i) \tag{5.48}$$

subject to the following constraints:

$$\sum_{i=1}^{M} p_{ki}(Y_i) \leq a_k \quad \text{for } k = 1, 2, 3, \ldots, n \tag{5.49}$$

where R_i is the ith stage or subsystem reliability; in addition, R_i is a function of Y_i, the number of stage i components.

Remaining notation was defined in Sections 5.3.4 and 5.3.5.

5.3.7 A Non–Series-Parallel Network Reliability Maximization

To optimize the reliability of such independent component networks, the problem is formulated [146]

$$\text{Max } R_{ss} = f(R_1, R_2, R_3, \ldots, R_M) \tag{5.49}$$

subject to the constraints

$$\sum_{i=1}^{M} p_{ki}(R_i) \leq a_k \qquad \text{for } k = 1, 2, 3, \ldots, n \tag{5.50}$$

Note that the system reliability, R_{ss}, is a function of R_i, which is the ith component reliability. Other notation was defined in Sections 5.3.4 to 5.3.6.

Literature relating to redundancy is presented in references 1 and 7 and on reliability allocation in references 22 and 148.

5.4 RELIABILITY OPTIMIZATION OF REDUNDANT CONFIGURATIONS WITH TWO FAILURE MODE COMPONENTS

This section presents a reliability optimization of three-state device networks. A device is said to have three states if it operates normally and fails either in an opened or shorted (closed) mode. Typical examples of three-state devices are a fluid flow valve and an electronic diode [173]. The following reliability optimization discussion covers identical and independent components series, parallel, series-parallel, and parallel-series networks.

5.4.1 Parallel Network

For a simple parallel configuration, all the elements must fail in the open mode or any one of the elements must stop functioning in a short mode to cause the system to fail completely. The parallel network reliability, R_p, expression from Eq. (3.51) and references 171 and 173 is

$$R_p = (1 - F_s)^k - F_o^k \tag{5.51}$$

where F_o is the component's open mode failure probability

F_s is the component's short (closed) mode failure probability

k is the number of identical and independent components

The optimum number of elements is obtained by differentiating Eq. (5.51) with respect to k and setting the resulting expression equal to zero

$$\frac{\partial R_p}{\partial k} = (1 - F_s)^k \ln(1 - F_s) - F_o^k \ln F_o = 0 \qquad (5.52)$$

Solving Eq. (5.52) for k, an optimum number of elements, k', expression results

$$k' = \frac{\ln\left[\dfrac{\ln F_o}{\ln(1 - F_s)}\right]}{\ln \theta_p} \qquad (5.53)$$

where $\theta_p \equiv (1 - F_s)/F_o$

EAMPLE 2

Suppose a number of identical and independent electronic diodes (components with two mutually exclusive failure modes) are to be connected in a parallel configuration. Each diode has open and short failure mode probabilities of 0.5 and 0.2, respectively. Find the optimum number of diodes, k', to be connected in a parallel configuration. Using Eq. (5.53) we get

$$k' = \frac{\ln\left[\dfrac{\ln 0.5}{\ln(1 - 0.2)}\right]}{\ln \theta_p} \qquad (5.54)$$

where $\theta_p = (1 - 0.2)/0.5$

From Eq. (5.54) the optimum value for k' is equal to 2.41. Therefore, two diodes should be connected in parallel to obtain optimum network reliability.

5.4.2 Series Network

The optimum number of identical and independent series elements, x', expression is obtained directly from Eq. (5.53) by the series to parallel duality

$$x' = \frac{\ln\left[\dfrac{\ln F_s}{\ln(1 - F_o)}\right]}{\ln \theta_s} \qquad (5.55)$$

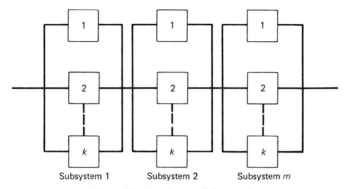

Figure 5.5. A series-parallel network.

where $\theta_s \equiv (1 - F_o)/F_s$

Derivations of Eq. (5.53) and (5.54) are presented in reference 171.

EXAMPLE 3

Suppose the same components as in Example 2 are to be connected in the series form. If their open and short mode probabilities are 0.2 and 0.2, respectively, find the optimum number of diodes, x', to be connected in series.

Utilizing Eq. (5.55) we get

$$x' = \frac{\ln\left[\dfrac{\ln 0.2}{\ln(1 - 0.2)}\right]}{\ln \theta_s} = 1.426$$

where $\theta_s \equiv (1 - 0.2)/0.2$

x' is equal to 1.426. Therefore, the optimum number of components in series configuration must not be more than one to achieve maximum reliability.

5.4.3 Series-Parallel Network

An independent and identical component series-parallel network is shown in Fig. 5.5. The identical and independent components series-parallel network shown in Fig. 5.5 can be represented by the equation given in references 171 and 173.

$$R_{sp} = (1 - F_o^k)^m - [1 - (1 - F_s)^k]^m \qquad (5.56)$$

where m is the number of identical subsystems

k is the number of identical components in each subsystem

F_o is the open mode failure probability of each component
F_s is the short mode failure probability of each component

To find the optimum values of k and m take the partial derivatives of Eq. (5.56) with respect to m and k and then set the resulting expressions equal to zero. Solve the resulting equations for k and m.

The resulting expressions for the partial derivatives of Eq. (5.56) with respect to m and k are

$$\frac{\partial R_{sp}}{\partial m} = (1 - F_o^k)^m \ln(1 - F_o^k)$$

$$- [1 - (1 - F_s)^k]^m \ln[1 - (1 - F_s)^k] \qquad (5.57)$$

and

$$\frac{\partial R_{sp}}{\partial k} = -m\{1 - F_o^k\}^{m-1} F_o^k \ln F_o$$

$$+ m[1 - (1 - F_s)^k]^{m-1}(1 - F_s)^k \ln(1 - F_s) \qquad (5.58)$$

Setting Eq. (5.57) equal to zero and rearranging we get

$$\left[\frac{1 - F_o^k}{1 - (1 - F_s)^k}\right]^m = \frac{\ln\{1 - (1 - F_s)^k\}}{\ln(1 - F_o^k)} \qquad (5.59)$$

Taking the logarithm of Eq. (5.59) and rearranging in the term of $m*$

$$m* = \frac{\ln\left\{\dfrac{\ln[1 - (1 - F_s)^k]}{\ln(1 - F_o^k)}\right\}}{\ln\left[\dfrac{1 - F_o^k}{1 - (1 - F_s)^k}\right]} \qquad (5.60)$$

Similarly, setting Eq. (5.58) equal to zero, rearranging, and then substituting in Eq. (5.60) for m, we get

$$\ln\left\{\frac{\ln[1 - (1 - F_s)^k]}{\ln(1 - F_o^k)}\right\} - \ln\left(\left\{\frac{(1 - F_s)^k \ln(1 - F_s)}{[1 - (1 - F_s)^k]}\right\}\right.$$

$$\left. \div \left[\frac{F_o^k \ln F_o}{(1 - F_o^k)}\right]\right) = 0 \qquad (5.61)$$

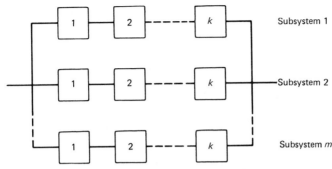

Figure 5.6. A parallel-series network.

The approximate optimum numerical values for m^* and k^* can be obtained from Eqs. (5.60) and (5.61).

5.4.4 Parallel-Series Network

Figure 5.6 represents an identical and independent component parallel-series network. Equations (5.62) to (5.64) are obtained directly from Eqs. (5.56) and (5.60) to (5.61) by using the series-parallel to parallel-series duality

$$R_{ps} = (1 - F_s)^m - [1 - (1 - F_o)^k]^m \qquad (5.62)$$

where R_{ps} is the reliability of the parallel-series network shown in Fig. 5.6

$$m^* = \frac{\ln\left\{\dfrac{\ln[1 - (1 - F_o)^k]}{\ln(1 - F_s^k)}\right\}}{\ln\left[\dfrac{1 - F_s^k}{1 - (1 - F_o)^k}\right]} \qquad (5.63)$$

$$\ln\left\{\frac{\ln[1 - (1 - F_o)^k]}{\ln(1 - F_s^k)}\right\} - \ln\left(\left\{\frac{(1 - F_o)^k \ln(1 - F_o)}{[1 - (1 - F_o)^k]}\right\}\right.$$
$$\left. \div \left[\frac{F_s^k \ln F_s}{1 - F_s^k}\right]\right) = 0 \qquad (5.64)$$

The optimum values for m and k can be approximated numerically by using Eqs. (5.63) and (5.64).

5.5 MECHANICAL COMPONENT RELIABILITY OPTIMIZATION

Stress-strength interference theory is used to determine the reliability of a mechanical component when the component stress-strength probability distributions are known. In the probabilistic design problem, the factors which determine the stress-strength distribution parameters can be controlled. This means finding the optimum parameter values, subject to constraints. These could be the design and resource constraints. According to reference 143, this leads to the study of the following two types of the problems:

1. At a specified component reliability, minimize the cost of design.
2. Maximize component reliability under resource constraints.

It is assumed here that the component stress and strength distributions are known and follow normal probability density functions. Limitations on the distribution and details on the following two models are given in reference 143. A well known equation to determine reliability, from Eq. (4.25) and reference 172, is

$$R_c = \int_{-\infty}^{\infty} f(s) \left[\int_{s}^{\infty} y(S) \, dS \right] ds \tag{5.65}$$

where R_c is the component reliability
$\quad f(s)$ is the component stress, s, probability density function
$\quad y(S)$ is the component strength, S, probability density function

5.5.1 Design Cost Minimization of a Component When Stress and Strength are Independent and Normally Distributed

The following component reliability, R_c, equation is developed by using Eq. (5.65) as in reference [172]

$$R_c = \frac{1}{\sqrt{2\pi}} \int_{\alpha}^{\infty} e^{-(x^2/2)} \, dx \tag{5.66}$$

where $\alpha = -[(\overline{m}_{Sth} - \overline{m}_{st})/(\sigma_{Sth}^2 + \sigma_{st}^2)^{1/2}]$
$\sigma_{Sth}, \sigma_{st}$ are the strength and stress standard deviations, respectively
$\overline{m}_{Sth}, \overline{m}_{st}$ are the strength and stress mean values, respectively

In Eq. (5.66) the component reliability value depends on α. In order to maximize reliability, the value of α must be as low as possible. When a component reliability is specified, the total cost, TK, minimization expressions may be written [172] as

$$\text{Min TK} = a(\mu_{Sth}) + b(\sigma_{Sth}) + c(\mu_{st}) + d(\sigma_{st}) \tag{5.67}$$

subject to the constraint:

$$[(\mu_{Sth} - \mu_{st})/(\sigma^2_{Sth} + \sigma^2_{st})^{1/2}] \geq x \qquad (5.68)$$

x is obtained for the desired level of reliability from normal tables
$a(\mu_{Sth})$ is a monotonically increasing mean strength cost function
$c(\mu_{st})$ is a monotonically decreasing mean stress cost function
$b(\sigma_{Sth})$ is the monotonically decreasing cost function of σ_{Sth}
$d(\sigma_{st})$ is the monotonically decreasing cost function of σ_{st}

The Lagrangian equation associated with Eqs. (5.67) and (5.68) is

$$f(\mu_{st}, \mu_{Sth}, \sigma_{Sth}, \sigma_{st}, \lambda) = a(\mu_{Sth}) + b(\sigma_{Sth}) + c(\mu_{st})$$
$$+ d(\sigma_{st}) + \lambda W \qquad (5.69)$$

where $W = \mu_{Sth} - \mu_{st} - x(\sigma^2_{Sth} + \sigma^2_{st})^{1/2}$
      λ is the Lagrange multiplier

In order to obtain local optimal solutions, differentiate (partially) Eq. (5.69) with respect to μ_{st}, μ_{Sth}, σ_{Sth}, σ_{st}, and λ, and then set the resulting expressions equal to zero. Solve these equations to find values for these five unknowns. These will be the local optima. To choose the global optimal solution, substitute these local solutions in the objective function. See reference 172 for details.

5.5.2 Component Reliability Maximization Subject to Resource Constraints When Stress and Strength are Independent and Normally Distributed

The following formulations are the dual of Eqs. (5.67) and (5.68)
     Maximize

$$x = \frac{\mu_{Sth} - \mu_{st}}{(\sigma^2_{Sth} + \sigma^2_{st})^{1/2}} \qquad (5.70)$$

subject to

$$a(\mu_{Sth}) + b(\sigma_{Sth}) + c(\mu_{st}) + d(\sigma_{st}) \leq R \qquad (5.71)$$

where R represents the quantity of available resources.

As for Eq. (5.69), the Lagrangian function for the problem is

$$f(\mu_{st}, \mu_{Sth}, \sigma_{Sth}, \sigma_{st}, \lambda) = \lambda T + \frac{\mu_{Sth} - \mu_{st}}{(\sigma_{Sth}^2 + \sigma_{st}^2)^{1/2}} \tag{5.72}$$

where $T = a(\mu_{Sth}) + b(\sigma_{Sth}) + c(\mu_{st}) + d(\sigma_{st}) - R$

To obtain the local optimal solutions, differentiate (partially) Eq. (5.72) with respect to μ_{st}, μ_{Sth}, σ_{Sth}, σ_{st}, and λ, and then set the resulting expressions equal to zero. Solve the five equations for five unknowns to find local optimal solutions. To choose the global optimal solution substitute the local optimal solutions in the objective function. For details see reference 172.

5.6 OPTIMUM MAINTENANCE POLICY

This section presents one optimum preventive maintenance (PM) policy. For a more detailed explanation and derivation, consult reference 174. This preventive policy suggests preventive maintenance should be performed after operating the equipment for x_o hours without a failure. The following assumptions were used to develop this optimum preventive maintenance policy:

1. When x_o is very large (infinite), PM is not scheduled.
2. Maintenance is performed immediately, if the equipment malfunctions or fails before x_o hours of operation. After the repair, preventive maintenance is rescheduled.
3. Equipment is as good as new after performing any type of maintenance or replacement.
4. Failures are statistically independent.
5. The hazard rate of the system or component under study is strictly increasing.

The well known system or component hazard rate expression of Eq. (3.17) is

$$z(x) = \frac{f(x)}{R(x)} = \frac{f(x)}{1 - F(x)} \tag{5.73}$$

where x denotes time
     $f(x)$ is the system or component failure probability density function
     $R(x)$ represents system or component reliability
     $F(x)$ is the cumulative distribution function

The optimum preventive maintenance period x_o is obtained from the following expression when it is satisfied

$$z(x_o)y - F(x_o) = \frac{x_s}{x_e - x_s} \qquad \text{if } x_e > x_s \qquad (5.74)$$

where $y = \int_0^{x_o} R(x) \, dx$

x_e denotes the expected time to carry out emergency maintenance
x_s denotes the expected time to carry out scheduled maintenance

When x_e equals x_s, x_o is very large or infinite. This indicates that one should never carry out the scheduled maintenance.

EXAMPLE 4

An equipment failure probability density is defined by the following Rayleigh distribution probability function [Eq. (2.41)]

$$f(x) = \frac{2}{n} e^{x^2 / n} \qquad (5.75)$$

where x is the time, and
       n is the scale parameter.

Find the expression for the optimum preventive maintenance period, x_o. Assume $n = 1$.

Integrating Eq. (5.75) over the range $[0, x]$ we get

$$F(x) = \int_0^x 2x \, e^{-x^2} \, dx = 1 - e^{-x^2} \qquad (5.76)$$

Equation (5.73) can be rewritten

$$R(x) + F(x) = 1$$

Therefore

$$R(x) = 1 - F(x) \qquad (5.77)$$

Substituting Eq. (5.76) in Eq. (5.77) we get

$$R(x) = e^{-x^2} \qquad (5.78)$$

Utilizing Eq. (5.78) to evaluate for y in Eq. (5.74), the following equation results [175]

$$y = \int_0^{x_0} e^{-x^2}\, dx = \sum_{i=0}^{\infty} \frac{(-1)^i x_0^{2i+1}}{i!\,(2i+1)} \tag{5.79}$$

To obtain the equipment hazard rate, $z(x)$, substitute Eqs. (5.75) and (5.78) in Eq. (5.73) to get

$$z(x) = \frac{2x\, e^{-x^2}}{e^{-x^2}} = 2x \tag{5.80}$$

To find the optimum preventive maintenance time, x_0 substitute Eqs. (5.76), (5.79), and (5.80) in Eq. (5.74) to get

$$2x_0 \left[\sum_{i=0}^{\infty} \frac{(-1)^i x_0^{2i+1}}{i!\,(2i+1)} \right] - (1 - e^{-x^2}) = \frac{x_s}{x_e - x_s} \qquad \text{if } x_e > x_s \tag{5.81}$$

For known values of x_e and x_s, the optimum value of preventive maintenance period, x_0, can be found from Eq. (5.81) when the equation is satisfied.

5.7 SUMMARY

This chapter is concerned with redundancy optimization, reliability optimization of redundant networks with two failure mode components, mechanical component reliability optimization, and optimum preventive maintenance policy. The author focuses more on the structure of the optimization concepts than the minute details. References 1 to 175 are provided for the reader who wants to delve deeper into this particular area. This bibliography is reasonably complete.

5.8 REFERENCES

1. K. K. Aggarwal, "Redundancy Optimization In General Systems," *IEEE Trans. on Reliability,* December 1976, pp. 330–332.
2. K. K. Aggarwal, "Minimum Cost Systems With Specified Reliability," *IEEE Trans. On Reliability,* August 1977.
3. K. K. Aggarwal, K. B. Misra, and J. S. Gupta, "A New Heuristic Method for Solving a Redundancy Optimization Criterion," *IEEE Trans. on Reliability,* Vol. R-24, April 1975, pp. 86–87.
4. K. K. Aggarwal, "Optimum Redundancy Allocation in Non-Series-Parallel Systems by Using Boolean Differences," *IEEE Trans. on Reliability,* Vol. R-28, 1979, pp. 79–80.
5. M. Alam and V. V. S. Sarma, "An Application of Optimal Control Theory to Repairman Problem With Machine Interference," *IEEE Trans. on Reliability,* June 1977.
6. R. N. Austin, "Optimized Operational Payloads for Manned Space Missions," *5th Reliability and Maintainability Conference,* 1966, pp. 202–214.

7. S. K. Banarjee, K. Rajamani, and S. S. Deshpande, "Optimal Redundancy Allocation for Non-Series-Parallel Networks," *IEEE Trans. on Reliability*, June 1976, pp. 115–117.

8. S. K. Banarjee and K. Rajamani, "Optimization of System Reliability Using a Parametric Approach," *IEEE Trans. on Reliability*, April 1973, pp. 35–39.

9. S. P. Bansal and S. Kumar, "Maximum Reliability Route Subject to K-Improvements in a Directed Network," *IEEE Trans. on Reliability*, June 1977.

10. R. E. Barlow and L. C. Hunter, "Criteria for Determining Optimal Redundancy," *IRE Trans. on Reliability and Quality Control*, April 1960, pp. 73–77.

11. R. E. Barlow, L. C. Hunter, and F. Proschan, "Optimum Redundancy When Components are Subject to Two Kinds of Failures," *J. SIAM*, Vol. 11, pp. 64–73, March 1963.

12. Y. Barness and H. Livni, "Reliability Optimization In the Design of Telephone Networks," *IEEE Trans. on Reliability*, December 1978.

13. P. W. Becker, "The Highest & Lowest Reliability Achievable With Redundancy," *IEEE Trans. on Reliability*, August 1977.

14. D. Behara and K. B. Misra, "Reliability Optimization Through Random Search Algorithm," *Microelectronics and Reliability*, Vol. 13, August 1974, pp. 295–297.

15. F. Beichelt and K. Fischer, "General Failure Model Applied to Preventive Maintenance Policies," *IEEE Trans. on Reliability*, April 1980.

16. R. E. Bellman and S. E. Drefus, "Dynamic Programming and the Reliability of Multi-Component Devices," *Operations Research*, Vol. 6, No. 2, pp. 200–206, March–April 1956.

17. G. Black and F. Proschan, "On Optimal Redundancy," *Operations Research*, Vol. 7, pp. 581–588, 1959.

18. M. Blitz, "Optimum Allocation of a Apares Budget," *Naval Research Logistic Quarterly*, Vol. 10, June 1963, pp. 175–191.

19. L. D. Bodin, "Optimization Procedures for the Analysis of Coherent Structures," *IEEE Trans. on Reliability*, August 1969, pp. 118–126.

20. A. M. Breipohl, "On Reliability Optimization A Preface," *IEEE Trans. on Reliability*, August 1977.

21. W. C. Broding, F. W. Diederich, and P. S. Parker, "Structural Optimization and Design Based on a Reliability Design Criterion," *J. Spacecraft*, Vol. 1, No. 1, pp. 56–61, January 1964.

22. R. M. Burton and G. T. Howard, "Optimal Design for System Reliability and Maintainability," *IEEE Trans. on Reliability*, May 1974, pp. 56–60.

23. R. M. Burton and G. T. Howard, "Optimal System Reliability for a Mixed Series and Parallel Structure," *J. Math. Anal.*, Vol. 28, Nov. 1969, pp. 370–382.

24. D. A. Butler, "The Use of Decomposition in the Optimal Design of Reliable Systems," *Operations Research*, Vol. 25, 1977, pp. 459–468.

25. P. K. W. Chan and T. Downs, "Optimization of Maintained Systems," *IEEE Trans. on Reliability*, April 1980.

26. S. C. Chay and M. Mazumdar, "Determination of Test Intervals In Certain Repairable Standby Protective Systems," *IEEE Trans. on Reliability*, August 1975.

27. S. S. Cheng, "Optimal Replacement Rate of Devices with Lognormal Failure Distribution," *IEEE Trans. on Reliability*, August 1977.

28. J. H. S. Chin, "Optimum Design for Reliability—The Group Reliability Approach," *IRE Wisconsin Convention Record*, Part 6, August 1969, pp. 23–29.

29. A. S. Cici and V. O. Muglia, "Computer Reliability Optimization System," *IEEE Trans. on Reliability*, August 1971, pp. 110–116.

30. N. D. Cox, "The Application of Reliability Theory to Process Design," *Annals of Reliability and Maintainability*, 1970, pp. 506–511.

31. N. Deo, "Optimization of Hardware Redundancy in Space Computers," *Annual Technical Conference Transactions,* ASQC, 1968, pp. 389–396.

32. Derby, Rasmuson, and Burdick, "A Risk-Based Approach to Advanced Reactor Design," *IEEE Trans. on Reliability,* August 1977.

33. J. Donelson, "Cost Model for Testing Program Based on Non-Homogenous Poison Failure Model," *IEEE Trans. on Reliability,* August 1977.

34. R. H. Dudley, K. K. Hekemian, and H. Laitin, "The Economics of Reliability," *Annals of Reliability and Maintainability,* 1970, pp. 451–470.

35. H. Everett, "Generalised Lagrange Multiplier Method for Solving Problems of Optimum Allocation of Resources," *Operations Research,* Vol. 11, pp. 399–417, May–June 1963.

36. L. T. Fan, C. S. Wang, and F. A. Tillman, C. L. Hwang, "Optimization of System Reliability," *IEEE Trans. on Reliability,* Vol. R-16, pp. 81–86, 1967.

37. M. Feldman and R. Gould, "Optimization of AEGIS Availability," *Proceedings Annual Symposium on Reliability,* 1971, pp. 213–218, Available from IEEE.

38. A. J. Federowicz and M. Mazumdar, "Use of Geometric Programming to Maximize Reliability Achieved by Redundancy," *Operations Research,* Vol. 16, pp. 948–956, September–October 1968.

39. O. T. Fleig, "Optimizing Reliability and Maintainability in the Airline Support Cycle," *Annals of Reliability and Maintainability,* 1971, pp. 11–14.

40. H. E. Frederic, "A Reliability Allocation Technique," *Proceedings 4th National Symposium on Reliability and Quality Control,* 1958, pp. 314–317.

41. L. C. Friar, P. M. Ghare, and K. L. Friar, "Optimization of System Reliability via Redundancy and/or Design Considerations," *IEEE Trans. on Reliability,* April 1980.

42. D. E. Fyffe, W. W. Hines, and N. K. Lee, "System Reliability Allocation and a Computational Algorithm," *IEEE Trans. on Reliability,* Vol. 17, pp. 64–69, June 1968.

43. P. M. Ghare and R. E. Taylor, "Optimal Redundancy for Reliability in Series Systems," *Operations Research,* Vol. 17, September–October 1969, pp. 838–847.

44. M. Goetz and R. H. Johnson, "Economically Optimum Receiving Inspection," *5th National Symposium on Reliability and Quality Control,* 1958, pp. 230–234.

45. L. C. Goheen, "On the Optimal Policy for the Machine Repair Problem When Failure and Repair Times Have Erlang Distributions," *Operations Research,* Vol. 25, pp. 484–492. 1977.

46. K. Gopal, K. K. Aggarwal, and J. S. Gupta, "Reliabiity Optimization in Systems with Many Failure Modes," *Microelectronics and Reliability,* Vol. 18, 1978, pp. 422–426.

47. K. Gopal, K. K. Aggarwal, and J. S. Gupta, "An Improved Algorithm for Reliability Optimization," *IEEE Trans. on Reliability,* December 1978.

48. K. Gopal, K. K. Aggarwal, and J. S. Gupta, "A New Method for Solving Reliability Optimization Problem," *IEEE Trans. on Reliability,* April 1980.

49. K. Gopal, K. K. Aggarwal, and J. S. Gupta, "A New Method for Reliability Optimization," *Microelectronics and Reliability,* Vol. 17, 1978, pp. 605–668.

50. K. Gopal, K. K. Aggarwal, and J. S. Gupta, "A New Approach to Reliability Optimization in General Modular Redundant Systems," *Microelectronics and Reliability,* Vol. 18, 1978, pp. 419–422.

51. K. Gordon, "Optimum Component Redundancy for Maximum System Reliability," *Operations Research,* April 1957, Vol. 5, pp. 229–247.

52. H. J. Greenberg, "An Application of a Lagrangian Penalty Function to Obtain Optimal Redundancy," *Technometrics,* Vol. 12, 1970, pp. 545–552.

53. S. K. Gupta and J. K. Sengupta, "Application of Reliability Programming to Production Planning," *International Journal of Systems Science,* Vol. 6, 1975, pp. 633–644.

54. P. R. Gyllenhal and J. E. Robinson, "A Reliability–Cost Optimization Procedure," *5th National Reliability and Quality Control Symposium,* 1959, pp. 43–54.

55. R. N. v. Hees and H. W. v.d. Meevendonk, "Optimal Reliability of Parallel Multi-Component Systems," *Operations Research,* Vol. 12, No. 1, pp. 16–26.

56. D. P. Herron, "Optimizing Trade-Offs of Reliability vs Weight," *IEEE Trans. on Reliability,* December 1963, pp. 50–54.

57. H. Hilton and M. Feirgen, "Minimum Weight Analysis Based on Structural Reliability," *J. Aerosp. Sci.,* Vol. 27, pp. 641–652. 1960.

58. G. C. Hunt and H. Ridley, "The Optimization of Earth-Orbiting Space Station Missions," *5th Reliability and Maintainability Conference,* 1966, pp. 193–201, Availability from the American Institute of Aeronautics and Astronautics.

59. R. F. Hynes, "Techniques in Optimizing Spare-Parts Provisions," *Annual Symposium on Reliability,* 1967, pp. 511–526.

60. K. N. Hyun and H. Okuno, "Optimization of the Redundant Allocation and Selection Problems in a Systems Reliability," *Keiei-Kagaku,* Vol. 18, pp. 132–147, 1974. Available from the Operations Research Society of Japan.

61. K. N. Hyun and H. Okuno, "Optimization of the Redundant Allocation in a System with Several Failure Modes of Zero-One Programming," *Proceedings of 7th Hawaii International Conference on Systems Science,* January 1974, pp. 189–191.

62. K. N. Hyun, "Reliability Optimization by 0–1 Programming for a System with Several Failure Modes," *IEEE Trans. on Reliability,* August 1975, pp. 206–210.

63. C. L. Hwang, K. C. Lai, and F. A. Tillman, "Optimization of System Reliability by the Sequential Unconstrained Minimization Technique," *IEEE Trans. on Reliability,* June 1975, pp. 133–135.

64. C. L. Hwang, F. A. Tillman, and W. Kuo, "Reliability Optimization Using Generalized Lagrangian Function and Reduced Gradient Methods," *IEEE Trans. on Reliability,* Vol. R-28, 1979, pp. 376–379.

65. C. L. Hwang, F. A. Tillman, W. K. Wei, and C. H. Lie, "Optimal Scheduled-Maintenance Policy Based on Multiple-Criteria Decision Making," *IEEE Trans. on Reliability,* Vol. R-28, 1979, pp. 394–399.

66. T. Inagaki, K. Inoue, and H. Akashi, "Improvement of Supervision Schedules for Protective Systems," *IEEE Trans. on Reliability,* June 1979.

67. T. Inagaki, K. Inoue, and H. Akashi, "Interactive Optimization of System Reliability Under Multiple Objectives," *IEEE Trans. On Reliability,* October 1978, pp. 264–267.

68. T. Inagaki, K. Inoue, and H. Akashi, "Optimal Redundancy Allocation Under Preventive Maintenance Schedule," *IEEE Trans. on Reliability,* April 1978, pp. 39–40.

69. K. Inoue, S. L. Gandhi, and E. J. Henley, "Optimal Reliability Design of Process Systems," *IEEE Trans. on Reliability,* April 1974, pp. 29–33.

70. R. K. Iyer and T. Downs, "A Variance Minimization Method of Reliability Design," *IEEE Trans. on Reliability,* June 1977, pp. 106–110.

71. R. K. Iyer, "A Moment Approach to Evaluation & Optimization of Complex System Reliability," *IEEE Trans. on Reliability,* August 1978, pp. 226–229.

72. N. Johnson, "On Optimizing Maintainability," *Microelectronics and Reliability,* Vol. 17, 1978, pp. 41–46.

73. R. Kalaba, "Design of Minimal-Weight-Structures for Given Reliability Cost," *J. Aerospace Sci.,* March 1962, pp. 355–356.

74. K. C. Kapur, "Optimization in Design By Reliability," *AIIE Transactions,* Vol. 7, No. 2, 1975, pp. 185–192.

75. K. C. Kapur, "Reliability Bounds in Probabilistic Design," *IEEE Trans. on Reliability,* Vol. R-24, 1975, pp. 193–195.

76. J. D. Kettelle, "Least-Cost Allocation of Reliability Investment," *Operations Research,* Vol. 10, March, April 1967, pp. 229–265.

77. D. N. Khandelwal, J. Sharma, L. M. Ray, "Optimal Periodic Maintenance Policy for Machines Subject to Deterioration and Random Breakdown," *IEEE Trans. on Reliability,* 1979, Vol. R-28, pp. 328–330.

78. D. N. Khandelwal, J. Sharma, L. M. Ray, "Optimal Overhaul Policy for a Machine," *IEEE Trans. on Reliability,* Vol. R-28, 1979, p. 327.

79. J. M. Kontonleon, "Optimum Redundancy of Repairable Modules," *IEEE Transactions on Reliability,* Vol. R-26, October 1977, p. 277.

80. J. M. Kontoleon, "Optimum Link Allocation of Fixed Topology Networks," *IEEE Trans. on Reliability,* June 1979, p. 145.

81. W. Kuo, C. L. Hwang, and F. A. Tillman, "A Note on the Heuristic Methods in Optimal System Reliability," *IEEE Trans. on Reliability,* December 1978, pp. 320–324.

82. B. K. Lambert, A. G. Walvekar, and J. P. Hirmas, "Optimal Redundancy and Availability in Multistage Systems," *IEEE Trans. on Reliability,* August 1971, pp. 181–185.

83. H. R. Lawrence and J. M. Vogel, "Some Thoughts on Reliability Estimation," *Proceedings of the IAS Aerospace Systems Reliability Symposium,* April 1962, pp. 61–66.

84. P. R. Leclercq, C. M. Marcovici, and S. A. Engins, "Complex Mission Worth Optimization by Redundancies," *Annual Reliability and Maintainability Symposium Proceedings,* 1972, pp. 354–360.

85. C. A. Locurto, W. T. Weir, and J. S. Youtcheff, "Reliability Test Optimization," *Proc. of the ISA Aerospace Systems Reliability Symposium,* 1962, pp. 87–97.

86. R. Luus, "Optimization of System Reliability by a New Non-Linear Integer Programming Procedure," *IEEE Trans. on Reliability,* April 1975, pp. 14–16.

87. S. Masanobu and J. N. Yang, "Optimum Structural Design Based on Reliability and Proof-Load Test," *8th Reliability and Maintainability Conference,* 1969, pp. 375–391.

88. D. W. McLeavy and J. A. McLeavy, "Optimization of System Reliability by Branch-and-Bound," *IEEE Trans. on Reliability,* Vol. R-25, December 1976, pp. 327–329.

89. D. W. McLeavy and J. A. McLeavy, "Analysis of Pseudo-Reliability of a Combat Tank System and Its Optimal Design," *IEEE Trans. on Reliability,* December 1976, pp. 327–329.

90. D. W. McLeavy and J. A. McLeavy, "Parallel Optimization Methods in Standby Reliability," Univ. of Connecticut, School of Business Administration, Bureau of Business Research, Working Paper No. 2, 12 pages, 1975.

91. L. D. Maxim and H. D. Weed, "Allocation of Test Effort for Minimum Variance of Reliability," *IEEE Trans. on Reliability,* Vol. R-26, June 1977, pp. 111–115.

92. M. Mazumdar, "An Optimum Procedure for Component Testing in the Demonstration of Series System Reliability," *IEEE Trans. on Reliability,* Vol. R-26, December 1977, pp. 342–344.

93. E. Menipaz, "Cost Optimization of Some Stochastic Maintenance Policies," *IEEE Trans. on Reliability,* Vol. R-28, June 1979, pp. 133–136.

94. H. Mine and H. Kawai, "Optimal Ordering and Replacement for a 1-Unit System," *IEEE Trans. on Reliability,* Vol. R-26, October 1977, pp. 273–276.

95. H. Mine and H. Kawai, "An Optimal Maintenance Policy for a 2-unit Parallel System with Degraded States," *IEEE Trans. on Reliability,* Vol. R-23, June 1974, pp. 81–85.

96. K. B. Misra, "A Simple Approach for Constrained Redundancy Optimization Problem," *IEEE Trans. on Reliability,* February 1972, pp. 30–35.

97. K. B. Misra and J. Sharma, "Reliability Optimization with Integar Constraint Coefficients," *Microelectronics and Reliability,* Vol. 12, 1973, pp. 431–433.

98. K. B. Misra, "A Method of Solving Redundancy Optimization Problems," *IEEE Trans. on Reliability,* August 1971, pp. 117–120.

99. K. B. Misra, "Reliability Design of a Maintained System," *Microelectronics and Reliability,* Vol. 13, 1974, pp. 495–500.

100. K. B. Misra and M. D. Ljubojevic, "Optimal Reliability Design of a System—A New Look," *IEEE Trans. on Reliability,* December 1973, pp. 255–258.

101. K. B. Misra, "Reliability Optimization of a Series-Parallel System," *IEEE Trans. on Reliability,* Nov. 1972, pp. 230–238.

102. K. B. Misra, "Dynamic Programming Formulation of Redundancy Allocation Problem," *Int. J. Math. Edu. Sci. Tech.,* Vol. 2, July–September 1971, pp. 207–215.

103. K. B. Misra, "Least-Square Approach to System Reliability Optimization," *Int. Jr. Contr.,* January 1973.

104. K. B. Misra and T. S. M. Rao, "Reliability Analysis of Redundant Networks Using Flow Graphs," *IEEE Trans. on Reliability,* February 1970, pp. 19–24.

105. K. B. Misra, "A Method of Redundancy Allocation," *Microelectronics and Reliability,* Vol. 12, pp. 389–393.

106. K. B. Misra, "A Fast Method of Redundancy Allocation," *Microelectronics and Reliability,* Vol. 12, pp. 385–387.

107. R. B. Misra and G. Agnihotri, "Peculiarities in Optimal Redundancy for Bridge Networks," *IEEE Trans. on Reliability,* April 1979, pp. 70–72.

108. K. Mizukami, "Optimum Redundancy for Maximum System Reliability by the Method of Convex and Integer Programming," *Operations Research,* Vol. 16, March–April 1968, pp. 392–406.

109. J. M. Mogg, "Constrained Optimum Test Configuration for Reliability Acceptance Test Incorporating Environmental Stresses," *IEEE Trans. on Reliability,* Vol. R-24, August, 1975.

110. J. M. Mogg, "Optimal Selection of Sequential Tests for Reliability," *IEEE Trans. on Reliability,* Vol. R-26, June 1977, pp. 116–118.

111. F. Moses and D. E. Kinser, "Optimum Structural Design with Failure Probability Constraints," *AIAA J.,* Vol. 5, No. 6, June 1967, pp. 1152–1158.

112. F. Moskowitz and J. B. McLean, "Some Reliability Aspects of System Design," *IRE Trans. on Reliability and Quality Control,* vol. RQc-8, September 1956, pp. 7–35.

113. E. J. Muth, "An Optimum Decision Rule for Repair vs Replacement," *IEEE Trans. on Reliability,* Vol. R-26, August 1977, p. 179.

114. L. B. Myers and N. L. Enrick, "Algorithmic Optimization of System Reliability," *Annual Technical Conference Transactions,* ASQC, 1968, pp. 455–460.

115. Y. Nakagawa and K. Nakashima, "A Heuristic Method for Determining Optimal Reliability Allocation," *IEEE Trans. on Reliability,* August 1977, pp. 156–161.

116. N. Nagakawa, "Optimum Policies When Preventive Maintenance is Imperfect," *IEEE Trans. on Reliability,* Vol. R-28, 1979, pp. 331–332.

117. T. Nakagawa and S. Osaki, "Optimum Preventive Maintenance Policies for a 2-Unit Redundant System," *IEEE Trans. on Reliability,* Vol. R-23, June 1974, pp. 86–90.

118. T. Nagagawa, "Optimum Preventive Maintenance Policies for Repairable Systems," *IEEE Trans. on Reliability,* Vol. R-26, August 1977, pp. 168–173.

119. N. Nagakawa and Y. Hattori, "Reliability Optimization with Multiple Properties and Integer Variables," *IEEE Trans. on Reliability,* Vol. R-28, 1979, pp. 73–78.

120. K. Nakashima and K. Yamato, "Optimal Design of a Series-Parallel System with Time Dependent Reliability," *IEEE Trans. on Reliability,* June 1977, Vol. R-26, pp. 119–120.

121. K. Narasimhalu and H. Sivaramakrishnan, "A Rapid Algorithm for Reliability Optimization of Parallel Redundant Systems," *IEEE Trans. on Reliability,* October 1978, pp. 261–263.

122. H. P. Nicely and C. B. Mayer, "Design Tools for the Optimization of Redundancy for a Planetary Vehicle," *7th Reliability and Maintainability Conference,* 1968, pp. 50–60.

123. K. S. Park, "Optimal Number of Minimal Repairs Before Replacement," *IEEE Trans. on Reliability,* Vol. R-28, June 1979.

124. B. Parhami, "Optimal Placement of Spare Modules in a Cascaded Chain," *IEEE Trans. on Reliability,* Vol. R-26, October 1977, pp. 280–282.
125. K. T. Plesser and T. O. Field, "Cost-Optimized Burn-In Duration for Repairable Electronic Systems," *IEEE Trans. on Reliability,* August 1977, pp. 195–197.
126. H. W. Price, "A Reliability Economic Decision Method," *6th National Symposium on Reliability and Quality Control,* 1959, pp. 88–92.
127. F. Proschan and T. A. Bray, "Optimum Redundancy under Multiple Constraints," *Operations Research,* Vol. 13, September–October 1965, pp. 800–814.
128. F. Proschan, "Optimal System Supply," *Naval Research Logistics Quarterly,* Vol. 7, December 1960, pp. 609–646.
129. W. J. Reich, W. A. Flannery, and D. A. Miller, "Reliability-Maintainability Cost Trade-Off Via Dynamic and Linear Programming," *Annals of Reliability and Maintainability,* 1966, pp. 310–329.
130. M. Sakawa, "Multi-Objective Optimization by the Surrogate Worth Trade-Off Method," *IEEE Trans. on Reliability,* Vol. R-27, December 1978, pp. 311–314.
131. M. Sasaki, "A Simplified Method of Obtaining Highest System Reliability," *8th National Symposium on Reliability and Quality Control,* 1962, pp. 489–502.
132. Sasaki, Kaburaki, and Yanagi, "System Availability & Optimum Spare Units," *IEEE Trans. on Reliability,* Vol. R-26, August 1977, pp. 182–188.
133. M. Sasaki, "An Easy Allotment- Method Achieving Maximum System Reliability," *Proc. 9th National Symposium on Reliability and Quality Control,* 1963, pp. 109–124.
134. R. L. Schaeffer, "Optimum Age Replacement In the Bivariate Exponential Case," *IEEE Trans. on Reliability,* Vol. R-24, August 1975. pp. 214–215.
135. V. Selman and N. T. Grisamore, "Optimum System Analysis by Linear Programming," *1966 Annual Symposium on Reliability,* pp. 696–703.
136. J. Sharma and K. V. Venkateswaran, "A Direct-Method for Maximizing the System Reliability," *IEEE Trans. on Reliability,* Nov. 1971, pp. 256–259.
137. B. V. Sheela, "Optimization of System Reliability by Sequential Weight Increasing Factor Technique," *IEEE Trans. on Reliability,* December 1977, pp. 339–341.
138. H. K. V. Shetty and D. P. Sengupta, "Reliability Optimization Using SLUMT," *IEEE Trans. on Reliability,* April 1975, pp. 80–82.
139. M. Shinozuka and J.-N. Yang, "Optimum Structural Design Based on Reliability and Proof-Load Test," *Annals of Assurance Sciences, Reliability & Maintainability Conference,* 1966, pp. 375–391.
140. H. Sivaramakrishnan and A. D. Narasimhalu, "Correction to Redundancy Optimization in General Systems," *IEEE Trans. on Reliability,* Vol. R-26, December 1977, p. 345.
141. Y. V. Subramanyan, "Some Observations on Optimization of System Reliability by SWIFT," *IEEE Trans. on Reliability,* Vol. R-29, April 1980, p. 38.
142. H. Switzky, "Minimum Weight Design with Structural Reliability," *AIAA 5th Annual Structures and Materials Conference,* New York, 1964, pp. 316–322.
143. S. I. Taraman and K. C. Kapur, "Optimization Considerations in Design Reliability by Stress-Strength Interference Theory, *IEEE Trans. on Reliability,* June 1975, pp. 136–138.
144. R. E. Taylor, Optimum Resource Allocation for Reliability in Series System, Ph.D. Dissertation, VPI & SU, 1970.
145. F. A. Tillman, C. L. Hwang, and W. Kuo, "Determining Component Reliability and Redundancy for Optimum System Reliability," *IEEE Trans. on Reliability,* August 1977, pp. 162–165.
146. F. A. Tillman, C. L. Hwang, and W. Kuo, "Optimization Techniques for System Reliability with Redundancy A Review," *IEEE Trans. on Reliability,* Vol. R-26, August 1977, pp. 148–155.
147. F. A. Tillman, "Optimization by Integer Programming of Constrained Reliability Problem with Several Modes of Failure," *IEEE Trans. on Reliability,* May 1969, pp. 47–53.

148. F. A. Tillman, C. L. Hwang, L. T. Fau, and K. C. Lai, "Optimal Reliability of a Complex System," *IEEE Trans. on Reliability,* Vol. R-19, August 1970, pp. 95–100.
149. E. A. Tinsans, R. T. McNichols, and S. L. Berry, "Availability Allocation Using a Family of Hyperbolic Cost Functions," *IEEE Trans. on Reliability,* Vol. R-24, 1975, p. 335.
150. W. E. Tipton, "Measuring and Optimizing Domant Weapon System Availability," *Annals of Assurance Sciences, Reliability and Maintainability Conference,* 1968, pp. 221–228.
151. O. Tosun, "Dynamic Decision Elements for 3-Unit System," *IEEE Trans. on Reliability,* Vol. R-26, Dec. 1977, pp. 335–338.
152. I. A. Ushakov, "A Heuristic Method of Optimization of the Redundancy of Multi-Function Systems," *Engineering Cybernetics,* Vol. 10, No. 4, 1972, pp. 612–613.
153. L. R. Webster, "Choosing Optimum System Configurations," *Proceedings of the National Symposium on Reliability and Quality Control,* 1964, pp. 345–359.
154. L. R. Webster, "Optimum System Reliability and Cost Effectiveness," *Annual Symposium on Reliability,* 1967, pp. 489–500.
155. K. Weir, "Unit Cost Data to Optimize Reliability," *Technical Conf. Trans.,* 1968, pp. 491–497.

Optimization Techniques

156. J. Abadie and J. Guigou, "Numerical Experiments with the GRG method," in *Integer and Nonlinear Programming* J. Abadie (ed.), North-Holland Amsterdam, 1970.
157. R. Bellman, *Dynamic Programming,* Princeton University Press, Princeton, NJ, 1957.
158. J. Bracken and G. P. McCormick, *Selected Applications of Nonlinear Programming,* Wiley, New York, 1968.
159. M. M. Denn, *Optimization by Variational Methods,* McGraw-Hill, New York, 1969.
160. R. J. Duffin, E. L. Peterson, and C. Zener, *Geometric Programming,* Wiley, New York (1967).
161. L. T. Fan, C. L. Hwang, and F. A. Tillman, "A Sequential Simplex Pattern Search Solution to Production Planning Problems," *American Institute of Industrial Engineers Transactions,* Vol. 1, 1969, pp. 267–273.
162. R. Fletcher, *Optimization,* Academic, New York, 1969.
163. A. V. Fiacco and G. P. McCormick, *Nonlinear Programming· Sequential Unconstrained Minimization Techniques,* Wiley, New York, 1969.
164. R. S. Garfinkel and G. L. Nemhauser, *Integer Programming,* Wiley, New York, 1972.
165. G. Hadley, *Nonlinear and Dynamic Programming,* Addison-Wesley, Reading, MA, 1964.
166. N. W. Kuhn and A. W. Tucker, "Nonlinear Programming," *Proceedings Second Berkeley Symposium on Mathematical Statistics and Probability,* J. Neyman (ed.), University of California Press, Berkeley, CA, pp. 481–492, 1951.
167. J. D. Roode, *Generalized Lagrange Function and Mathematical Programming, in Optimization,* R. Fletcher (ed.), Academic, New York, 1969.
168. H. A. Taha, *Integer Programming—Theory, Applications and Computations,* Academic, New York, 1975.
169. D. J. Wilde, *Optimum Seeking Methods,* Prentice-Hall, Englewood Cliffs, NJ, 1964.
170. C. Zener, *Engineering Design by Geometric Programming,* Wiley-Interscience, New York, 1971.

Miscellaneous

171. B. L. von Alven (ed.), *Reliability Engineering,* Prentice-Hall, Englewood Cliffs, NJ, 1964.
172. K. C. Kapur, L. R. Lamberson, *Reliability in Engineering Design,* Wiley, New York, 1977.

173. B. S. Dhillon, "The Analysis of the Reliability of Multi-State Device Networks," Ph.D. Dissertation, 1975, Available from the National Library of Canada, Ottawa, Ontario. Canada.

174. R. E. Barlow and F. Proschan, "Optimum Preventive Maintenance Policies," *Operations Research*, Vol. 8, 1960, pp. 90–100.

175. I. S. Gradshteyn and I. M. Ryzhik, *Table of Integrals, Series, and Products,* Academic, New York, 1980.

6
Engineering Reliability
Growth Models

6.1 INTRODUCTION

There are several reliability techniques that can be used to predict product reliability. Most of these methods predict reliability at a single point in time. To obtain more accurate product reliability results we should also consider the time variations introduced by design modifications or changes in maintenance procedures of a product. Reliability growth modeling is an analytical tool to analyse and control the reliability changes of a product over time. Furthermore, a reliability growth model can also be used to determine the time required to develop a marketable product. In addition, reliability growth models are used to monitor the progress of the development program as well as to predict the time required to achieve the reliability target when programs are modified. Reference 16 defines reliability growth as a continuing decrease in a product's actual hazard rate that will approach the inherent value targeted during the conceptual and development phase of the product's design. According to reference 46, a product's reliability grows due to the following reasons:

1. Correcting the weaknesses or errors in manufacturing methods, design, etc.
2. Operating each newly developed product to identify and eliminate bad components.
3. "Settling down" in product servicing, use, and manufacturing as well as increased operator skill and familiarization. The time for settling down may be reduced by good choice and planning and, of course, the right conditions.

6.1.1. A Brief Review

The early serious thinking on reliability growth occurred in the late fifties. A popular reliability growth model was first postulated in 1962 by Duane [27]. This was first published as a General Electric report and later in reference 27. According to Duane, cumulative hours and the cumulative failure rate plot as a straight line on log-log paper. He supported this by obtaining several data sets during product development test phases. His model can be applied during a reliability improvement program whenever corrective actions are introduced.

Since 1962 several technical papers have been published on the various aspects of reliability growth modeling. References 6, 8, 18, 24, 57, and 59 present developments on the Duane model. In particular, papers published in references 24 and 57 are concerned with relating the Duane model to the Weibull hazard function. In addition, papers published in references 5, 21, and 47 deal with estimating the model parameter confidence limits.

Some papers concerned with reliability growth management planning are presented in references 18, 62, and 74. Reliability growth optimization is covered in reference 67 and some reliability growth models are presented in references 1, 2, 82, 14, 33, and 43. An extensive bibliography is presented at the end of this chapter. This bibliography was mainly obtained from conference proceedings and journals. It also includes books and reports that cover reliability growth. The period covered is from 1959 to 1980.

Figure 6.1 shows the block diagram of a typical reliability growth feedback system. This figure shows the test-redesign-retest cycle. This criteria forms the integral part of most developmental programs. Figure 6.1 clearly shows that the testing is not the only factor in the reliability growth. Growth also

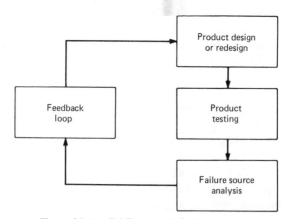

Figure 6.1 A reliability growth feedback system.

results from redesigns which are the result of the failure analyses. In other words, the failure sources discovered through testing are eliminated through redesign of the product.

The subsequent sections of this chapter are on reliability growth management, reliability growth models, growth model parameter estimation, and software reliability growth modeling.

6.2 RELIABILITY GROWTH MANAGEMENT

A typical reliability growth management program discussion is presented in reference 48. This section is based on this article. Reliability growth management (RGM) is defined as the application of management planning control methodology to the growth of product reliability during the test-redesign-retest phase (TRRP) of a product development program. A properly designed reliability growth management may help a program manager by answering questions such as: How well the product development program is progressing and how much resources and time are required to meet the end specified reliability target?

Systematically, it may be said that a reliability growth management program [48] may be useful at least in the following four ways in the product development program:

1. Projecting the product reliability trends.
2. Evaluating the time and resources requirements during test phases. In addition, determining the consequence of changes and limitations in the development program.
3. To assess the reliability growth progress against the specified reliability growth plan.
4. Preparing reliability growth time-phased and planned profiles.

One should note that the reliability growth management program is a subset of the overall set of reliability program management controls. A typical reliability growth management system is shown in Fig. 6.2. Note that the feedback system of Fig. 6.1 is a subsystem of the reliability growth management system shown in Fig. 6.2.

6.2.1. Reliability Growth Management Basic Tools

There are four types of reliability growth models which can be used as basic reliability growth management tools.

1. Assessment model: This model is used during the assessment period for estimating the reliability parameter values when value growth occurs.

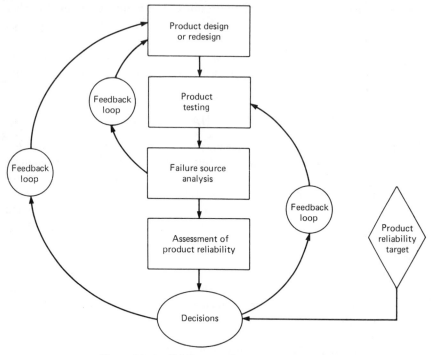

Figure 6.2 A reliability growth management system.

2. Generic model: This model represents a general pattern of growth in reliability plotted against time. The generic model is based on historical data for a group of products developed under a certain reliability program. A typical type of generic model is Duane's model [27].

3. Targeted model: This model is derived from the generic model by knowing certain values for discontinuities, slopes, starting points, and end points. These values are unique to the program under execution. Furthermore, these values may be functions of system complexity, state-of-the-art, etc.

4. Projection model: This model is used to forecast future reliability parameter values on the basis of past and future characteristics of the ongoing program as well as on past reliability growth.

Interested readers should consult references 45 and 48 for a detailed description of these four types of models.

6.3. RELIABILITY GROWTH MODELS

This section presents ten reliability growth models.

6.3.1 Duane Model

As mentioned in Section 6.1, this model was first postulated by Duane in 1962. He postulated that a cumulative hours versus cumulative failure rate plot would produce a straight line on log-log paper, when a continued reliability effort is maintained. He supported this by obtaining five data sets from product development test phases. This model is quite commonly used for the tasks such as reliability growth monitoring, program planning, resources planning, and so on.

Duane [27] defined the cumulative failure rate of his model as

$$\lambda_c = \frac{F}{T} = \alpha T^{-\beta} \tag{6.1}$$

where λ_c is the cumulative failure rate
     T is the total test hours
     F are the failures during T
     $\alpha; \beta$ are unknown parameters; α determined by circumstances such as product complexity, design objective, and design margin; β is the growth rate

The parameter estimating procedure of this model is presented in Section 6.4.1.

6.3.2. Weiss Model

This model is presented in references 43 and 78. The model depends upon the occurrence of product failures and corrective actions. When $(j - 1)$ corrective actions on a product have been completed then that product failure rate can be defined as

$$\lambda_j = (\theta + j)/\gamma j \tag{6.2}$$

where θ and γ are the unknown parameters.
     λ_j is a constant between failures (because this model is not time dependent)

For further readings on this model consult reference 78.

6.3.3. A Generalized Model

This model is used when the testing is conducted in M stages. A certain number of assumptions regarding this model are outlined in references 2 and 35.

The reliability equation for the model is

$$R_j = R_\infty - \beta f(j) \qquad \text{for } \beta > 0 \tag{6.3}$$

R_j is system's actual reliability at the jth stage of testing

R_∞ is the limiting reliability of a system for a very large value of j (or $j \to \infty$)

β is the rate of growth modifying parameter, i.e., β quantifies the amount of growth between stages one to infinity

$f(j)$ represents a positively decreasing function; which depends upon j, and in addition, characterizes the reliability growth

6.3.4. Hyperbolic Model

This is the special case of the model presented in Section 6.3.3, i.e., when $f(j) = 1/j$. This special case is taken from reference 2. From Eq. (6.3), we may write

$$R_j = R_\infty - \frac{\beta}{j} \tag{6.4}$$

The assumptions associated with this model are presented in reference 2. Parameter estimating equations for the model are presented in Section 6.4.4.

6.3.5. Modified Exponential Model

This model is used for one-shot devices such as small rockets [9]. It can also be utilized for test program planning before testing. As for models presented in Sections 6.3.3 and 6.3.4, the testing is conducted in m stages. The model reliability equation is

$$R_j = R_\infty - F\alpha^{j-1} \tag{6.5}$$

where F is the initial unreliability prior to testing

$\quad\quad\quad \alpha$ is called a growth factor; and is given by the ratio of the end of each stage unreliability to the unreliability value at the beginning of the stage

$\quad\quad\quad j$ is the stage number

$\quad\quad\quad R_j$ is the jth test reliability

$\quad\quad\quad R_\infty$ is the ultimate inherent reliability for a large value of j

At $\alpha = 0$, Eq. (6.5) indicates that there is no unreliability and $\alpha = 1$ means no gain in reliability. Equation (6.6) results from Eq. (6.5), where in terms of α, if we let $j = m$, where the m is number of stages or tests

$$\alpha = \left[\frac{R_\infty - R_m}{F} \right]^{1/(1-m)} \tag{6.6}$$

Plots of Eq. (6.5) are shown in Fig. 6.3(a), (b), and (c) for the various values of F, j, and α.

EXAMPLE 1

Suppose a customer requires the demonstrated reliability of some sort of a one-shot device (e.g., small rockets) to be 0.99 with 95 percent confidence. The customer wishes to know the planned reliability growth test program, which shows points to be observed on the manufacturer's planned curve. In other words, the reliability growth points are to be observed for each test or stage. Furthermore, the same customer has specified the following tests or stages:

1. Prototype
2. Qualification
3. Production
4. Demonstration

In addition, assume an earlier prototype exhibited 0.3 reliability when it was first tested.

This problem is solved in the following steps:

1. Determine number of trials needed for the reliability demonstration test. To determine number of trials needed for a specified reliability demonstration test (specified reliability with confidence limit), the following equation is used:

$$\sum_{i=0}^{Q} \binom{m}{i} (1-R)^i R^{m-i} \le \alpha \tag{6.7}$$

where Q is the acceptable number of failures in m trials
m is the number of trials
α is the consumer's risk index
R is the specified demonstration reliability with a consumer risk of α

From reference 9 and Eq. (6.7), we need 473 trials with one failure allowed to meet our requirement of 0.99 reliability with 95 percent confidence. Therefore, the planned observed reliability with 1 allowed failure in 473 tests will be given by

$$= \frac{473 - 1}{473} = \frac{472}{473} = 0.99788$$

2. Obtain the initiating reliability value of the growth curve. This is given in the problem statement and is equal to 0.3. Thus

$$F = 1 - 0.3 = 0.7$$

3. Estimate the value of α by using $R_\infty = 1$ and $m = 4$ (number of tests or stages). Substituting the results of steps 1 and 2 in Eq. (6.6) we get

$$\alpha = \left[\frac{1 - 0.99788}{0.70}\right]^{1/3} = 0.14468$$

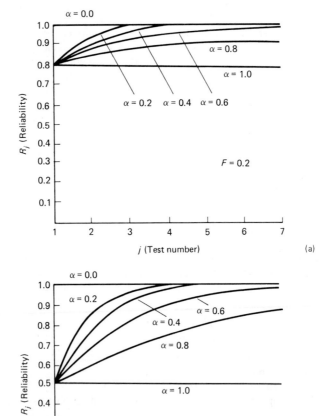

(a)

(b)

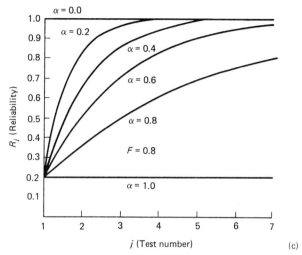

Figure 6.3 (a) Reliability growth curves of expression (6.5) for $F = 0.2$; (b) Reliability growth curves of expression (6.5) for $F = 0.5$; and (c) Reliability growth curves of expression (6.5) for $F = 0.8$.

This means that for reliability growth from 0.3 to 0.99788 in four tests or stages requires an α value $= 0.14468$.

4. Evaluate Eq. (6.5) for

$$R_\infty = 1.0, \ \alpha = 0.14468, \ F = 0.7 \text{ and } j = 1, 2, 3, 4.$$

The resultant values of Eq. (6.5) for $j = 1, 2, 3, 4$ are tabulated in Table 6.1.

Table 6.1. Tabulated Values of R_j to Be Observed

j	$R_j = R_\infty - F\alpha^{j-1}$
1	$R_1 = 0.3$
2	$R_2 = 0.899$
3	$R_3 = 0.985$
4	$R_4 = 0.99788$

A plot of the computed values of R_j versus j, in Table 6.1, is shown in Fig. 6.4. For each test, the planned reliability growth curve shows the reliability values to be observed if the specified reliability target is to be met.

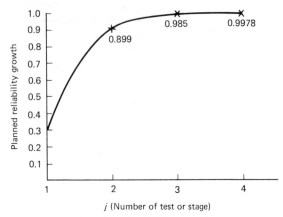

Figure 6.4 Planned reliability growth curve.

6.3.6 Generalized Hyperbolic Model

This model is the generalized version of the model presented in Section 6.3.4 letting $R_\infty = 1$ and raising the (right-hand side) j to the power n. Equation (6.4) is modified [1] to

$$R_j = 1 - \frac{\beta}{j^n} \tag{6.8}$$

where β and n are the parameters.

Equation (6.8) is the simplified version of the equation developed in reference 72. Note that at $n = 1$, Eq. (6.8) reduces to Eq. (6.4) as shown in reference 2.

6.3.7 Gompertz Model

This model is taken from reference 76. The Gompertz equation is quite useful when projecting the reliability of equipment with nonlinear reliability rate of change.

The Gompertz reliability growth curve is

$$R = \alpha A^{\gamma^t} \tag{6.9}$$

where $0 < \gamma < 1, 0 < A < 1$

α represents the upper limit reliability, R, as time $t \to \infty$.

One numerical example and a procedure to estimate the parameters of this model are presented in Section 6.4.3.

6.3.8 Exponential Model

This exponential model is presented in references 1, 14, and 35. The reliability growth equation of the model is

$$R = 1 - \alpha e^{-\beta t} \qquad (6.10)$$

where $0 < \alpha \leq e^{-\beta t}, \beta > 0$

α represents a constant which indicates the amount of reliability growth

t represents number of tests or time

β is a constant which measures the rate of reliability growth

At $t = 0$ and $\alpha = 1$, the reliability growth curve goes through zero. A plot of Eq. (6.10) is shown in Fig. 6.5 with $\beta = 0.5$, for the various values of t and α.

6.3.9 Golovin Model

This model is simply presented here to show its relationship [18] to the Duane model. This model is similar to the model presented in Section 6.3.6 and is based on test situations such as go/no-go. Golovin used his model to predict the reliability growth of rocket engines.

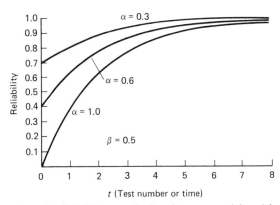

Figure 6.5 Reliability growth plot of an exponential model.

This mathematical model is

$$R = 1 - \gamma/(t + \theta)^\beta \qquad (6.11)$$

β, θ, γ are the unknown parameters or constants
t denotes number of tests

The drawbacks of this model are presented in reference 18.
To derive the relationship to the Duane model let

$$F = 1 - R = \gamma/(t + \theta)^\beta \qquad (6.12)$$

where F is the failure probability.

To predict the total number of failures in t tests, integrate Eq. (6.12) over the time interval $[0, t + \theta]$

$$F(t + \theta) = \gamma \int_0^{t+\theta} \frac{1}{(t + \theta)^\beta} d(t + \theta)$$

$$= \gamma(t + \theta)^{1-\beta}/(1 - \beta) \qquad (6.13)$$

To obtain the cumulative failure rate, λ_c, divide Eq. (6.13) by $(t + \theta)$

$$\lambda_c = \frac{\gamma(t + \theta)^{1-\beta}}{(1 - \beta)(t + \theta)} = \gamma(t + \theta)^{-\beta}/(1 - \beta) \qquad (6.14)$$

If we let $T = t + \theta$ and $\alpha = \gamma/(1 - \beta)$ then we can write Eq. (6.14) as

$$\lambda_c = \alpha T^{-\beta} \qquad (6.15)$$

Equation (6.15) is same as Eq. (6.1) which was postulated by Duane. The main difference between the Golovin and Duane models is in the plotting of the data points. Golovin plots the variables' instantaneous values while Duane plots cumulative values. Another difference is that Duane uses the plain variable T instead of Golovin's $(t + \theta)$. Further discussion on the subject is given in reference 18.

6.3.10 Weibull Growth Model

This model was first postulated by Crow in reference 24. He observed that the cumulative failure rate, λ_c, of the Duane model by definition is

$$\lambda_c = \frac{F(T)}{T} \tag{6.16}$$

where $F(T)$ represents expected number of failures during time interval T.

Therefore, one can rewrite Eq. (6.16)

$$F(T) = \lambda_c T \tag{6.17}$$

By substituting Eq. (6.1) in Eq. (6.17) we get

$$F(T) = \alpha\, T^{1-\beta} \tag{6.18}$$

To obtain the instantaneous failure rate, $\lambda(T)$, differentiate Eq. (6.18) with respect to T and get

$$\lambda(T) = \alpha(1 - \beta)T^{-\beta} \tag{6.19}$$

If we let $(1 - \beta) = b$, the Eq. (6.19) becomes

$$\lambda(T) = \alpha b T^{b-1} \tag{6.20}$$

Equation (6.20) represents the Weibull hazard rate. This model is based upon the Duane model. The parameter estimating equations are presented in Section 6.4.2.

6.3.11 Other Reliability Growth Models

Some of the reliability growth models are presented in section 6.3. Other reliability growth models not presented in this chapter may be found in references 5, 8, 14, 20, 24, and 35 or in the bibliography at the end of this chapter.

6.4 GROWTH MODELS PARAMETER ESTIMATES

The parameter estimates for some of the models discussed in Section 6.3 are presented in this section.

6.4.1 Duane Model Parameter Estimates

To estimate the parameter values for Duane's model, we take the logarithm of Eq. (6.1) to get

$$\log \lambda_c = \log \alpha - \beta \log T \tag{6.21}$$

Equation (6.21) is a straight line equation. Therefore, α and β can easily be estimated from the plot of the cumulative failure rate against cumulative operating hours. β is the slope of the straight line, (growth rate). At $T = 1$, α is equal to the corresponding cumulative failure rate.

Unweighted Least-Squares Fit. The unweighted least-squares technique can be used to find the best fitted straight line parameters α and β [47, 21] by using Eq. (6.22) and (6.23)

$$\log \alpha = \frac{\sum\limits_{j=1}^{m} \log n_j - \beta \left(\sum\limits_{j=1}^{m} \log t_j \right)}{m} \tag{6.22}$$

and

$$\beta = \frac{m \sum\limits_{j=1}^{m} (\log t_j \log n_j) - \left(\sum\limits_{j=1}^{m} \log t_j \sum\limits_{j=1}^{m} \log n_j \right)}{m \sum\limits_{j=1}^{m} (\log t_j)^2 - \left(\sum\limits_{j=1}^{m} \log t_j \right)^2} \tag{6.23}$$

where m denotes the total number of failures occurred during testing
$\quad t_j$ denotes jth failure time to failure
$\quad n_j$ represents the cumulative mean time between failures at time t_j

Weighted Least-Squares Fit. If the plotted data points are not independent, then proportional weighting the cumulative number of failures at each point is a reasonable way to improve accuracy of the estimates. This technique assigns greater weight to the preceding data point (the most recent one) [47, 79]. This method is based on the assumption that each data point is plotted m number of times at that point. If we assume that each failure is weighted by its failure number, m_j, the Eqs. (6.22) and (6.23) are modified to

$$\log \alpha = \frac{\Sigma \, m_j \log n_j - \beta \, \Sigma \, m_j \log t_j}{\Sigma \, m_j} \tag{6.24}$$

$$\beta = \frac{\Sigma \, m_j (\Sigma \, m_j \log t_j \log n_j) - (\Sigma \, m_j \log t_j)(\Sigma \, m_j \log n_j)}{\Sigma \, m_j \, \Sigma (m_j (\log t_j)^2) - (\Sigma \, m_j \log t_j)^2} \tag{6.25}$$

where $\Sigma \, m_j$ are total number of assumed points

6.4.2 Weibull Growth Model Parameter Estimates

To estimate the Weibull model parameters, α and b, of Eq. (6.20), the maximum likelihood (ML) is utilized

$$\hat{\alpha} = \frac{M}{T^{\hat{b}}} \tag{6.26}$$

and

$$\hat{b} = M \sum_{j=1}^{M-1} \log (T/T_j) \tag{6.27}$$

where M is the number of failures during testing
T_j is time of the jth failure
T is the termination time of the test

The derivations of Eqs. (6.26) and (6.27) are presented in reference 26.

6.4.3 Gompertz Model Parameter Estimates

The method presented in this section is useful to estimate reliability growth and its upper limit. It is taken from reference [76]. The general Gompertz reliability equation from Section 6.3.7 [Eq. (6.9)] is presented here

$$R = \alpha A^{\gamma^t} \tag{6.28}$$

Taking logarithm of Eq. (6.28) we get

$$\log R = \log \alpha + \gamma^t \log A \tag{6.29}$$

In Eq. (6.29), the failure data are substituted and the values of t are chosen at equal intervals. To obtain resulting expressions for α, A, and γ, Eq. (6.29) is divided into three parts with m number of data points for each part. If we add the data points or members of each part, the following three equations result

$$I = \sum_{R_0}^{R_{m-1}} \log R = m \log \alpha + \log A \sum_{0}^{m-1} \gamma^t \tag{6.30}$$

$$J = \sum_{R_m}^{R_{2m-1}} \log R = m \log \alpha + \log A \sum_{m}^{2m-1} \gamma^t \tag{6.31}$$

$$K = \sum_{R_{2m}}^{R_{3m-1}} \log R = m \log \alpha + \log A \sum_{2m}^{3m-1} \gamma^t \qquad (6.32)$$

If we subtract Eq. (6.31) from Eq. (6.30), and Eq. (6.32) from Eq. (6.31), the following expressions result

$$I - J = \log A \sum_{0}^{m-1} \gamma^t - \log A \sum_{m}^{2m-1} \gamma^t \qquad (6.33)$$

$$J - K = \log A \sum_{m}^{2m-1} \gamma^t - \log A \sum_{2m}^{3m-1} \gamma^t \qquad (6.34)$$

Dividing Eq. (6.33) by Eq. (6.34) and rearranging, we get the resulting expression for γ

$$\gamma = \left(\frac{J-K}{I-J}\right)^{1/m} \qquad (6.35)$$

Similarly, the following equations for parameters α and A are obtained

$$\log \alpha = \frac{I}{m} - \frac{(I-J)}{(1-\gamma^m)m} \qquad (6.36)$$

$$\log A = \frac{(I-J)(1-\gamma)}{(1-\gamma^m)(1-\gamma^m)} \qquad (6.37)$$

The following example demonstrates how to estimate parameters α, γ, and A.

EXAMPLE 2

It is required that an equipment under development should have assessed reliability of 0.95 when its 12-month development period ends. Using the known reliability data for 5 months presented in Table 6.2, predict the equipment reliability at the end of a 12-month period as well as its achievable upper limit if no major modification is carried out.

As shown in Table 6.2, the group size is equal to 2 ($m = 2$), i.e., each group has only two data points. To obtain estimates for γ, α, and A we substitute the results of Table 6.2 in Eqs. (6.35) to (6.37) to get

$$\gamma = \left(\frac{3.658 - 3.804}{3.393 - 3.658}\right)^{1/2} = 0.74225$$

$$\log \alpha = \frac{3.393}{2} - \frac{3.393 - 3.658}{[1 - (0.74225)^2]2} = 1.9916$$

Table 6.2. Known Reliability Data for Five Months.

PARTS OR GROUPS	GROUP SIZE, m	TIME (MONTHS)	EQUIPMENT RELIABILITY, R		SUMMATION OF EACH GROUP RELIABILITY
			R (%)	LOG R	Σ LOG R
1	2	0	45	1.653	$I = 3.393$
		1	55	1.74	
2	2	2	65	1.813	$J = 3.658$
		3	70	1.845	
3	2	4	75	1.875	$K = 3.804$
		5	85	1.929	

$\therefore \alpha = 98.07$ percent; this the upper limit of the equipment reliability. Similarly,

$$\log A = \frac{(3.393 - 3.658)(1 - 0.74225)}{[1 - (0.74225)^2][1 - (0.74225)^2]} = -0.3386$$

$$\therefore A = 0.4585$$

Thus substituting for α, γ, and A in Equation (6.28) we get

$$R = (98.07)(0.4585)^{(0.74225)\,t} \tag{6.38}$$

To obtain the equipment reliability after 12 months, set $t = 12$ in Eq. (6.38)

$$R = (98.07)(0.4585)^{(0.74225)^{12}}$$

$$= 95.95\%$$

As it can be easily seen that the assessed reliability after a 12-month period will be 95.95 percent which will fulfill the required reliability target of 95 percent. Other similar examples are presented in reference 76.

6.4.4 Hyperbolic Model Parameter Estimates

To estimate R_∞ and β for Eq. (6.4) we use the least-squares method [2]. In order to use the least-squares technique, one has to minimize the summation of squares, F, of the deviations of the successful trials, t_{sj}, to the, m_j, number of tests ratio from its expected value R_j with respect to parameters in question

$$F = \sum_{j=1}^{M} [(t_{sj}/m_j) - R_j]^2 \tag{6.39}$$

where M is the number of stages the test program is conducted.

By substituting for R_j from Eq. (6.4), and differentiating Eq. (6.39) with respect to parameters β and R_∞ and setting these derivative equal to zero, the following expressions result

$$\Sigma\, 2\left(t_{sj}/jm_j - \frac{R_\infty}{j} + \frac{\beta}{j^2}\right) = 0 \tag{6.40}$$

$$\Sigma\,(-2)\left[\frac{t_{sj}}{m_j} - R_\infty + \frac{\beta}{j}\right] = 0 \tag{6.41}$$

By solving the linear Eqs. (6.40) and (6.41) we get

$$\hat{\beta} = [\Sigma\, 1/j\, \Sigma\, H_j - M\, \Sigma\, (H_j/j)]/(M\, \Sigma\, 1/j^2) - (\Sigma\, 1/j)^2 \tag{6.42}$$

where $H_j = t_{sj}/m_j$
$$\hat{R}_\alpha = \Sigma\, 1/j^2\, \Sigma\, H_j - \Sigma\, 1/j\, \Sigma\, (H_j/j)/(M\, \Sigma\, 1/j^2) - (\Sigma\, 1/j)^2 \tag{6.43}$$

Interested readers requiring further detailed information on the parameter estimation of this model should consult reference [2]. This reference presents two estimating procedures: (1) maximum likelihood; and (2) least squares.

6.5 SOFTWARE RELIABILITY GROWTH MODELING

Because of the failure of computer programs in recent years, this subject has received a widespread attention. Software failures may be one of the prime sources of failures in complex computer systems. Software reliability is equally important to the hardware and human reliability if a complex system has to operate successfully. A section on the software reliability is included in Chapter 4. Readers interested in software reliability growth modeling should consult reference [75]. This paper discusses several software reliability models and provides an extensive bibliography on the subject.

6.6 SUMMARY

This chapter presents a brief review of the subject, a section on reliability growth management, equations for 10 selective reliability growth models, and parameter estimation equations for four models as well as two solved numerical examples. An extensive list of references is presented in the following section for those readers who would like to gain further knowledge on the subject.

6.7 REFERENCES

Books

1. B. L. Amstadter, *Reliability Mathematics: Fundamentals; Practices; Procedures,* McGraw-Hill, New York, 1971.
2. D. K. Lloyd and M. Lipow, *Reliability: Management Methods and Mathematics,* Prentice-Hall, Englewood Cliffs, NJ, 1962.
3. R. H. Myers, K. Wong, and H. M. Gordy, *Reliability Engineering for Electronic Systems,* J Wiley, New York, 1964.

Articles

4. D. R. Barr, "A Class of General Reliability Growth Model, Operations Research," Vol. 18, pp. 52–65, 1970.
5. R. E. Barlow and E. M. Scheuer, "Reliability Growth During a Development Testing Program," *Technometrics,* Vol. 8, No. 1, pp. 53–60, 1966.
6. D. E. Beachler and W. A. Chapman, "Reliability Proving for Commercial Products," *Annual Reliability and Maintainability Symposium,* pp. 89–94, 1977.
7. A. G. Bezat and L. L. Montague, "The Effect of Endless Burn-In on Reliability Growth Projections," *Annual Reliability and Maintainability Symposium,* pp. 392–397, 1979.
8. A. Bezat, V. Norquist, and L. L. Montague, "Growth Modeling Improves Reliability Predictions," *Annual Reliability and Maintainability Symposium,* pp. 317–322, 1975.
9. A. J. Bonis, "Reliability Growth Curves for One Shot Devices," *Annual Reliability and Maintainability Symposium,* pp. 181–185, 1977.
10. T. C. Bowling, "Relationships of Hazard Rate Distributions to Reliability Growth and Confidence Computators," *Annals of Reliability and Maintainability,* pp. 705–718, 1968.
11. A. J. Bonis, "Operating Characteristics Curves for Reliability Measurements," *IRE Transactions on Reliability and Quality Control,* pp. 1–7, 1962.
12. R. W. Bonnett and R. D. Oglesby, "Patriot Ground Equipment Reliability Growth," *Reliability Growth Management, Testing and Modelling,* pp. 23–27, 1978.
13. J. E. Bresenham, "Reliability Growth Models," *Annual Technical Conference Transactions,* ASQC, pp. 179–187, 1966.
14. J. E. Bresenham, "Reliability Growth Models," Technical Report No. 74, AD 605 993, NTIS, Springfield, VA, 22161, August 1964.
15. J. K. Byers, "Reliability Growth Apportionment," *IEEE Trans. on Reliability,* Vol. R-26, No. 4, pp. 242–244, 1977.
16. J. M. Clarke, "No-Growth Growth Curve," *Annual Reliability and Maintainability Symposium,* pp. 407–412, 1979.
17. J. M. Clarke and W. P. Cougan, "A Recent Real-Life Case History," *Reliability Growth Management, Testing and Modelling,* pp. 10–17, 1978.
18. E. O. Codier, "Reliability Growth in Real Life," *Annual Symposium on Reliability,* pp. 458–469, 1968.
19. E. O. Codier, "Reliability Prediction—Help or Hoax?" *Proc. 1969 Annual Symposium on Reliability,* pp. 383–390. Available from the IEEE.
20. W. J. Corcoran, H. Weingarten, and P. W. Zehna, "Estimating Reliability After Corrective Action," *Management Science,* Vol. 10, No. 4, pp. 786–795, 1964.
21. T. D. Cox and J. Kelly, "Reliability Growth Management of SATCOM Terminals," *Annual Reliability and Maintainability Symposium,* pp. 218–233, 1976.
22. W. P. Cougan and W. P. Kindig, "A Real-Life MTBF Growth Program for a Deployed Radar," *Annual Reliability and Maintainability Symposium,* pp. 121–127, 1979.

23. L. H. Crow, "On the AMSAA Reliability Growth Model," *Reliability Growth Management, Testing and Modelling,* pp. 28–31, 1978. Available from the Institute of Environmental Sciences, U.S.A.

24. L. H. Crow, "On Tracking Reliability Growth," *Annual Reliability and Maintainability Symposium,* pp. 438–443, 1975.

25. R. C. Dahiya, "Estimation of Reliability After Corrective Action," *IEEE Trans. on Reliability,* Vol. R-26, No. 5, pp. 348–351, 1977.

26. "Development Guide for Reliability, Part 4," *Reliability Measurements,* AMCP 706–198, 1976—ADA 027371. Available from the NTIS, Springfield, VA 22161.

27. J. T. Duane, "Learning Curve Approach to Reliability Monitoring," *IEEE Trans. on Aerospace,* pp. 563–566, 1964.

28. D. R. Earles, "Reliability Growth Prediction During the Initial Design Analysis," *Seventh National Symposium on Reliability and Quality Control,* pp. 380–393, 1961. Available from the IEEE.

29. F. A. Eble and E. W. Richards, "Rapid RMA Assessment—The Painless Plot," *Annual Reliability and Maintainability Symposium,* pp. 398–402, 1979.

30. J. C. Evans and S. I. Legreid, "Reliability Growth on B-52 FLIR System," *Reliability Growth Management, Testing and Modelling,* pp. 18–22, 1978. Available from the Institute of Environmental Sciences, U.S.A.

31. F. W. Fertig and V. K. Murthy, "Models for Reliability Growth During Burn-in: Theory and Applications," *Annual Reliability and Maintainability Symposium,* pp. 504–509, 1978. Available from the IEEE.

32. R. W. Fink, "Screening for Reliability Growth," *Annual Symposium on Reliability,* pp. 316–320, 1971. Available from the IEEE.

33. J. M. Finkelstein, "Starting and Limiting Values for Reliability Growth," *IEEE Trans. on Reliability,* Vol. R-28, No. 2, pp. 111–114, 1979.

34. J. E. Green, "The Problems of Reliability Growth and Demonstration with Military Electronics," *Proceedings of the NATO Conference on Reliability Evaluation,* pp. VII-C-1-VII-C-14, 1972.

35. A. J. Gross and M. Kamins, "Reliability Assessment in the Presence of Reliability Growth," *Symposium on Reliability,* pp. 406–416, 1968.

36. A. J. Gross, "The Application of Exponential Smoothing to Reliability Assessment," *Technometrics,* Vol. 13, No. 4, pp. 877–883, November 1973.

37. J. E. Green, "The Problems of Reliability Growth and Demonstration with Military Electronics," *Microelectronics and Reliability,* Vol. 12, pp. 513–520, 1973.

38. D. O. Hamilton and W. G. Ness, "Aircraft Reliability Growth Characteristics," *Proc. 1969 Annual Symposium on Reliability,* pp. 465–471.

39. G. R. Herd, "Comparisons of Reliability Growth Experience," *Reliability Growth Management, Testing and Modelling,* pp. 65–69, 1978. Available from the Institute of Environmental Sciences, U.S.A.

40. J. B. Hovis and D. O. Fiéni, "Reliability Growth Planning for Airborne Radar RIW? GMTBF Requirements," *Reliability Growth Management, Testing and Modelling,* pp. 60–64, 1978. Available from the Institute of Environmental Sciences, U.S.A.

41. J. Imai, H. Karasawa, and H. Machida, "Reliability Learning Model: Application to Color TV," *Microelectronics and Reliability,* Vol. 19, pp. 73–80, 1979.

42. E. T. Ireton, "Evolutionary Growth of Helicopter Maintainability, Reliability and Safety," *Seventh Reliability and Maintainability Conference,* pp. 167–176, 1968.

43. T. Jayachandran and L. R. Moore, "A Comparison of Reliability Growth Models," *IEEE Trans. on Reliability,* Vol. R-25, No. 1, pp. 49–51, 1976.

44. F. Kreuze, "Growth Curves—A Practical Management Tool," *Annual Reliability and Maintainability Symposium,* pp. 430–436, 1972. Available from the IEEE.

45. W. A. Lilius, "Reliability Growth Planning and Control," *Proceedings Annual Reliability and Maintainability Symposium,* pp. 267–270, 1978.
46. P. H. Mead, "Reliability Growth of Electronic Equipment," *Microelectronics and Reliability,* Vol. 14, pp. 439–443, 1975.
47. P. H. Mead, "Duane Growth Model: Estimation of Final M.T.B.F. with Confidence Limits Using a Hand Calculator," *Annual Reliability and Maintainability Symposium,* pp. 269–274, 1977.
48. C. Mead, T. Cox, and J. Lavery, "Reliability Growth Management in USAMC," *Annual Reliability and Maintainability Symposium,* pp. 432–437, 1975.
49. P. Mead, "The Role of Testing and Growth Techniques in Enhancing Reliability," *IREE* (British), Vol. 48, No. 7/8, July/Aug/1978.
50. C. J. Napolitano, "Assessing Reliability Growth Potential," *Proceedings Eleventh National Symposium on Reliability and Quality Control,* pp. 320–323, 1965.
51. D. G. Newman, "Reliability Growth Through the Air Force Reliability Improvement Warranty," *Reliability Growth Management, Testing and Modelling,* pp. 46–51, 1978. Available from the Institute of Environmental Sciences, U.S.A.
52. L. Nenoff, "Failure Evaluation as a Tool for Systems Reliability Growth," *23rd Annual Technical Meeting,* pp. 222–227, 1977. Available from the Institute of Environmental Sciences, U.S.A.
53. L. M. Nevola, "The Reliability Growth Aspects of MIL-STD-781D," *Reliability Growth Management, Testing and Modelling,* pp. 70–73, 1978. Available from the Institute of Environmental Sciences, U.S.A.
54. H. P. Norris and A. R. Timins, "Failure Rate Analysis of Goddard Space Flight Center Spacecraft Performance During Orbital Life," *Annual Reliability and Maintainability Symposium,* pp. 120–125, 1978.
55. D. E. Olsen, "A Confidence Interval for the Barlow-Schever Reliability Growth Model," *IEEE Trans. on Reliability,* Vol. R-27, No. 5, pp. 308–310, 1978.
56. D. E. Olsen, "Estimating Reliability Growth," *IEEE Trans. on Reliability,* Vol. R-26, No. 1, pp. 50–53, 1977.
57. E. J. Peacore, "Reliability Developments—AWACS," *Annual Reliability and Maintainability Symposium,* pp. 383–389, 1975.
58. F. Proschan, "Reliability Growth," *Proceedings of the Nineteenth Conference on the Design of Experiments in Army Research,* Development and Testing, Rock Island, IL, October 24–26, 1973.
59. A. Pollack and R. A. Nulk, "Reliability and Choosing Number of Prototype," *Annual Reliability and Maintainability Symposium,* pp. 84–90, 1974.
60. S. M. Pollock, "A Bayesian Reliability Growth Model," *IEEE Trans. on Reliability,* Vol. R-17, No. 4, December 1978.
61. P. S. K. Prasad, "Reliability Prediction and Growth Studies—How Accurate and Useful?," *Microelectronics and Reliability,* Vol. 13, pp. 195–202, 1974.
62. *Reliability Growth Management, Testing and Modelling,* IES Environmental Reliability Project Group, Feb. 27–28, 1978, Washington, D.C. Available from the Institute of Environmental Sciences, U.S.A.
63. R. R. Read, "A Remark on the Barlow-Schever Reliability Growth Estimation Scheme," *Technometrics,* Vol. 13, No. 1, pp. 199–200, 1971.
64. C. L. Richardson, "Is Reliability Growth Being Mismanaged?," *Reliability Growth Management, Testing and Modelling,* pp. 74–76, 1978. Available from the Institute of Environmental Sciences, U.S.A.
65. M. M. Roffman, "Reliability Growth Versus Technical Innovation," *Modeling and Simulation Conference,* pp. 531–534, 1975. Available from the Instrument Society of America.

66. M. H. Saltz, "Methods for Evaluating Reliability Growth and Ultimate Reliability During Development of a Complex System," *Fifth National Symposium on Reliability and Quality Control,* pp. 89–97, 1959. Available from the IEEE.

67. P. Sen, "Optimization of a Reliability Growth Problem," *Microelectronics and Reliability,* Vol. 13, pp. 87–89, 1974.

68. G. L. Sriwastav, "A Reliability Growth Model," *IEEE Trans. on Reliability,* Vol. R-27, No. 8, pp. 306–307, 1978.

69. D. J. Simkins, "A Reliability Growth Management Approach," *Annual Reliability and Maintainability Symposium,* pp. 356–360, 1979.

70. D. J. Simkins, "Triple Tracking Growth," *Annual Reliability and Maintainability Symposium,* pp. 158–163, 1974.

71. C. N. Stoll and W. S. Oliveri, "Reliability Growth—Actual Versus Predicted," *Annual Reliability and Maintainability Symposium,* pp. 391–395, 1974.

72. J. Sogorka and J. Peterson, "Dynamic Characteristics of Reliability Growth and its Implications," *Proc. Joint Mil-Ind. Guided Missile Res. Technol., 5th,* Chicago, 1959.

73. A. F. M. Smith, "A Bayesian Note on Reliability Growth During a Development Testing Program," *IEEE Trans. on Reliability,* Vol. R-26, No. 5, pp. 346–347, 1977.

74. B. H. Swett, "Reliability Growth Management," *Reliability Growth Management, Testing and Modelling,* 4, pp. 4–9, 1978.

75. W. E. Thompson, "Software Reliability Growth Modelling," *Reliability Growth Management, Testing and Modelling,* pp. 32–41, 1978. Available from the Institute of Environmental Sciences, U.S.A.

76. E. P. Virene, "Reliability Growth and its Upper Limit," *Annual Symposium on Reliability,* pp. 265–270, 1968.

77. W. E. Wallace, "Initial Results of the Fleet Reliability Assessment Program (FRAP)," *Reliability Growth Management, Testing and Modelling,* pp. 52–59, 1978. Available from the Institute of Environmental Sciences, U.S.A.   /

78. H. K. Weiss, "Estimation of Reliability Growth in a Complex System with a Poisson-Type Failure," *Operations Research,* Vol. 4, pp. 532–545, 1956.

79. L. A. Weaver and G. A. Rilgore, "Comparison of Burn-in and Life Test Growth Models," *ASQC Technical Conference Transactions,* pp. 53–57, 1978.

80. M. C. Weinrich and A. J. Gross, "The Barlow-Schever Reliability Growth Model from a Bayesian Viewpoint," *Technometrics,* Vol. 20, No. 3, pp. 249–254, August 1978.

81. I. E. Willard, "Reliability Growth Curves," *Annual Technical Conference Transactions,* ASQC, pp. 321–328, 1979.

82. M. Zelen (ed.), "Statistical Theory of Reliability," University of Wisconsin, Madison, pp. 149–166, 1963.

83. M. G. Zsak, "Coradcom's Reliability Growth Policy," *Reliability Growth Management, Testing and Modelling,* pp. 42–45, 1978. Available from the Institute of Environmental Sciences, U.S.A.

7
System Safety Engineering

7.1 INTRODUCTION

System safety has been receiving attention from the United States Department of Defence ever since the early fifties. In the late fifties, system safety began to be identified as a separate discipline. According to reference 9, the first military document entitled "System Safety Engineering for the Development of United States Air Force (USAF) Ballistic Missiles" was published in 1962. In 1963 the USAF published MIL-STD-38130 [82]. This document was revised in 1966. The revised edition became known as MIL-STD-38130A [83]. This document became a requirement to be followed by all the Department of Defence equipment or systems contractors. It was superseded by MIL-STD-882, entitled "System Safety Program for Systems and Associated Subsystems and Equipment: Requirements For," in 1969 [80]. Another document, entitled *System Safety Design Handbook,* was published by the United States Air Force Systems Command in 1967 [14]. Important publications on the subject are reference 1 to 130.

The following are the definitions of safety and system safety:

Safety. This may be defined [80] as freedom from those conditions that can cause damage to or loss of equipment or injury or death to human beings.

System Safety. System safety [80] is defined as the optimum level of safety subject to resource and operational effectiveness constraints attained by applying engineering and system safety management principles throughout the life cycle of a system.

This chapter discusses the following areas of system safety:

1. System safety functions
2. Product safety program approach
3. System safety analytical techniques
4. Legal aspects of safety.

For those readers who want to delve further in the subject, a bibliography on the subject is presented at the end of this chapter.

7.2 SYSTEM SAFETY FUNCTIONS

The following list presents some of the system safety functions:

1. Establishing requirements for basic design to prevent accidents
2. Participate in the hazard analysis throughout the entire system design
3. Design review participation
4. Emergency procedures determination
5. Maintaining accident/safety records
6. Recommending for the need of test, study, or research
7. Liaison with outside safety bodies
8. Providing education and training on the subject of safety
9. Developing plans for accident investigation
10. Taking part in investigations of accidents
11. Communicating to others the information to prevent accidents
12. Follow-up all actions which were the result of accident investigations
13. Prepare a system safety management plan

Although the above list of system safety functions is not all-inclusive, it clearly shows the wide system safety interest spectrum. System safety is related to relatively new disciplines such as reliability, value engineering, human factors, maintainability, logistics/product support, and quality assurance. Figure 7.1 shows the relatively new disciplines for system effectiveness. The definition of the system effectiveness is the probability that a system, within a specified time, can successfully meet an operational demand when used under stated conditions.

7.3 PRODUCT SAFETY PROGRAM APPROACH

Today, engineering managers are faced with the increasing costs of hardware and system analysis. In some aerospace organizations [9], 10 to 15 percent of the total program dollars are spent on the product assurance areas. We

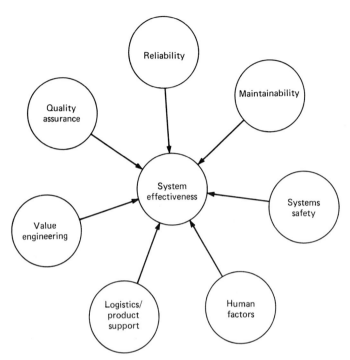

Figure 7.1 New disciplines for system effectiveness.

may ask the question, how does the system safety integrate into the overall product development cycle? This cycle may be divided [9] into the following four phases:

1. Conceptual phase
2. Design and development phase
3. Manufacturing phase
4. Operational phase

7.3.1 Conceptual Phase

During the conceptual phase, the following system safety tasks should be accomplished:

1. Prepare the product safety plan. The plan should state the procedures to establish and execute the system safety program. The system safety plan is a basic management document which is used by program managers to measure the progress of the system safety effort during the program life cycle. Some of the areas to be included in this plan are:

   a. The establishment and maintenance of types of records
   b. The type of system safety analysis to be performed
   c. The organization of the safety management and its relationship to other program functions
   d. The identification of hazards to be controlled and minimized to a level of acceptability

2. Participation in meetings for concept formulation and to ensure that all the associated hazards are considered.
3. Establishing a file of system safety information and documentation.
4. Preliminary level hazard analysis preparation.

7.3.2 Design and Development Phase

During this phase, the system safety effectiveness is determined. At this stage, it is the responsibility of the system safety engineer to ensure that all hazards were identified in the conceptual phase and are eliminated or controlled and minimized.

During the design and development, the following system safety efforts should be taken care of:

1. Implementation of the system safety plan
2. Circulation of the design criteria of safety to be incorporated in the system design
3. Inclusion of safety requirements into the specification for subcontractors
4. Documentation of decisions regarding system safety
5. Participation in design meetings to ensure system safety
6. Preliminary hazard analysis update and its possible expansion to establish a complete system safety hazard analysis

7.3.3 Manufacturing Phase

Some of the system safety efforts which are to be accomplished during the manufacturing phase are:

1. Products, processes, and test procedures are to be evaluated and reviewed from the safety point of views
2. System safety verification tests and inspections are to be outlined for the use by the quality assurance personnel.
3. Accomplish the Operational System Safety Analysis (OSSA) which will be useful to the system operation and maintenance personnel. These analy-

ses could already have been included in the conceptual and design and development phases. However, the Operational System Safety Analysis should be separated into one report.

4. Examine the system design or other changes to ensure system safety.
5. Establish the product failures feedback system to examine failures from the standpoint of hazardous conditions.
6. Establish a monitoring program to ensure that all actions were performed as outlined in the system safety hazard analysis specification.

7.3.4 Operational Phase

This is a system safety verification and monitoring phase. If the system safety effort was performed effectively in the other three phases, then this phase should be hazard free.

Some of the system safety efforts to be accomplished during this phase are:

1. Audit maintenance, operating and emergency procedures. In addition, evaluate their adequacy as conceived in the product design.
2. Retain the failure feedback system as established in the manufacturing phase. Utilize this feedback system to make decisions regarding the future product safety.
3. To ensure safety, evaluate the operational equipment modifications and changes from the safety point of views.
4. Accidents and incidents to be analyzed and reviewed. Follow-up on results to eliminate reoccurrences.
5. Audit the operational system safety to ensure the maintenance of the desired level of safety.

7.4 SYSTEM SAFETY ANALYTICAL TECHNIQUES

There are several analytical techniques developed in system safety engineering. The aim of these techniques is toward the hazard identification and control. Some of the main objectives of [91] system safety analysis are

1. To demonstrate by means of analysis, that the safety specifications, criteria, or objectives are fully satisfied.
2. To provide engineering professionals and decision makers with the necessary information and data to perform trade-off studies to achieve optimum system safety.
3. To assist the safety analyst to better understand and evaluate the system safety by performing the analysis.

7.4.1 System Safety Analyses Elements

As shown in Fig. 7.2, identification, evaluation, and communication are the basic three key elements of system safety analyses. These three elements must be balanced within any analytical approach, in order to integrate the safety considerations effectively into the system engineering activity. These elements are presented in the following sections:

(a) Identification.   Obviously, potential hazard identification is the first step in any form of system safety analysis. It must be understood that hazard identification alone will not be enough so that it will be properly controlled. Therefore, one has to go one step further to ensure that it is adequately evaluated and communicated to those concerned with the decision-making responsibilities.

In complex systems, the potential hazards are uncovered by looking at the same aspect from different directions by following a systematic approach.

(b) Evaluation.   It must be understood that all hazardous events are not of equal importance. In other words, the resulting effects of hazards could be minimal or catastrophic. The hazard levels should be categorized by using MIL-STD-882 [80, 81].

The other aspect of the hazard evaluation deals with determining the likelihood of occurrence of the hazardous event. The likelihood of occurrence of an event could be stated in qualitative or quantitative form. Qualitatively, it could be classified as improbable, unlikely, or possible, whereas, quantitatively, it is given in numeric terms (i.e., rate or probability of occurrence).

The third or the final aspect of evaluation is concerned with assigning the priority order to hazardous events.

(c) Communications.   Communicating the end results of system safety analysis to other concerned bodies is equal in importance to evaluation and identification, because the corrective measures have to be taken to implement the

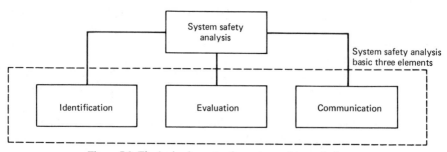

Figure 7.2 The basic elements of a system safety analysis.

evaluation phase findings. To communicate safety analysis results to others, the basic requirements are timeliness, accurateness, and concisiveness.

7.4.2 System Safety Analysis Classification

System safety analysis techniques [91] may be categorized into one of three basic classifications

1. Conceptual system safety analysis
   This category includes two techniques
   a. Preliminary hazard analysis (PHA)
   b. Preliminary safety matrix analysis (PSMA)
2. System design and development safety analysis
   Under this classification are
   a. Subsystem hazard analysis
   b. System hazard analysis
   c. Safety logic model analysis
   d. Fault hazard analysis (FHA)
   e. Fault tree analysis (FTA)

The fault hazard and fault tree analysis techniques will be discussed later in the chapter. Readers interested in subsystem (design safety matrix) hazard, system hazard, and safety logic model analysis techniques should consult reference 91.

3. Functional system safety analysis
   Two techniques that fall into this category are
   a. Operating hazard analysis (OHA)
   b. Operational safety matrix analysis (OSMA)

Details of some of the techniques of the above three main classifications (1, 2, and 3) are presented in the following sections.

7.4.3 Conceptual System Safety Analysis

Conceptual system safety analysis techniques are used to perform an initial safety evaluation analysis of a system during the conceptual phase. The conceptual safety analysis provides the basis of subsequent system safety tasks, criteria, and requirements. This type of analysis should include the following items as applicable

1. Hazardous components identification
2. Product life cycle hazard operations, testing, and emergency condition identification
3. Support equipment, training, routine, and special safety related facility identification, and so on.

The safety analyst performing the conceptual system safety analysis should have access to the reference input information such as system functional flow diagrams, drawings, data, design sketches, and so on. Two conceptual system safety analysis techniques (PHA and PSMA) are discussed in the following sections.

Preliminary Hazard Analysis Technique (PHA). This is relatively an unstructured technique because of the unavailability of definitive functional flow diagrams, drawings, and so on, during the product conceptual phase. The safety analyst should structure his or her own understanding in such a way that the objective of the conceptual phase safety analysis is fulfilled.

Preliminary Safety Matrix Analysis Technique (PSMA). This widely used procedure uses a matrix form to document and perform the preliminary hazard analysis. There are many types of preliminary hazard analysis matrix formats which can be used to accomplish a systematic and comprehensive assessment. These formats are quite similar, therefore, one should tailor the format according to the need of the system in question.

A typical matrix is shown in Fig. 7.3. The steps to be used when performing the preliminary safety matrix analysis are

1. Subdivide the system (Column 1) into
   a. Subsystems or items
   b. Functions or operations
   c. Life cycle stages such as storage, repair, use, testing, and so on
   Make sure that all these steps are properly accomplished for the entire system.
2. Identify each item's hazardous aspects for each mode (Columns 2, 3). Accomplish this for the entire system.
3. Categorize each hazardous event's potential effects (Column 4). Each hazardous event's effects could be divided into four categories
   a. Catastrophic
   b. Critical
   c. Marginal
   d. Negligible

Subject: *PRELIMINARY SAFETY MATRIX ANALYSIS*

Col. 1	Col. 2	Col. 3	Col. 4	Col. 5	Col. 6
Function/ Element	Mode	Identification of Hazardous Aspects	Classification of Hazard	Safety Provision Needs	Priorty of Cor- rective Action

System:	*Sub- system*	*Date Prepared by*	*Page No. ___ of ___*
			Revision

Figure 7.3 A typical preliminary safety matrix format.

4. Outline the safety provisions needed to control or eliminate each hazardous event (Column 5). These safety provisions may include alternate design procedures, incorporation of fail-safe devices, inclusion of hazardous event warning devices, and so on.
5. Evaluate and assign the priority ranking of corrective action (Column 6) required for the Column 5. A suggested list of priority classification is
   a. *Urgent:* This signifies that the hazarous event requires immediate management attention.
   b. *Critical:* This dictates the special attention of management.
   c. *Routine:* Corrective action should be carried through routine channels.
   d. *Special:* This dictates special follow-up action on a certain corrective action which may otherwise be a problem if it carried out through the routine channels.

For more detailed information on this technique, consult reference 91.

7.4.4 System Design and Development Safety Analysis

These analyses are performed during the equipment design and development stage. Some of the objectives of the analysis are

1. To verify that the established safety related criteria is satisfied for a given product design procedure
2. To identify hazardous design features
3. To identify product design changes that will enhance safety

In order to perform an effective safety analysis at the design and development stage, the safety analyst must have access to the following information

1. System/subsystem/item design specifications and drawings
2. Product functional and operating environment details
3. Details of maintenance, handling, and servicing equipment
4. Interface drawings
5. Documentation of earlier safety related analyses and studies

Two system design and development safety analysis techniques are presented in the following sections (for other techniques, the reader should consult reference 91).

Fault Hazard Analysis Technique. This technique is used to define the effects of component failure modes. In addition, this technique is utilized to categorize the component failure modes effect on equipment. The form shown in Fig. 7.4 is used to perform a Fault Hazard Analysis. This form includes some of the most important items to be considered. The following sections describe each item of the form briefly.

System:		Analyst:			Date:	
(a) Major system components	(b) Failure modes of components	(c) Operational modes of the system	(d) Major component failure rates	(e) Secondary component failure factors	(f) Categorization of hazard effects	(g) Remarks

Figure 7.4 A fault hazard analysis form.

a. Major items or components of the system: These are normally defined by the analysts according to their judgement which is influenced by various factors. A suggested list of major components is mechanical devices, chemical systems, safety devices, electrical systems, wiring, electronic logic circuits, etc.

b. Failure modes of a component: This requires the listing of all failure modes of each major component under study.

c. Operational modes of the system: This requires the analysis of the failure modes of each major component in order to investigate the effects on all operational modes of the system.

d. Major component failure rate: This information is obtained from the field failure data or other sources for each failure mode of each major component.

e. Secondary component failure factors: There could be a situation when a major component has abnormal or out-of-tolerance inputs. An analyst should take into consideration the transient effects, temperature limits, chemical effects, power reversal effects, vibration limits, etc., on a system when the major components are to operate as part of a system.

f. Categorization of hazard effects: In this column, the hazard effect should be categorized as safe, marginal, critical, or catastrophic.

g. Remarks: One may include in this column any additional information needed for clarification or verification presented in other columns. Furthermore, this column could be used for any other desired remarks.

This technique is described in detail in reference 91.

Fault Tree Analysis Techniques. This technique has been discussed in Chapter 4 of this book. It is used to evaluate reliability and safety of complex systems. The fault tree method was first developed at the Bell Laboratories to analyze the Minuteman Launch Control System. More information on the technique may be found in references 131 and 132. Most of the fault tree symbols are presented in Chapter 4. Some of them are repreated here for completeness.

OR Gate. This symbol represents an output event that occurs if any one or more of the n input events occur.

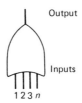

AND Gate. This is the dual of the OR gate. An AND gate denotes that an output event only occurs if all of the n input events occur.

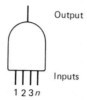

Resultant Event. This event is represented by a rectangle. The rectangle represents an event which results from the combination of fault events through the input of a logic gate.

Basic Fault Event. This event is denoted by a circle. It represents the failure of an elementary component or a basic fault event. The event parameters such as the probability of occurrence, the failure rate, and the repair rates are obtained from the field failure data or other reliable sources.

Incomplete Event. This event is denoted by a diamond. It denotes a fault event whose cause has not been fully determined, either due to lack of interest or to lack of information or data.

Transfer-in and Transfer-out.   These situations are denoted by a triangle. A line from the top of the triangle denotes transfer-in, whereas a line from the side represents transfer-out.

Conditional Event.   This event is represented by an ellipse. This symbol indicates any condition or restriction to a logic gate.

Other fault tree symbols may be found in reference 131.

EXAMPLE 1

An OR gate situation is represented by the hypothetical example shown in Fig. 7.5. The top event, "Fire alarm system without power," will occur if either the power line or the commercial power fails.

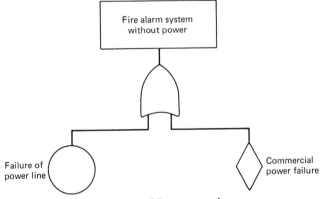

Figure 7.5 OR gate example.

EXAMPLE 2

An AND gate situation is represented by the hypothetical example shown in Fig. 7.6. The top event, "Fire without fire alarm," can only occur if the fire alarm is disabled and there is a fire. Furthermore, the fire alarm must be disabled before the fire.

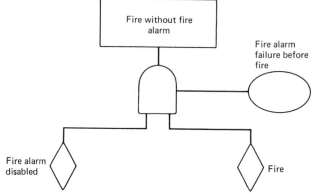

Figure 7.6 AND gate example.

EXAMPLE 3

A simple example of a fault tree is shown in Fig. 7.7. Note that this fault tree contains the logic gates of Fig. 7.5 and 7.6. The top event of the tree is "House fire without fire alarm."

The main idea here is to investigate the occurrence of the top event. The intermediate and basic events are shown in Fig. 7.7. These events contribute to the occurrence of the top event.

Probability Evaluation of Fault Trees. In this section, we will deal with the probability evaluation of fault trees. The OR and AND gate top (output) event probability expressions for the statistically independent input events are given below.

OR Gate

$$P(a + b + c + \cdots) \simeq P(a) + P(b) + P(c) + \cdots \qquad (7.1)$$

where $P(a + b + c + \cdots)$ is the output event's occurrence probability
$P(a)$ is the probability of occurrence of input event a
$P(b)$ is the probability of occurrence of input event b
$P(c)$ is the probability of occurrence of input event c

Equation (7.1) is true for very small input events probabilities. An OR gate representing Eq. (7.1) is shown in Fig. 7.8.

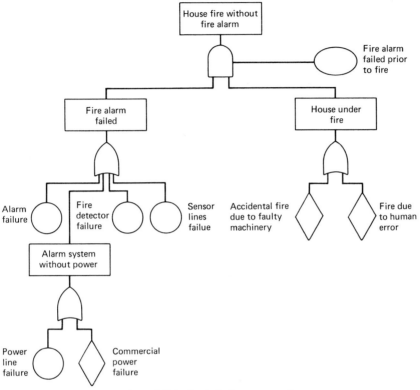

Figure 7.7 A simple fault tree example.

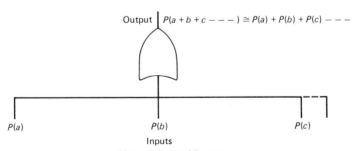

Figure 7.8 An OR gate.

AND Gate

$$P(a \cdot b \cdot c \cdots) = P(a) \cdot P(b) \cdot P(c) \cdots \tag{7.2}$$

where $P(a)$, $P(b)$, $P(c)$, . . . are the probability of occurrence of input events a, b, c, . . . , respectively.

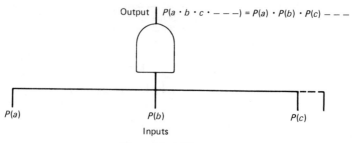

Figure 7.9 AND gate.

An AND gate representing Eq. (7.2) is shown in Fig. 7.9.

The above expressions are also presented in Chapter 4 and reference 131.

EXAMPLE 4

For the probability failure data in Fig. 7.10, evaluate the top event occurrence probability.

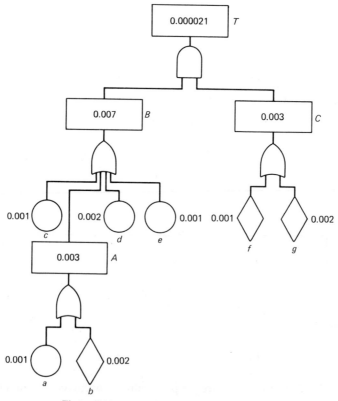

Figure 7.10 A probability evaluation fault tree.

The letters A, B, C, and T in Fig. 7.10 represent the resultant fault events, whereas the letters a, c, d, e, and b, f, g denote basic and incomplete fault events, respectively.

Proceeding step by step, the following results are calculated from Fig. 7.10.

Using Eq. (7.1), the event A probability is

$$P(A) \simeq P(a) + P(b) = 0.001 + 0.002 = 0.003 \qquad (7.3)$$

The event B probability is

$$P(B) \simeq P(c) + P(A) + P(d) + P(e) \qquad (7.4)$$

where P(A) is given by Eq. (7.3).

$$\therefore P(B) \simeq 0.001 + 0.003 + 0.002 + 0.001 = 0.007 \qquad (7.5)$$

Similarly, the event C probability is

$$P(C) \simeq P(f) + P(g) \simeq 0.001 + 0.002 = 0.003 \qquad (7.6)$$

To obtain the top event probability expression for $P(T)$, use Eq. (7.2) and then substitute in the numerical results of Eqs. (7.5) and (7.6):

$$P(T) = P(B) \cdot P(C) = 0.007 \times 0.003 = 0.000021 \qquad (7.7)$$

The probability of occurrence of the top event is 0.000021.

For those readers who wish to probe deeper into the fault tree technique, refer to references 131 and 132.

7.4.5 Functional System Safety Analyses

The main objective of these analyses is to identify and determine the safety aspects associated with personnel, environment, equipment, and approaches involved during the system life cycle. These analyses may be performed on activities such as maintenance, support, testing, installation, training, operations, and so on. The end results of such analyses provide input for the need

1. To include system safety devices and consider design modifications in order to eliminate potential hazards.
2. To have special handling, storage, transportation, and servicing procedures.

3. To have special inspections and other procedures for the system operating and maintenance instructions.

The Functional System Safety Analysis (FSSA) must start at the development stage of the system. The analysis techniques associated with (FSSA) are

1. Operating Safety Matrix Analysis
2. Operating Hazard Analysis

These techniques are discussed in detail in reference 91.

7.5 LEGAL ASPECTS OF SAFETY

Perhaps the recent emphasis on the system safety may be due to the recent court decisions in product liability suits. The term product liability is widely used by the product manufacturers, users, and others. This term is used when a product performance is encountered in such a way that the product's integrity is questioned in a court.

The annual number of court cases involving product liability were around [9] 50,000 in 1963 in the United States. At the beginning of the last decade, the number of court cases approached 500,000. The number of product liability suits predicted for the mid-seventies was around one million.

The following are among the reasons for consumers' filing court cases against manufacturers:

1. The direct cause of injury was the defective product.
2. The occurrence of the dangerous condition was the result of defective design or manufacturing.
3. The existence of a defect when the product was sold by the seller.

In the past few years, the negligence of the following product actions has been established by liability suits:

1. Operational instructions misleading or incomplete.
2. Inappropriate tests to find defects.
3. Failure to warn of possible dangers to consumers. These could be due to the use or misuse of the product.
4. Inability to obtain information regarding the consequences of the product use or misuse.

Some of the areas normally investigated to build a product liability suit are

1. False product advertising claims
2. Procedures used for testing
3. Selection of materials
4. Design defectiveness
5. Murphy's law
6. Departure from specifications, manufacturing procedures, blueprints, etc.
7. Quality of materials used
8. Product safety procedure inadequacy

7.6 SUMMARY

This chapter has discussed the various areas of system safety. Furthermore, this chapter emphasizes the structure of the concepts and procedures rather than the minute details, because it is impossible to cover the subject of system safety engineering effectively in a single chapter. An extensive list of references will be found at the end of this chapter.

7.7 REFERENCES

Books

1. D. B. Brown, *Systems Analysis and Design for Safety,* Prentice Hall, Inc., Englewood Cliffs, NJ, 1976.
2. R. A. Cole, *Industrial Safety Techniques,* West Publishing Company, Sydney, Australia, 1975.
3. R. De Reamer, *Modern Safety Practices,* Wiley, New York, 1958.
4. W. Hammer, *Handbook of System and Product Safety,* Englewood Cliffs, NJ, Prentice Hall, 1972.
5. W. Hammer, *Occupational Safety Management and Engineering,* Prentice-Hall, Englewood Cliffs, NJ, 1975.
6. W. Handley, *Industrial Safety Handbook,* McGraw-Hill, London, 1969.
7. J. Kolb and S. S. Ross, *Product Safety and Liability, A Desk Reference,* McGraw-Hill, New York, 1980.
8. S. W. Malasky, *System Safety: Planning, Engineering, and Management,* Hayden, Rochelle Park, NJ 1974.
9. W. P. Rodgers, *Introduction to System Safety Engineering,* Wiley-Interscience, New York, 1971.
10. A. D. Swain, *The Human Element in System Safety: A Guide for Modern Management,* Industrial and Commercial Techniques, London, 1974.
11. W. E. Tarrants (ed.), *A Select Bibliography of Reference Materials in Safety Engineering and Related Fields,* Published by ASME, 1967.
12. Versicherungs Allianz, *Handbook of Loss Prevention,* Springer-Verlag, New York, 1978.

Articles and Reports

13. L. A. Adkins, and M. C. Thatro, "Reliability Requirements for Safe All-Weather Landings," *Annals of Assurance Sciences, Seventh Reliability and Maintainability Conference,* 1968, pp. 328–335.

14. AFSC. DH 1–6 AFSC, Design Handbook, *System Safety,* July, 1967.

15. AFSCM 127–1, *Safety Management,* January, 1967.

16. AMC Pamphlet 385–23, *Management System Safety,* July, 1969.

17. K. K. Arora and H. C. Edfors, "Automated Production System Safety," *Proceedings Annual Reliability and Maintainability Symposium,* 1975, pp. 531–537.

18. L. W. Ball and R. G. Penny, "Safety Achievement Through Systems Safety Engineering," *Annals of Reliability and Maintainability,* 1971, pp. 278–285.

19. L. W. Ball and R. G. Penny, "Safety Inputs to Development Program Plans," *Annals of Reliability and Maintainability,* 1971, pp. 258–267.

20. J. A. Barton, "The State of Art of Safety Analysis," LTV Aerospace Corporation, Presented at the *Aviation Contractors' Safety Representatives' Conference,* Norfolk, Virginia, 1965.

21. J. A. Barton, "Operational Safety Analysis Techniques," *Annals of Reliability and Maintainability,* 1965, pp. 299–302.

22. G. A. Berman, "Dare You Watch Your Color TV Without a Fire Extinguisher?," *Proceedings Annual Reliability and Maintainability Conference,* 1975, pp. 528–530.

23. C. B. Boehmer, "Safety Aspects of Personal Rapid Transit," *Proceedings Annual Reliability and Maintainability Symposium,* 1974, pp. 210–216.

24. P. H. Bologer, "Evaluation of Safety Programs," *Annals of Reliability and Maintainability,* 1970, pp. 614–621.

25. Booz-Allen Applied Research, "Vehicle-in-Use Systems Safety Analysis," Final Report, Contract No. FH-11–7316 for the National Highway Traffic Safety Administration, Washington, D.C., September 1970.

26. E. S. Brown, "Systems Safety and Human Factors: Some Necessary Relationships," *Proceedings Annual Reliability and Maintainability Symposium,* 1974, pp. 197–200.

27. S. Canale, "System Safety Measurement and Control," *Annals of Reliability and Maintainability,* 1966, pp. 482–489.

28. S. Canale, "Safety Measurement Concept for Naval Weapons System," Contract No. 178–67-C-0036, U.S. Naval Weapons Lab, Norfolk, Virginia, 1967.

29. S. Canale, "Hazard Risk Measurement and Optimization," *Government-Industry Systems Safety Conference,* 1968, published by the NASA Safety Office, Washington, D.C.

30. E. Christie, "Analysis of the Results of Diagnostic Tests of Automobiles and Their Relation to Used Car Safety," *Proceedings Annual Symposium on Reliability,* 1971, pp. 145–149.

31. N. E. Classon, "Results for Systems Safety," *Reliability and Maintainability Conference,* 1969, p. 315.

32. N. E. Classon, "Pre-Accident Investigation," *Government-Industry Systems Safety Conference Proceedings,* 1968, published by the NASA Safety Office, Washington, D.C.

33. J. de S. Coutinho, "Failure-Effect Analysis," *Transactions of the New York Academy of Sciences,* Vol. 26, No. 5, March 1964, pp. 564–584.

34. R. D. Craig, "Application of Safety Disciplines to SRAM Program," *Proceedings 1973 Annual Reliability and Maintainability Symposium,* pp. 392–402.

35. G. E. Cranston and G. F. Ruff, "System Safety Hazard Analyses," *Government-Industry Systems Safety Conference Proceedings,* 1968, published by the NASA Safety Office, Washington, D.C.

36. D. M. Critchiow, "System Safety Applications to United States Air Force," *Proceedings Annual Reliability and Maintainability Symposium,* 1972.

37. R. N. Danzeisen, J. A. Matayka, and D. W. Weiss, "A Reliability and Safety Analysis

of Automotive Vehicles," *Proceedings of the 1971 Annual Symposium on Reliability and Maintainability,* Washington, D.C.

38. E. S. Dean, "Engineering Risk Reduction in Satellite Programs," *Proceedings Annual Reliability and Maintainability Symposium,* 1979.

39. F. J. Denny, "A System Engineering Approach to Safety for a Naval Operational Unit," *Annals of Reliability and Maintainability,* 1970, pp. 445–448.

40. F. J. Denny, "Application of System Safety to Commercial Products," *10th Reliability and Maintainability Conference,* Vol. 10, 1971.

41. W. R. Downs and D. B. Boies, "Detection and Minimization of Chemically Induced Ignition Hazards of Glycol Fluids," *Annals of Assurance Sciences, Reliability and Maintainability Conference,* 1969, pp. 316–324.

42. P. E. V. Dries, "Rational Risk Assessment for Defence System Safety," *Proceedings Annual Reliability and Maintainability Symposium,* 1979.

43. F. R. Farmer, "Reactor Safety and Siting: A Proposed Risk Criterion," *Nuclear Safety,* 8(6), Nov.–Dec. 1967, pp. 539–548.

44. R J. Feutz and J. Tracy, "Fault Tree for Safety," The Boeing Company, D6–5713, January 3, 1966, Seattle, WA.

45. R. J. Feutz and T. A. Waldeck, "The Application of Fault Tree Analysis to Dynamic Systems," *Systems Safety Symposium,* June 1965, University of Washington, Seattle.

46. J. B. Fussel, D. P. Wagner, J. S. Arendt, and J. J. Rooney, "Improving System Safety Through Risk Assessment," *Proceedings Annual Reliability and Maintainability Conference,* 1979.

47. F. X. Gavigan and J. D. Griffith, "The Application of Probabilistic Methods to Safety Research and Development and Design Choices, Nuclear Systems Reliability Engineering and Risk Assessment," SIAM, 1977, pp. 22–41.

48. L. S. Gephart and R. H. Keegan, "Flight Test Safety Analysis of the All-Weather Landing System (AWLS) Program," *Proceedings Annual Reliability and Maintainability Symposium,* 1974, pp. 436–439.

49. R. B. Gordon, "Defining the System Mission," *Annals of Reliability and Maintainability,* Vol. 6, 1967, pp. 536–541.

50. R. B. Gordon, "Application of Systems Analysis in Occupational Hazard Situations," *Proceedings 1969 Professional Conference,* ASSE, Park Ridge, IL.

51. R. B. Gordon and P. Woodbury, "Product Safety Assurance Through Systems Safety Techniques," *Annals of Reliability and Maintainability,* 1970.

52. K. Green and W. Cinibulk, "Quantitative Safety Analysis," *Proceedings Annual Reliability and Maintainability Symposium,* 1972.

53. D. F. Haasl, "Advanced Concepts in Fault Trees," *Systems Safety Symposium,* University of Washington, Seattle, WA 1965.

54. W. Hammer, "Numerical Evaluation of Accident Potentials," *Annals of Reliability and Maintainability,* Vol. 5, 1966, pp. 494–504.

55. J. Hrizina, "Single-Point Failure Analysis in Systems Safety Engineering," *Professional Safety,* March 1980, pp. 20–26.

56. K. Inone and H. Daito, "Safety Analysis of Automobile Brake Systems by Fault Tree Methods," *HOPE International, JSME Symposium,* 1977, pp. 213–220. Available from the JSME, Tokyo.

57. M. L. Jones, "Systems Safety Management by Hazard Evaluation and Risk Control," *Third International Systems Safety Conference,* Washington, D.C., October 1977, pp. 1013–1031.

58. J. R. Jordon and R. L. Buchanana, "Systems Safety—A Quantitative Fallout from Reliability Analysis," *Proceedings of the Sixth Reliability and Maintainability Conference,* July 1967.

59. K. Kanda, "Computerization of System/Product Fault Trees," *8th Reliability and Maintainability Conference,* 1969.

60. K. Kanda, "Concept of System Safety Mathematics," *Systems Safety Symposium,* Seattle, WA, June 1965.

61. K. Kanda, "Sub-System Safety Analysis Techniques," *Annals of Reliability and Maintainability,* Vol. 4, 1965.

62. D. M. Kelly, "Flight Safety Prediction Technique: Development Program," Phase II-B, Document No. 754–01–985 ARINC Research Corporation, Santa Ana, CA.

63. G. H. Kinchin, "Plant Reliability and Safety," Safety and Reliability Directorate, UKAEA (1978), U.K.

64. C. L. Kitchens, "The Probability of 2, Redundancy and Reliability," *Professional Safety,* March 1980, pp. 27–30.

65. H. J. Kolodner, "The Quantification of Safety," *Journal of the American Society of Safety Engineers,* March 1965.

66. H. J. Kolodner, "Correlation of Systems Safety to System Reliability," *Eleventh National Symposium on Reliability and Quality Control,* 1965, pp. 347–360.

67. T. J. Lebel, "The RAIDS Data Systems as a Tool of Systems Safety Engineering," *Annals of Reliability and Maintainability,* Vol. 6, 1967, pp. 648–652.

68. "Launch Control Safety Study, Vol. I & II," Bell Laboratories, September 15, 1962, NJ.

69. A. M. Lester, "Airline Safety Record," *18th Annual International Air Safety Seminar,* November 7–10, 1965, pp. 16–20.

70. B. K. O. Lundberg, "Speed and Safety in Civil Aviation, Part II—Safety," *The Aeronautical Research Institute of Sweden,* Stockholm, 1963, Report 65.

71. B. K. O. Lundberg, "The Allotment-of-Probability-Shares, APS Method," *The Aeronautical Research Institute of Sweden (FFA),* FFA Memo PE-18, April 26, 1966.

72. W. F. Jutzweit, "Safety in Poseidon Technical Manuals," *Proceedings Annual Reliability and Maintainability Symposium,* 1973, pp. 391–396.

73. D. V. MacCollum, "Reliability as a Quantitative Safety Factor," *ASSE Journal,* May 1969.

74. E. D. MacKenzie, "On Stage for System Safety," *ASSE Journal,* October 1968.

75. J. A. Matayeka, "Maintainability and Safety of Transit Buses," *Proceedings Annual Reliability and Maintainability Symposium,* 1974, pp. 217–225.

76. G. B. McIntire, "Development of a System Safety Design Handbook," Martin Marietta Corporation, Denver, Colorado, AIAA paper No. 67–849, *AIAA 4th Annual Meeting and Technical Display,* Anaheim, CA, October 1969.

77. B. B. McIntosh and C. E. Corneu, "The Probabilistic Prediction of Human Behaviour for Assessment of Human Error Contribution to System Safety Analysis," McDonnell-Douglas Astronautics Company, Huntington Beach, CA, 1965.

78. J. F. Medford, "System Safety on the Fleet Satellite Communication Program," *Proceedings of the Annual Reliability and Maintainability Conference,* 1974, pp. 179–185.

79. J. M. Michels, "Computer Evaluation of the Safety Fault Tree Model," *Systems Safety Symposium,* Univ. of Washington, Seattle, WA, June 1965.

80. MIL-STD-882, Systems Safety Program for System and Associated Subsystem and Equipment—Requirements for, July 15, 1962.

81. MIL-STD-882-A, Systems Safety Program Requirements, June 28, 1977.

82. MIL-STD-38130 (A) Proposed, Safety Engineering of Systems and Associated Sub-systems and Equipment—General Requirements for, Sept. 30, 1963.

83. MIL-S-38130 (USAF), Safety Engineering of System and Associated Subsystems and Equipment—General Requirements for, March 1966.

84. MIL-STD-58077, Safety Engineering of Aircraft System, Associated Subsystem and Equipment—General Requirements for, October 26, 1967.
85. MIL-STD-23069, Safety Requirements Minimum for Air Launched Guided Missiles, October 31, 1965.
86. C. O. Miller, "Hazard Analysis and Identification in System Safety Engineering," *Annals of Reliability and Maintainability Conference,* 1968, pp. 336–343.
87. C. O. Miller, "The Role of Systems Safety in Aerospace Management," Institute of Aerospace Safety and Management, University of Southern California, Los Angeles, August 1966.
88. L. C. Montgomery, "Improving Aerospace System Safety Programs Through an Intercompany Information Exchange," *Annals of Reliability and Maintainability,* Vol. 6, 1967, pp. 542–546.
89. G. B. Mumma, "System Safety, An Engineering Discipline," *7th Reliability and Maintainability Conference,* 1968, pp. 321–327.
90. G. B. Mumma, "An Information Analysis Center for Safety Documentation," *NASA System Safety Symposium,* Goddard Space Flight Center, May 23, 1968.
91. G. B. Mumma and R. B. Gordon, Chairmen, "System Safety Analytical Techniques," Safety Engineering Bulletin, No. 3, May 1971. Available from the Electronic Industries Association, Washington, D.C.
92. P. M. Nagel, "The Monte Carlo Method to Compute Fault-Tree Probabilities," Presented at the *System Safety Symposium,* Univ. of Washington, Seattle, WA, June 1965.
93. R. L. Newman and R. G. Snyder, "General Aviation System Safety Engineering," *Third International System Safety Conference,* Washington, D.C., October 1977.
94. L. W. Newton and J. F. Krey, AH-56A (AAFSS), "Safety Engineering," *Annals of Reliability and Maintainability,* Vol. 6, 1967, pp. 529–535.
95. NHB 5300.4 (ID-1), "Safety, Reliability, Maintainability, and Quality Provisions for the Space-Shuttle Program," Reliability and Quality Assurance Publication, NASA, August 1974.
96. NHB-1700–1, Vol. 1, "NASA Safety Manual, Basic Safety Requirements," NASA, July 1969.
97. NHB-1700–1, Vol. 3, "NASA Safety Manual, Systems Safety," NASA, March 1970.
98. Office of the Manned Space Flight (NASA), Safety Program Directive No. 1, System Safety Requirement for Manned Space Flight, January 1969.
99. R. E. Olson, "Remarks on Safety Criteria," Presented at the Institute of Aerospace Safety and Management, University of Southern California, Los Angeles, October 5, 1967.
100. M. Onda, "Safety Design in the Comprehensive Automobile Traffic Control System," *HOPE International, JSME Symposium,* 1977, pp. 221–228. Available from the JSME, Tokyo.
101. J. R. Penland, "A Formulation for Risk Assessment and Allocation," *Proceedings Annual Reliability and Maintainability Symposium,* 1975, pp. 1–5.
102. G. A. Peters and F. S. Hall, "To Cut Down Accidents Design for Safety," *Product Engineering,* September 1965, pp. 125–128.
103. G. A. Peters, "Human Error: Analysis and Control," *ASSE Journal,* Technical Section, Vol. XI, No. 1, January 1966, pp. 9–15.
104. G. A. Peters and F. S. Hall, "Missile System Safety; An Evaluation of Test Data," Report R-5135, Rocketdyne, Canoga Park, CA., March 1963 (AD 418 644, NASA, N-63–15902).
105. G. Peterson and R. S. Babin, "Integrated Reliability and Safety Analysis of the DC-10 AWLS," *Proceedings Annual Reliability and Maintainability Symposium,* 1973, pp. 403–409.

106. N. B. Petter, "Fault-Tree Analysis as a Tool for System Safety Engineering," North American Aviation Astronautics, Anaheim, CA, Report X5-1002/319, January 14, 1965.
107. L. L. Philipson and M. S. Schaeffer, "Hazardous Materials Shipment Data in Risk Analysis," *Proceedings Annual Reliability and Maintainability Symposium,* 1975, pp. 519–527.
108. E. W. Pickrel and T. A. McDonald, "Quantification of Human Performance in Large Complex Systems," *Human Factors,* December 1964.
109. W. C. Pitts, "Summary Report of Reliability-Safety Analysis Methodology for Manned Space Vehicles," Chance Vought Aircraft Inc., Dallas, July 1960.
110. "Reactor Safety Study—An Assessment of Accident Risks in U.S. Commercial Nuclear Power Plants," Wash.-1400, USNRC, Washington, D.C., 1975.
111. J. L. Recht, "Systems Safety Analysis: Error Rates and Costs," *National Safety News,* June 1966.
112. J. L. Recht, "Systems Safety Analysis: Failure Mode and Effect," *National Safety News,* February 1966.
113. J. L. Recht, "Systems Safety Analysis: An Introduction," *National Safety News,* February 1966.
114. N. P. Rodgers, "Integration of System Safety in Total Product Integrity," *10th Annual West Coast Reliability Symposium,* 1969, pp. 217–223. Available from Western Periodicals Company, West Hollywood, CA.
115. P. Rubel, "Cultivating the Logic Tree for Reactor Safety," *Proceedings Annual Reliability and Maintainability Symposium,* 1975, pp. 13–17.
116. G. F. Ruff, "The Role of Systems Safety Engineering and its Relation to Reliability," *Technical Conference Transactions,* ASQC, Los Angeles, 1965.
117. "Safety and Failure of Components," *Proceedings 1969–70,* Vol. 184, Part 3B, Institution of Mechanical Engineers (British). Available from the IME, 1, Birdcage Walk, Westminster, London.
118. "Safety and Operational Guidelines for Undersea Vehicles," *Marine Technology Society,* 1968, Washington, D.C.
119. "Safety and Reliability of Metal Structures," *Conference Proceedings,* Pittsburgh, 1972. Available from the American Society of Civil Engineers, New York.
120. B. Sayers, "Safety and Risk in a Chemical Plant, A Case History," *Proceedings Annual Reliability and Maintainability Conference,* 1979, pp. 174–178.
121. J. L. Simpson, "Operational Safety for the TITAN II, Update Program," *6th Annual SAG/ASME/AIAA Reliability and Maintainability Conference,* July 1967.
122. W. Steiglitz, "Numerical Safety Goals—Are They Predictable?," *5th Reliability and Maintainability Conference,* 1966, pp. 490–493.
123. A Stresau, "Safety-Reliability: Fundamentals of System Safety Engineering," *ASSE Journal,* January 1966, pp. 16–18.
124. System Safety Program for Space and Missile Systems, MIL-STD-1574 (USAF), March 15, 1977.
125. "System Safety in a World of Diminishing Resources," *Third International System Safety Conference,* Washington, D.C., October 17–21, 1977.
126. "System Safety Implementation in the Reliability Program," *9th Annual West Coast Reliability Symposium,* pp. 105–123.
127. T. A. Waldeck, and R. B. McMurdo, "A System Safety Mathematical Model for Commercial Jet Airlines," *AIAA 4th Annual Meeting and Technical Display,* Anaheim, CA (AIAA Paper No. 67–910), October 1967.
128. T. A. Waldeck and R. B. McMurdo, "Systems Safety Engineering Analysis Techniques," The Boeing Company, Document D2-84303-1, Seattle, WA, February 1968.

129. H. W. Wynholds, "Quantitative System Safety Analysis Technique—Evaluation and Proposal Study," Lockheed, Document No. LR 22854, Burbank, CA, June 1970.
130. H. W. Wynholds, "A General Quantitative System/Mission Safety Evaluation Model," *Annals of Reliability and Maintainability*, Vol. 10, 1971, pp. 268–277.

Miscellaneous

131. B. S. Dhillon and C. Singh, *Engineering Reliability: New Techniques and Applications*, Wiley, New York, 1981.
132. B. S. Dhillon and C. Singh, "Bibliography of Literature on Fault Trees," *Microelectronics and Reliability*, Vol. 17, 1978, pp. 501–503.

8
Failure Data Analysis

8.1 INTRODUCTION

Failure data analysis is the backbone of any reliability analysis. A wrong assumption of the failure time distribution of an item can lead to an incorrect system reliability decision.

When sufficient field or laboratory failure data is available, the next logical step is to fit the most likely statistical distribution to the data and estimate values for its parameters. This chapter covers the following topics:

1. Failure data banks
2. Nonrepairable item failure data analysis techniques
3. Repairable component failure data analysis

For further reading, references 1 to 102 are presented at the end of this chapter.

8.2 FAILURE DATA BANKS

The data contained in these banks is used to perform failure data analysis [78]. The following are the data sources of an equipment failure data bank [46]:

1. Field failure data
2. Development (prototype equipment) data
3. In-house acceptance test data
4. Field repair data
5. Quality and manufacturing data

6. Qualification test data
7. Field demonstration data
8. Warranty associated data

There are various types of failure data banks. Most of the time we are concerned with obtaining the failure data for electrical, electronic, and mechanical items. These data banks may be associated with one or more application areas such as electrical, electronic, and mechanical components/systems. For example, aerospace, military, electric power generation, and transportation, equipment manufacturers and users have their own failure data banks. Each data bank has its own objective in collecting failure data. For example, the generation equipment failure data bank is concerned with collecting data for power equipment and components used in the power generation area.

Each data source has its reasons for the data collection. The following are the objectives [78] of the field data collection and analysis

1. Providing assistance for corrective actions
2. Determining past corrective actions success
3. Determining the actual field failure rate

The following are some of the usage areas of the analyzed data:

1. Setting reliability targets for a system/equipment to be developed
2. Design trade-off studies
3. Reliability prediction
4. Provision of spares
5. Life cycle cost modeling
6. To choose a reliability demonstration technique
7. Procurement
8. Advancing knowledge in failure patterns

The main aim of this section is to discuss briefly the failure data banks. It is not the author's intention to discuss them in depth. For further reading, if necessary, the reader should consult references 6, 8, 14, 19, 20, 33, 34, 36, 48, 56, 59, 62, 66, 76, 77, 78, 89, 92, and 96.

8.3 NONREPAIRABLE ITEMS FAILURE DATA ANALYSIS TECHNIQUES

This section presents two failure data analysis techniques. Both techniques can be easily applied by engineers to estimate distribution parameters.

8.3.1 Incomplete Failure Data Hazard Plotting Technique

This is a graphical technique which is used to estimate distribution parameters (point estimates) [71, 73, 75, 74]. The technique is popular because engineers are used to plotting data. Some of the advantages of this technique are

1. Fits the data to a straight line
2. Results obtained through plotting are convincing
3. It is easy to visualize the theoretical distribution that fits the field failure data.
4. It can be understood easily by the less mathematically inclined.
5. It can be used to analyze data composed of both complete and incomplete observations.

Complete and incomplete observations are defined as follows.

Complete Data.   These are data generated when the failure times of all units of a sample are known.

Incomplete Data.   This is also known as censored data. When the failure data are composed of the running times of unfailed units and the failure times of failed units, the data are said to be incomplete. The *censoring times* are the running times of unfailed units. The failure data is known as the *multi-censored* when the censoring times of the unfailed units are different. Another term, *singly censored,* is used when the censoring times of all the unfailed units are the same and greater than the failed units' failure times. The multi-censored data occurrence is due to (1) removing items or terminating their use before they fail, (2) some extraneous cause caused units to fail, and (3) collecting data from operating units. In real life the multi-censored data occur in unit life testing work [74].

Hazard Plotting Theory.   The basis for this theory is a distribution hazard function. The following three definitions were first given in Chapter 2 for the cumulative probability distribution, $F(t)$, the hazard rate, $h(t)$, and the cumulative hazard functions, $h_c(t)$, respectively

$$F(t) = \int_0^t f(t)\, dt \tag{8.1}$$

where $f(t)$ is the distribution probability density function
$\quad\ t \quad$ is time

$$h(t) = \frac{f(t)}{R(t)} = \frac{f(t)}{1 - F(t)} \qquad \text{since } R(t) + F(t) = 1 \tag{8.2}$$

where $R(t)$ is the reliability function, and

$$h_c(t) = \int_0^t h(t)\, dt \tag{8.3}$$

Equations (8.1) to (8.3) are used to develop a straight line expression to estimate parameters for the following distributions.

Exponential Distribution. From Chapter 2, the probability density function for the distribution is

$$f(t) = \lambda e^{-\lambda t} \qquad t \geq 0 \tag{8.4}$$

where λ is the constant failure rate
    t is time

By substituting Eq. (8.4) in Eq. (8.1), and then integrating the resulting expression, we get

$$F(t) = 1 - e^{-\lambda t} \tag{8.5}$$

By using Eqs. (8.4) and (8.5) in Eq. (8.2), the following hazard function expression results

$$h(t) = \lambda \tag{8.6}$$

To obtain the cumulative hazard function, substitute Eq. (8.6) in Eq. (8.3), and then integrate:

$$h_c(t) = \lambda t \tag{8.7}$$

Using Eq. (8.7) to express the time to failure, t, as a function of h_c, gives the result

$$t(h_c) = \theta h_c \tag{8.8}$$

where $\theta = 1/\lambda$ is the mean time to failure.
    Equation (8.8) is the equation of a straight line. Therefore, to estimate θ, one can plot time to failure t against the cumulative hazard h_c on a graph as shown in Fig. 8.1. The plot can be used to estimate θ. At $h_c = 1$ on the data plot, the corresponding value of t is the estimated value of θ. The slope of the line is also equal to θ.

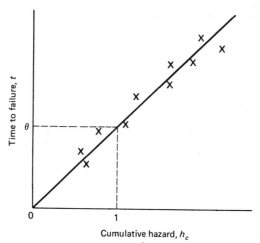

Figure 8.1 A hypothetical plot of Eq. (8.8).

Weibull Distribution. The Weibull probability density function, $f(t)$, is

$$f(t) = \frac{b}{\theta^b} t^{b-1} e^{-(t/\theta)^b} \qquad t \geq 0 \qquad (8.9)$$

where b is the shape parameter
θ is the scale parameter
t is time

The Weibull cumulative distribution function, $F(t)$, is obtained by substituting Eq. (8.10) in Eq. (8.1) and evaluating its integral

$$F(t) = 1 - e^{-(t/\theta)^b} \qquad t \geq 0 \qquad (8.10)$$

By substituting Eqs. (8.9) and (8.10) in Eq. (8.2) and then substituting the resulting equation in Eq. (8.3) and evaluating its integral, the cumulative hazard expression results

$$h_c(t) = (t/\theta)^b \qquad (8.11)$$

By expressing the time to failure t as a function of h_c, Eq. (8.11) is rewritten

$$t(h_c) = \theta(h_c)^{1/b} \qquad (8.12)$$

Taking logarithms (base 10), Eq. (8.12) becomes

$$\log t = \frac{1}{b} \log h_c + \log \theta \tag{8.13}$$

The plot of the above equation can be used to estimate parameters b and θ because $\log t$ against $\log h_c$ plots linearly. In other words, Eq. (8.13) is the equation of a straight line. The value of the parameter b is given by the reciprocal of the slope of the straight line. The condition $h_c = 1$ is used on the straight line plot to estimate θ. At the condition ($h_c = 1$), the corresponding value of t equals θ. Graphically, this fact is used to obtain the value of θ.

Extreme Value Distribution. The smallest extreme value distribution [73, 74] cumulative function is

$$F(t) = 1 - e^{-e^{(t - \alpha)/\beta}} \qquad -\infty < t < \infty \tag{8.14}$$

where β denotes the scale parameter
$\quad \alpha$ denotes the location parameter
$\quad t$ is time

To obtain the probability density function, $f(t)$, differentiate Eq. (8.14) with respect to time t. Similarly, as for the exponential and Weibull distributions, the extreme value distribution cumulative hazard function, $h_c(t)$, is

$$h_c(t) = e^{(t - \alpha)/\beta} \tag{8.15}$$

By expressing t as a function of h_c, Eq. (8.15) is rewritten

$$t(h_c) = \alpha + \beta \ln h_c \tag{8.16}$$

Equation (8.16) is the equation of a straight line because $t(h_c)$ against $\ln h_c$ plots linearly. The slope of the line is equal to β. Graphically, $h_c = 1$ is used to estimate α because at this point, the corresponding value of t equals α.

Normal Distribution. The cumulative distribution function [73, 74] of the normal distribution is

$$F(t) = \Phi\left(\frac{t - m}{\sigma}\right) \qquad -\infty < t < \infty \qquad (8.17)$$

where m is the mean
   σ is standard deviation
   t is time

The probability density function, $f(t)$, may be obtained by differentiating Eq. (8.17) with respect to time. As for the exponential and Weibull distributions, the normal cumulative hazard function is

$$h_c(t) = -\ln\left[1 - \Phi\left(\frac{t - m}{\sigma}\right)\right] \qquad (8.18)$$

Rearranging Eq. (8.18) to have t as a function of h_c results in

$$t(h_c) = \sigma\Phi^{-1}(1 - e^{-h_c}) + m \qquad (8.19)$$

The plot of the above expression can be used to estimate m and σ. Note that $t(h_c)$ against $\Phi^{-1}(1 - e^{-h_c})$ plots linearly. The slope of the line is equal to σ. The point where the straight line intersects the t axis denotes the value of m.

Log Normal Distribution.   The cumulative distribution function [73, 74] is given by

$$F(t) = \Phi\left(\frac{\log t - m}{\sigma}\right) \qquad t > 0 \qquad (8.20)$$

where t is time
   m is the logarithmic time to failure mean
   σ is the logarithmic time to failure standard deviation

The distribution probability density function, $f(t)$, may be obtained by differentiating Eq. (8.20) with respect to time. As for the exponential and Weibull distributions, the log normal cumulative hazard function is

$$h_c(t) = -\ln\left[1 - \Phi\left(\frac{\log t - m}{\sigma}\right)\right] \qquad (8.21)$$

Rearranging Eq. (8.21) to have log t as a function of h_c, the resulting expression is

$$\log t(h_c) = \sigma \Phi^{-1}(1 - e^{-h_c}) + m \qquad (8.22)$$

The parameters m and σ are estimated from the plot of Eq. (8.22). The plot of log t against $\Phi^{-1}(1 - e^{-h_c})$ is a straight line. The standard deviation σ is given by the slope of the line. The point where the straight line intersects the log t axis is the value of m.

Bathtub Distribution. The cumulative distribution function [31] is given by

$$F(t) = 1 - \exp[-(e^{(\beta t)^b} - 1)] \qquad \text{for } \beta,\ b > 0;\ t \geq 0 \qquad (8.23)$$

where b is the shape parameter
β is the scale parameter
t is time

Special cases of the distribution are

$b = 1$: Extreme value distribution
$b = 0.5$: Bathtub hazard rate curve distribution

The distribution probability density function, $f(t)$, may be obtained by differentiating Eq. (8.23) with respect to time. As for the exponential function, the cumulative hazard function, $h_c(t)$, is given by

$$h_c(t) = e^{(\beta t)^b} - 1 \qquad (8.24)$$

By twice taking logarithms of $(h_c + 1)$ in Eq. (8.24), Eq. (8.25) results

$$\ln t = \frac{1}{b} \ln z - \ln \beta \qquad (8.25)$$

where $z \equiv \ln (h_c + 1)$

Using Eq. (8.25), ln z against ln t plots as a straight line. The slope of the line is $1/b$ and the intercept is $-\ln \beta$. Therefore, b and β can be estimated graphically from the field failure data.

Hazard Rate Distribution.   The cumulative distribution function [30] is

$$F(t) = 1 - e^{-\ln(\lambda t^n + 1)} \qquad n \geq 1; \qquad \lambda > 0, t \geq 0 \qquad (8.26)$$

where λ is the scale parameter
  n is the shape parameter
  t is time

To obtain the probability density function, $f(t)$, differentiate Eq. (8.26) with respect to time. The distribution cumulative hazard function is

$$h_c(t) = \ln(\lambda t^n + 1) \qquad (8.27)$$

This expression is obtained in a similar way to the Weibull cumulative hazard function. Equation (8.27) for the time to failure t as a function of h_c becomes

$$\frac{1}{n} \ln \frac{1}{\lambda} + \frac{1}{n} \ln(e^{h_c} - 1) = \ln t \qquad (8.28)$$

Plotting $\ln(e^{h_c} - 1)$ against $\ln t$ results in a straight line. The slope of the line is $1/n$ and the intercept is $(1/n)\ln(1/\lambda)$. Therefore, λ and n can be estimated graphically from the failure data.

EXAMPLE 1

The failure times of 35 and running time of one (denoted by an asterisk) identical units are given in Table 8.1 (Column 2). Find the statistical distribution fit to the failure data and estimate its parameter values.

Hazard Plotting Steps.   These steps are as follows:

1. Without paying any attention to whether the failure data is composed of units running (censoring) or failure times, order the times from smallest to largest. Mark the running (censoring) times with an asterisk. If, in some cases, the running and failure times are equal, then both the times should be well mixed on the ordered list of smallest to largest times. In our example in column 2 of Table 8.1, the unit failure and running times are ordered from smallest to largest. The censoring time is marked with an asterisk.

2. Calculate each failure time's corresponding hazard value. A single failure time hazard value is given by $1/N$, where N denotes the number of items or units whose running or failure times are greater than or equal to that failure. The conditional probability of failure at a failure time is known as the hazard value. In other words, $1/N$ out of total number of N units that operated up to that time and then failed.

To obtain the hazard value for each failure time, number the ordered list of times backwards. For example, the first (lowest) time should be labeled N, the second $(N - 1)$, the third $(N - 2)$, and the last 1.

In Table 8.1, there are 36 units data given in column 1. For example, the values of N, $N - 1$, $N - 2$, ... are shown in parentheses in column 1, and the corresponding

Table 8.1. Failure Data and Hazard Calculations

COLUMN 1 ITEM NO.	COLUMN 2 FAILURE TIME (HOURS)	COLUMN 3 HAZARD	COLUMN 4 CUMULATIVE HAZARD (h_c)	COLUMN 5 ln z where $z \equiv \ln(h_c + 1)$	COLUMN 6 ln t
1 (36)	58	0.028	0.028	−3.589	4.06
2 (35)	70	0.029	0.059	−2.859	4.25
3 (34)	90	0.029	0.086	−2.495	4.5
4 (33)	105	0.03	0.116	−2.21	4.65
5 (32)	113	0.031	0.147	−1.987	4.73
6 (31)	121	0.032	0.179	−1.804	5
7 (30)	153	0.033	0.212	−1.649	5.03
8 (29)	159	0.035	0.247	−1.511	5.07
9 (28)	224	0.036	0.283	−1.3895	5.41
10 (27)	421	0.037	0.32	−1.2815	6.04
11 (26)	570	0.039	0.359	−1.1817	6.34
12 (25)	596	0.04	0.399	−1.0914	6.39
13 (24)	618	0.042	0.441	−1.0069	6.43
14 (23)	834	0.044	0.485	−0.9278	6.73
15 (22)	1,019	0.046	0.531	−0.8535	6.93
16 (21)	1,104	0.048	0.579	−0.7835	7.23
17 (20)	1,497	0.05	0.629	−0.7175	7.31
18 (19)	2,027	0.053	0.682	−0.654	7.61
19 (18)	2,234	0.056	0.738	−0.5929	7.71
20 (17)	2,372	0.059	0.797	−0.5349	7.77
21 (16)	2,433	0.063	0.86	−0.4771	7.8
22 (15)	2,505	0.067	0.927	−0.4217	7.83
23 (14)	2,690	0.071	0.998	−0.3680	7.9
24 (13)	2,877	0.077	1.075	−0.3148	7.96
25 (12)	2,879	0.083	1.158	−0.2624	7.97
26 (11)	3,166	0.091	1.249	−0.2101	8.06
27 (10)	3,455	0.1	1.349	−0.1578	8.15
28 (9)	3,551	0.111	1.46	−0.1052	8.18
29 (8)	4,378	0.125	1.585	−0.0516	8.38
30 (7)	4,872	0.143	1.728	0.0036	8.49
31 (6)	5,085	0.167	1.895	0.061	8.53
32 (5)	5,272	0.2	2.095	0.1220	8.57
33 (4)	5,341	0.25	2.345	0.1885	8.58
34 (3)	8,952	0.333	2.678	0.2642	9.1
35 (2)	9,188*	—	—	—	—
36 (1)	11,399	1	3.678	0.434	9.34

* Still running

values are 36, 35, 34, ... Following are the examples of each failure hazard calculation.

a. For the smallest failure time, the hazard value is

$$\frac{1}{N} = \frac{1}{36} = 0.028$$

b. For the fifth smallest failure time, the hazard value is

$$\frac{1}{N-4} = \frac{1}{36-4} = \frac{1}{32} = 0.031$$

Column 3 of Table 8.1 shows the hazard values for all the 36 data points. The cumulative hazard values are given in column 4.

Following are the two examples of each failure cumulative hazard calculation:

a. For the fifth smallest failure time, the cumulative hazard value is $= 0.116 + 0.031 = 0.147$
b. For the largest failure time, the cumulative hazard value is $= 2.678 + 1 = 3.678$

3. Select a statistical distribution and use the failure time and corresponding cumulative hazard data to construct a plot. From our past experience, we select the bathtub distribution. To obtain a straight line plot for the distribution we have to plot $\ln z$ against $\ln t$ as expressed in Eq. (8.25). The processed data for $\ln z$ and $\ln t$ is given in columns 5 and 6 of Table 8.1, respectively.

4. Plot each cumulative hazard value against its corresponding failure time on a graph. In our case, we plotted the $\ln z$ values against the $\ln t$ values of column 5 and 6 of Table 8.1 as shown in Fig. 8.2. The reader should note that only selected

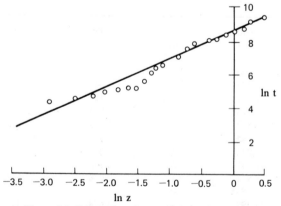

Figure 8.2 Failure data fit to a bathtub distribution.

values of columns 5 and 6 are shown on the plot because they are considered to be enough to show the trend of the data.

5. Check the plotted data to see if it follows a reasonably straight line. If it does, then draw a best fit straight line. If it does not, then try plotting the data for another theoretical distribution. In our case, the failure data fit to the bathtub distribution was adequate, as shown in Fig. 8.2.

6. Estimate the distribution parameters. In our case, utilizing Fig. 8.2 and Eq. (8.25) the slope of the line is

$$\frac{1}{b} = 1.6667$$

$$\therefore b = 0.6$$

and the intercept is

$$-\ln \beta = 8.4$$

$$\therefore \beta = e^{-8.4} = 0.000225$$

Finally, we emphasize that the end results obtained using this technique are valid only when the failure times of unfailed units are statistically independent of their censoring times, if the unfailed units were operated to failure. Uses of the hazard plots are outlined in reference 74.

8.3.2 Maximum Likelihood Estimation Technique

This is a commonly used method to estimate distribution parameters (point estimates) [99–101]. The basis for the technique is to assume a random sample $t_1, t_2, t_3, \ldots, t_m$ of size m from a population with a probability density function, $f(t; \lambda)$. Here, λ is the unknown parameter of the distribution. The joint density function of the m random variables of the random sample is known as the likelihood function. This is defined by

$$L = f(t_1; \lambda) \cdot f(t_2; \lambda) \cdot \cdots \cdot f(t_m; \lambda) \tag{8.29}$$

The value of λ that maximizes the log L or the L is called the maximum-likelihood estimator (MLE) of λ. Often, but not always, the value of λ is estimated by solving the equation

$$\frac{\partial \log L}{\partial \lambda} = 0 \tag{8.30}$$

The estimated value of λ is denoted by $\hat{\lambda}$. The following are the important properties [99] of the MLE:

1. For the large sample size m, the MLE is the most efficient estimate.
2. The invariance property is possessed by the MLE.
3. When a sufficient estimator exists, it (the MLE) is a sufficient estimator.
4. For the large sample size m, the variance is

$$\text{var } \lambda = -\left(\frac{\partial^2 \log L}{\partial \hat{\lambda}^2}\right)^{-1} \tag{8.31}$$

The maximum likelihood method is applied in the following examples to various distributions.

EXAMPLE 2   EXPONENTIAL DISTRIBUTION

The probability density function of the distribution is

$$f(t; \lambda) = \lambda e^{-\lambda t} \qquad t \geq 0 \tag{8.32}$$

where λ is the constant failure rate
$\quad$ t is time

Develop an expression for $\hat{\lambda}$ using the MLE procedure. By substituting Eq. (8.32) in Eq. (8.29), the resulting expression is

$$L = \lambda^m e^{-\sum_{j=1}^{m} \lambda t_j} \tag{8.33}$$

By taking logarithms of Eq. (8.33), the expression becomes

$$\log L = m \log \lambda - \sum_{j=1}^{m} \lambda t_j \tag{8.34}$$

Differentiating Eq. (8.34) with respect to λ, and setting the resulting expression equal to zero

$$\frac{d \log L}{d\lambda} = \frac{m}{\lambda} - \sum_{j=1}^{m} t_i = 0 \tag{8.35}$$

Therefore

$$\hat{\lambda} = m / \sum_{j=1}^{m} t_i \tag{8.36}$$

The following expression is obtained from Eq (8.34):

$$\frac{d^2 \log L}{d\lambda^2} = -\frac{m}{\lambda^2} \tag{8.37}$$

For large m, substituting Eq. (8.37) in Eq. (8.31), the following results [99]

$$\text{Var } \hat{\lambda} = \frac{\lambda^2}{m}$$

$$\simeq \frac{\hat{\lambda}^2}{m} \tag{8.38}$$

EXAMPLE 3 WEIBULL DISTRIBUTION

The probability density function of the distribution is

$$f(t) = b\beta t^{b-1} e^{-\beta t^b}$$
$$\text{for } t > 0,\ \alpha > 0,\ \lambda > 0 \tag{8.39}$$

where t is time
   b is the shape parameter
   β is the scale parameter

By substituting Eq. (8.39) in Eq. (8.29) and taking logarithms, the resulting equation [101] is

$$\log L = m \log \beta + m \log b + (b-1) \sum_{j=1}^{m} \log t_j - \beta \sum_{j=1}^{m} t_j^b \tag{8.40}$$

Differentiating Eq. (8.40) with respect to β, and setting the resulting expression equal to zero

$$\frac{\partial \log L}{\partial \beta} = \frac{m}{\beta} - \sum_{j=1}^{m} t_j^b = 0 \tag{8.41}$$

Similarly

$$\frac{\partial \log L}{\partial b} = \sum_{j=1}^{m} \log t_i - \beta \sum_{j=1}^{m} t_j^b \log t_j + \frac{m}{b} = 0 \tag{8.42}$$

Equations (8.41) and (8.42) are rewritten to

$$\hat{\beta} = m / \sum_{j=1}^{m} t_j^b \tag{8.43}$$

and

$$\beta = m/A \qquad (8.44)$$

where

$$A \equiv \beta \sum_{j=1}^{m} t_j^\beta \log t_j - \sum_{j=1}^{m} \log t_j$$

Equations (8.43) and (8.44) can be solved by using an iterative process [101].

EXAMPLE 4   GAMMA DISTRIBUTION

The probability density function [101] is

$$f(t) = \beta^b t^{b-1} e^{-\beta t}/\Gamma(b) \qquad \text{for } t > 0, \beta, b > 0 \qquad (8.45)$$

where β, b are the distribution parameters
      t is time

By substituting Eq. (8.45) in Eq. (8.29), the resulting expression is

$$L = \beta^{bm} \left(\prod_{j=1}^{m} t_j^{(b-1)} \right) e^{-\sum_{j=1}^{m} \beta t_j} (\Gamma(b))^{-m} \qquad (8.46)$$

Taking logarithms of Eq. (8.46)

$$\log L = bm \log \beta + (b-1) \sum_{j=1}^{m} \log t_j - \sum_{j=1}^{m} \beta t_j - m \log \Gamma(b) \qquad (8.47)$$

Differentiating Eq. (8.47) with respect to β and setting the resulting expression equal to zero, we get

$$\frac{\partial \log L}{\partial \beta} = \frac{bm}{\beta} - \sum_{j=1}^{m} t_j = 0 \qquad (8.48)$$

Similarly

$$\frac{\partial \log L}{\partial b} = m \log \beta + \sum_{j=1}^{m} \log t_j - m \, \psi(b) = 0 \qquad (8.49)$$

where

$$\psi(b) \equiv \frac{\partial [\log \Gamma(b)]}{\partial b} = \frac{\Gamma'(b)}{\Gamma(b)}$$

$\psi(b)$ is known as the Psi (Digamma) function [102]. The prime denotes the derivative with respect to b. The following approximate formula to calculate the vlaue of $\psi(b)$ is given in reference [101] for $b \geq 2$

$$\psi(b) \simeq [24(b - 0.5)^2]^{-1} + \log(b - 0.5) \qquad (8.50)$$

To compute the value of $\psi(b)$, a general formula is given in reference [102]. The values of β and b may be estimated by using an iterative process [101].

EXAMPLE 5 NORMAL DISTRIBUTION

The normal probability density is

$$f(t) = \frac{1}{\sigma\sqrt{2\pi}} e^{-\left[\frac{(t - \bar{m})^2}{2\sigma^2}\right]} \qquad \text{for} -\infty < t < \infty,\ -\infty\ \bar{m} < \infty,\ \sigma^2 > 0 \quad (8.51)$$

where t is time
σ and $\bar{m}$ are the parameters

By substituting Eq. (8.51) in Eq. (8.29) and taking logarithms, the likelihood function is

$$\log L = -\frac{m}{2} \log \sqrt{2\pi} - \frac{1}{2\sigma^2} \sum_{j=1}^{m} (t_i - \bar{m})^2 - \frac{m}{2} \log \sigma^2 \qquad (8.52)$$

Differentiating Eq. (8.52) with respect to σ^2 and setting the resulting expression equal to zero, we get

$$\frac{1}{\sigma^2} \sum_{j=1}^{m} (t_j - \bar{m})^2 - m = 0 \qquad (8.53)$$

Rearranging the above equation yields

$$\sigma^2 = \frac{1}{m} \sum_{j=1}^{m} (t_j - \hat{m})^2 \qquad (8.54)$$

Similarly

$$\bar{m} = \sum_{j=1}^{m} t_j/m \qquad (8.55)$$

EXAMPLE 6 BATHTUB DISTRIBUTION

From Eq. (8.23), the bathtub probability density function is

$$f(t) = b\beta\,(\beta t)^{b-1} \cdot e^{-[e^{(\beta t)^b} - (\beta t)^b - 1]} \qquad (8.56)$$

The resulting maximum-likelihood equations [31] are

$$\ln L = n \ln b + bn \ln \beta + n \ln e + (b - 1) \Sigma_i A_i - \Sigma_i e^{B_i} + \Sigma_i B_i \qquad (8.57)$$

where $A_i \equiv \ln t_i$, $B_i \equiv (\beta t_i)^b$, $C_i \equiv \ln \beta t_i$, Σ_i is the sum over i from 1 to n. n is the number of data points.

$$\frac{\partial \ln L}{\partial b} = \frac{n}{b} + n \ln \beta + \Sigma_i A_i - \Sigma_i B_i C_i e^{B_i} + \Sigma_i B_i C_i = 0 \qquad (8.58)$$

$$\frac{\partial \ln L}{\partial \beta} = \frac{bn}{\beta} - \frac{b}{\beta} \Sigma_i B_i e^{B_i} + \frac{b}{\beta} \Sigma_i B_i = 0 \qquad (8.59)$$

$$\frac{\partial^2 \ln L}{\partial b^2} = -\frac{n}{b^2} - \Sigma_i [C_i^2 B_i \{B_i + 1\} e^{B_i}] + \Sigma_i C_i^2 B_i \qquad (8.60)$$

$$\frac{\partial^2 \ln L}{\partial \beta^2} = \frac{bn}{\beta^2} - \frac{b}{\beta^2} \Sigma_i e^{B_i}[(b - 1)B_i + B_i^2\, b] + \frac{b(b - 1)}{\beta^2} \Sigma_i B_i \qquad (8.61)$$

Using Eqs. (8.58) to (8.59), parameters β and b can be estimated by trial and error. For large n, to estimate the parameter variance, substitute Eqs. (8.60) and (8.61) in Eq. (8.31) separately

$$\text{var } \hat{b} = -\left(\frac{\partial^2 \ln L}{\partial \hat{b}^2}\right)^{-1} \qquad (8.62)$$

$$\text{var } \beta = -\left(\frac{\partial^2 \ln L}{\partial \hat{\beta}^2}\right)^{-1} \qquad (8.63)$$

EXAMPLE 7 HAZARD RATE DISTRIBUTION

The distribution probability density [31] function is

$$f(t) = \lambda \frac{(b + 1)[\ln(\lambda t + 1)]^b}{(\lambda t + 1)} e^{-[\ln(\lambda t + 1)]^{b+1}} \qquad (8.64)$$

$$\text{for } b \geq 0, \lambda > 0, t \geq 0$$

where b is the shape parameter
λ is the scale parameter
t is time

To estimate b and λ, the maximum likelihood (ML) method is used. The resulting ML equations [31] are

$$\ln L = k \ln \lambda + k \ln(b + 1) + b \Sigma_i \ln\{\ln T_i\} - \Sigma_i \ln T_i - \Sigma_i [\ln T_i]^{b+1} \qquad (8.65)$$

where Σ_i implies the sum over i from 1 to k. k is the number of data points. And

$$T_i \equiv (\lambda t_i + 1)$$

$$\frac{\partial \ln L}{\partial \lambda} = \frac{k}{\lambda} + b \, \Sigma_i \, t_i/[(\ln T_i) \, T_i] - \Sigma_i \, t_i/T_i - (b+1) \, \Sigma_i t_i (\ln T_i)^b/T_i = 0 \qquad (8.66)$$

$$\frac{\partial \ln L}{\partial b} = \frac{k}{b+1} + \Sigma_i \ln(\ln T_i) - \Sigma_i \ln(\ln T_i) \, (\ln T_i)^{b+1} = 0 \qquad (8.67)$$

$$\frac{\partial^2 \ln L}{\partial \lambda^2} = -\frac{k}{\lambda^2} - b\Sigma_i \left\{ \frac{t_i^2 + t_i^2 \ln T_i}{(T_i \ln T_i)^2} \right\} + \Sigma_i \left(\frac{t_i}{T_i} \right)^2$$

$$- (b+1) \, \Sigma_i \left[\frac{t_i^2 \, (\ln T_i)^b}{T_i^2} \right] \left(\frac{b}{\ln T_i} - 1 \right) \qquad (8.68)$$

$$\frac{\partial^2 \ln L}{\partial b^2} = -\frac{k}{(b+1)^2} - \Sigma_i [\ln(\ln T_i)]^2 \, (\ln T_i)^{b+1} \qquad (8.69)$$

The parameters λ and b can be estimated from Eqs. (8.66) and (8.67) by trial and error. As for the bathtub distribution, for large k, the λ and b variance can be estimated using

$$\text{var } \hat{\lambda} = -\left(\frac{\partial^2 \ln L}{\partial \hat{\lambda}^2} \right)^{-1} \qquad (8.70)$$

and

$$\text{var } \hat{b} = -\left(\frac{\partial^2 \ln L}{\partial \hat{b}^2} \right)^{-1} \qquad (8.71)$$

Estimating Variances and Covariances.   The following procedure to estimate variances and covariances by solving one example of the bathtub distribution is presented. The following partial derivatives are obtained by utilizing Eq. (8.57)

$$L_{bb} = \frac{\partial^2 \ln L}{\partial b^2} = -\frac{n}{b^2} - \Sigma_i [C_i^2 B_i \, \{B_i + 1\} e^{B_i}] + \Sigma_i C_i^2 B_i \qquad (8.72)$$

$$L_{\beta\beta} = \frac{\partial^2 \ln L}{\partial \beta^2} = -\frac{bn}{\beta^2} - \frac{b}{\beta^2} \Sigma_i \, e^{B_i}[(b-1)B_i + B_i^2 b] + \frac{b(b-1)}{\beta^2} \Sigma_i B_i \quad (8.73)$$

$$L_{b\beta} = L_{\beta b} = \frac{\partial^2 \ln L}{\partial \beta \, \partial b} = \frac{n}{\beta} - \frac{1}{\beta} \Sigma_i B_i \, e^{B_i} [1 + b \ln \beta + bA_i + bB_i C_i]$$

$$+ \frac{1}{\beta} \Sigma_i B_i \{1 + b \ln \beta + bA_i\} \qquad (8.74)$$

Equations (8.72) to (8.74) are rewritten to form a matrix [101]

$$C = \begin{bmatrix} L_{bb} & L_{\beta b} \\ L_{\beta b} & L_{\beta\beta} \end{bmatrix}$$

(8.75)

The matrix D is defined by

$$D = \begin{bmatrix} \text{var } b & \text{cov } (b, \hat{\beta}) \\ \text{cov } (b, \hat{\beta}) & \text{var } \hat{\beta} \end{bmatrix}$$

(8.76)

The negative inverse of matrix C is equal to D.

$$D = -C^{-1}$$

(8.77)

The inverse of C is given by

$$C^{-1} = \frac{1}{\det C} \begin{bmatrix} L_{\beta\beta} & -L_{\beta b} \\ -L_{\beta b} & L_{bb} \end{bmatrix}$$

(8.78)

where $\det C = L_{\beta\beta}L_{bb} - (L_{\beta b})^2$

Utilizing Eqs. (8.76) to (8.78) we get

$$\text{var } b = -\frac{L_{\beta\beta}}{\det C}$$

(8.79)

$$\text{var } \hat{\beta} = -\frac{L_{bb}}{\det C}$$

(8.80)

$$\text{cov}(b, \hat{\beta}) = \frac{L_{\beta b}}{\det C}$$

(8.81)

EXAMPLE 8

Using the data given in example 1 in Eqs. (8.79) to (8.81), we get

$$\text{var } b = 7.5737 \times 10^{-3}$$
$$\text{var } \hat{\beta} = 1.1639847 \times 10^{-9}$$
$$\text{cov}(b, \hat{\beta}) = -6 \times 10^{-7}$$

The coefficient of correlation, $\rho(b, \hat{\beta})$, is given [101] by

$$\rho(\hat{b}, \hat{\beta}) = \{\text{cov}(\hat{b}, \hat{\beta})\}\{\text{var } \hat{b} \cdot \text{var } \hat{\beta}\}^{-1/2} \tag{8.82}$$

where $-1 \leq \rho \leq 1$.

By substituting the numerical values of the variances and covariances in Eq. (8.82), we get

$$\rho(\hat{b}, \hat{\beta}) = -0.20208$$

Interval Estimates. Once a distribution mean and variance are known, then we can establish the probability interval for the mean and variance using the Chebyshev inequality. This inequality evaluates the probability $P(\ )$ that a distribution random variable is in an interval. The Chebyshev inequality [99] is

$$P(|t - m| \geq k\sigma) \leq \frac{1}{k^2} \qquad k > 0 \tag{8.83}$$

where m is the mean
       σ^2 is the variance
       t is the distribution random variable
       k is any positive number

For large m(sample size), the distribution of a maximum likelihood estimator follows approximately the normal distribution function. Thus, the parameter confidence intervals may be obtained from the following:
For a 68 percent confidence the actual value of t is between

$$\hat{t} - \sigma_{\hat{t}} \leq t \leq \hat{t} + \sigma_{\hat{t}} \tag{8.84}$$

For a 95 percent confidence the actual value of t is between

$$\hat{t} - 2\sigma_{\hat{t}} \leq t \leq \hat{t} + 2\sigma_{\hat{t}} \tag{8.85}$$

For a 97 percent confidence, the actual value of t is between

$$\hat{t} - 3\sigma_{\hat{t}} \leq t \leq \hat{t} + 3\sigma_{\hat{t}} \tag{8.86}$$

EXAMPLE 9

Calculate the confidence interval of parameters $\hat{b}$ and $\hat{\beta}$ estimated in Example 8 using Eqs. (8.84) to (8.86) for large m(sample size). The resulting calculations are given in Table 8.2

Table 8.2. Interval Estimates for b and β.

PARAMETER	PROBABILITY (%)	INTERVAL ESTIMATES FOR b AND β
b	68	$0.532972 \leq b \leq 0.7070$
	95	$0.4460 \ \ \leq b \leq 0.7941$
	97	$0.3589 \ \ \leq b \leq 0.8811$
β	68	$0.0001709 \leq \beta \leq 0.000239$
	95	$0.0001368 \leq \beta \leq 0.0002732$
	97	$0.0001026 \leq \beta \leq 0.000307$

8.4 FAILURE DATA ANALYSIS OF REPAIRABLE SYSTEMS

This section includes a brief discussion on failure data analysis of repairable systems. The failure data analysis techniques such as hazard plotting and maximum likelihood estimation presented in Section 8.3 can be applied directly to repairable systems under certain assumptions. These assumptions are

1. After each repair, the system is as good as new.
2. Due to (1), the time interval between each failure is independently and identically distributed.

Now we can easily obtain the failure data for a repairable system. The number, n, of failure times of a system after each repair represent the failure times of n identical systems put on a test until their failure. Failure data analysis of repairable systems is also discussed in references 7, 9, 10, 11, 12, 16, 25, 55, 61, 83, and 84.

8.5 SUMMARY

The following topics are presented in this chapter:

1. Failure data banks
2. Failure data analysis of nonrepairable systems
3. Failure data analysis of repairable systems

Several examples are presented to demonstrate how to use the data analysis techniques.

Today there is no shortage of data analysis techniques, but obtaining good real-life failure data is still a problem. These techniques can only be useful when the input data is reliable. Otherwise, results obtained using these techniques will not be true. A list of 102 useful references is provided for further reading on the subject.

8.6 REFERENCES

Books

1. A. J. Gross and V. A. Clark, *Survival Distribution: Reliability Applications in the Biomedical Sciences,* Wiley, New York, 1975.
2. J. R. King, *Probability Charts for Decision Making,* Industrial Press, New York, 1970.
3. S. L. Meyer, *Data Analysis For Scientists and Engineers,* Wiley, New York, 1975.
4. F. Mosteller and J. W. Tukey, *Data Analysis and Regression: A Second Course in Statistics,* Addison-Wesley, Reading, MA, 1977.
5. J. W. Tukey, *Exploratory Data Analysis,* Addison-Wesley, Reading, MA, 1977.
6. A. Ullman, *Reliability Data Banks, 2nd Seminar on Reliability Data Banks,* Stockholm, Sweden, 1977. Available from the Military Electronics Laboratory, FOA3, 10450, Stockholm 80, Sweden.

Articles

7. S. Abe, "Field Data Analysis Via Markov Renewal Life Models," *Annual Reliability and Maintainability Symposium Proceedings,* 1975, pp. 562–567.
8. M. Anderson, L. Rauseén, and G. Karlén, "The Swedish Thermal Power Reliability Data System," *EOQC-EFNMS, Second Seminar on Reliability Data Banks,* Stockholm, March 1977. Available from the Military Electronics Laboratory, FOA3, 10450, Stockholm 80, Sweden.
9. H. E. Ascher, "Evaluation of Repairable System Reliability Using the Bad-AS Old Concept," *IEEE Trans. on Reliability,* Vol. 17, June 1968, pp. 103–110.
10. H. E. Ascher and H. Feingold, "The Aircraft Air Conditioner Data Revisited," *Annual Reliability and Maintainability Symposium Proceedings,* 1979, pp. 153–159.
11. H. E. Ascher and H. Feingold, "Is There Repair After Failure," *Proceedings Annual Reliability and Maintainability Symposium,* 1978, pp. 190–197.
12. H. Ascher and H. Feingold, "Bad-As-Old Analysis of System Failure Data," *Annual Reliability and Maintainability Proceedings,* 1969, pp. 49–62.
13. H. Ascher, "Distribution-Free Estimation of System Reliability," *Proceedings Annual Reliability and Maintainability Symposium,* 1980, pp. 374–378.
14. A. Avena and T. Gardini, " 'Experience Banks' and 'Single Systems' Some Recent Experiences In Steel Making Industry," *Second European Seminar On the Reliability Data Banks,* Stockholm, Sweden, March 1977. Available from the Military Electronics Laboratory, FOA3, 10450, Stockholm 80, Sweden.
15. R. E. Barlow, "Analysis of Retrospective Failure Data Using Computer Graphics," *Annual Reliability and Maintainability Symposium Proceedings,* 1978, pp. 103–111.
16. W. M. Bassin, "Increasing Hazard Functions and Overhaul Policy," *Annual Symposium On Reliability,* 1969, pp. 173–180.
17. I. Bazovsky, "Distribution of Downtimes," *Reliability and Maintainability Symposium,* 1974, pp. 470–480.
18. S. B. Bennett, A. L. Ross, and P. Z. Zemanick, *Failure Prevention and Reliability Conference Proceedings,* 1977, Available from the ASME.
19. B. Bergman, "Total Time on Teso Plotting for Failure Data Analyses," *EOQC-EFNMS, Second Seminar on Reliability Data Banks,* Stockholm, March 1977. Available from the Military Electronics Laboratory, FOA3, 10450, Stockholm 80, Sweden.
20. K. Boesebeck and P. Homke, "Failure Data Collection and Analysis in the Federal Republic of Germany," *The Energy Technology Conference,* Houston, TX, September 1977. Published by ASME, New York.

21. A. J. Bonis, "Reliability Demonstration Requirements: Tables and Charts for Bernoulli Trails," *IEEE Trans. on Reliability,* February 1971, pp. 33–36.
22. T. L. Burnett and B. A. Wales, "System Reliability Confidence Limits," *Seventh National Symposium on Reliability and Quality Control,* 1961, pp. 118–128.
23. T. L. Burnett, "Truncation of Sequential Life Tests," *National Symposium on Reliability,* 1962, pp. 7–13.
24. S. H. Bush, "Review of Reliability of Piping in Light Water Reactors," *The Energy Technology Conference on Failure Data and Failure Analysis,* September 1977, Houston, TX. Available from ASME, New York.
25. J. M. Cozzolino, "The Optimal Burn-in Testing of Repairable Equipment," Technical Report No. 23, MIT Operations Research Center, Cambridge, MA, October 1966.
26. D. E. Crawford, "Analysis of Incomplete Life Test Data on Motorettes," *Insulation/Circuits,* Vol. 16, 1970, pp. 43–48.
27. L. Crow, "Reliability Analysis for Complex Repairable Systems," *Reliability and Biometry,* SIAM, Philadelphia, 1974, pp. 370–410.
28. D. Davis, "An Analysis of Some Failure Data," *Journal of the American Statistical Association,* Vol. 43, 1952, pp. 113–150.
29. "Development of Prediction Techniques," in *System Reliability Prediction by Function, Vol. 1.* RADC-TDR-63-300. 1963, Rome Air Development Centre, Griffiss, AFB, N.Y.
30. B. S. Dhillon, "Statistical Functions to Represent Various Types of Hazard Rates," *Microelectronics and Reliability,* Vol. 20, 1980, pp. 581–584.
31. B. S. Dhillon, "Statistical Distributions," *IEEE Trans. on Reliability,* Vol. 30, December 1981, pp. 457–460.
32. R. H. Dudley, T. R. Chow, S. E. Van Vleck, and R. J. Pooch, "How to Get More Mileage Out of Your Data," *Annual Reliability and Maintainability Symposium Proceedings,* 1977, pp. 414–420.
33. R. J. Duphily and R. L. Long, "Enhancement of Electric Power Plant Reliability Data Systems," *Fourth Annual Reliability Engineering Conference for the Electric Power Industry,* June 1977, New York. Available from the IEEE.
34. A. R. Eames et al., "Operation of a Reliability Data Bank," *NATO Advanced Study Conference Generic Techniques in Systems Reliability Assessment,* 1976, Liverpool, England, pp. 99–115. North-Holland, Holland.
35. B. Epstein, "Estimation From Life Test Data," *IRE Trans. on Reliability and Quality Control,* April 1960, pp. 104–107.
36. Failure Data and Failure Analysis in the Power and Processing Industry, The Energy Technology Conference, Houston, TX, September 1977, proceedings published by ASME, New York.
37. P. I. Feder, "Graphical Techniques in Statistical Data Analysis Tools for Extracting Information from Data," *Technometrics,* May 1974, pp. 287–299.
38. J. W. Foster, W. T. Craddock, "Estimating Life Parameters from Burn-In Data," *Proceedings Annual Reliability and Maintainability Symposium,* 1974, pp. 206–209.
39. H. E. Franka and S. J. Dapkunas, "Failure Analysis and Failure Data Correction in the ERDA Coal Conversion System," *The Energy Technology Conference on Failure Data Analysis, The Energy Technology Conference,* Houston, TX, September 1977. Available from ASME, New York.
40. P. M. Ghare and Y. H. Kim, "An Approximate Function for the Hazard Rate Curve," *ASQC Technical Conference Proceedings,* Pittsburgh, 1970, pp. 367–371.
41. R. Gnanadesikan and M. B. Wilk, "Probability Plotting Methods for the Analysis of Data," *Biometrika,* Vol. 55, 1968, pp. 1–7.
42. R. Gnanadesikan and M. B. Wilk, "*Data Analytic Methods in Multi-Variate Statistical*

Analysis," Multi-Variate Analysis II, P. R. Krishniah (ed.), Academic, New York, pp. 593–638.

43. P. Gottfried, "The Interpretation of Statistically Designed Reliability and Maintainability Tests," *Annual Reliability and Maintainability Symposium,* 1977, pp. 203–205.

44. "Graphical Reliability Analysis Procedures," In *Failure Distribution Analysis Study, Vol. 2,* Computer Applications Inc., New York, August 1964.

45. G. H. Hahn and W. Nelson, "A Comparison of Methods for Analysing Censored Life Data to Estimate Relationship between Stress and Product Life," *IEEE Trans. on Reliability,* April 1974, pp. 1–11.

46. R. F. Hahn, "Data Collection Techniques," *Annual Reliability and Maintainability Symposium Proceedings,* 1972, pp. 38–43.

47. G. J. Hahn and W. B. Nelson, "Graphical Analysis of Incomplete Accelerated Life Test Data," *Insulation/Circuits,* Vol. 17, September 1971, pp. 79–84.

48. I. Hansson, "Experience from the Reliability Predictions by Means of the FTL Data Bank," *EOQC-EFNMS Second Seminar on Reliability Data Banks,* Stockholm, March 1977. Available from the Military Electronics Laboratory, F0A3, 10450, Stockholm 80, Sweden.

49. R. L. Haueter, "Current Program on Power Plant Availability Data Systems," *The Energy Technology Conference Proceedings,* September 1977, Houston, TX. Published by ASME, New York.

50. G. R. Herd, "Estimation of Reliability from Incomplete Data," *6th National Symposium on Reliability and Quality Control,* 1960, pp. 202–217.

51. G. R. Herd, "Estimation of Reliability Functions," *Proceedings of the 3rd National Symposium on Reliability and Quality Control,* 1957, pp. 113–122.

52. E. L. Kaplan and P. Mevi, "Non-Parametric Estimation from Incomplete Observations," *Journal of American Statistical Association,* Vol. 53, 1958.

53. J. H. K. Kao, "A Graphical Estimation of Mixed Weibull Parameters in Life Testing Electron Tubes," *Technometrics,* Vol. 1, 1959, pp. 389–407.

54. J. H. K. Kao, "A Summary of Some New Techniques in Failure Analysis," *Proceedings National Symposium on Reliability and Quality Control,* 1960, pp. 190–201.

55. K. Kivenko, "Electronic Equipment Burn-In for Repairable Equipment," *Journal of Quality Technology,* Vol. 5, No. 1, Jan. 1973, pp. 7–10.

56. M. Koch, "The Swedish Project: Information Systems for Reliability Data," *EOQC-EFNMS Second Seminar on Reliability Data Banks,* Stockholm, March 1977. Available from Military Electronics Laboratory F0A3, 10450, Stockholm 80, Sweden.

57. D. Kodlin, "A New Response Time Distribution," *Biometrics,* June 1967, pp. 227–239.

58. S. A. Lapp and G. J. Powers, "A Method for Generation of Fault Trees," *The Energy Technology Conference,* Houston, TX, September 1977. Published by ASME, New York.

59. P. H. Leclerc and S. A. Matra, "MISGRAD: Equipment Reliability Data Bank Usage for System Trade-Off," *EOQC-EFNMS, Second Seminar on Reliability Data Banks,* Stockholm, March 1977. Available from the Military Electronics Laboratory, F0A3, 10450, Stockholm 80, Sweden.

60. G. J. Lieberman, "Some Problems in Reliability Estimation," *Proceedings 3rd Annual Aerospace Reliability and Maintainability Conference,* 1964, pp. 136–140. Available from ASME, New York.

61. W. A. Lilius, "Graphical Analysis of Repairable Systems," *Proceedings Annual Reliability and Maintainability Symposium,* 1979, pp. 403–406.

62. C. Marcovici and S. A. Matra, "Data Collection for an Equipment Reliability Data Bank," *EOQC-EFNMS, Second Seminar on Reliability Data Banks,* Stockholm, March 1977. See reference 6.

63. W. Q. Meeker and W. Nelson, "Weibull Percentile Estimates and Confidence Limits from

Singly Censored Data by Maximum Likelihood," *IEEE Trans. on Reliability,* Vol. R-25, April 1976, pp. 20–24.

64. W. Mendenhall, "A Bibliography of Life Testing and Related Topics," *Biometrika,* Vol. 45, Parts 3 & 4, December 1958, pp. 521–543.

65. L. A. Morris, "Estimation of Hazard Rate from Incomplete Data," *Annual Symposium on Reliability,* 1970, pp. 337–342.

66. T. R. Moss, "Recent Developments in the SRS Data Base," *EOQC-EFNMS, Second Seminar on Reliability Data Banks,* Stockholm, 1977. See reference 6.

67. W. Nelson, "Analysis of Residuals from Censored Data," *Technometrics,* November 1973, pp. 697–715.

68. W. Nelson, "Graphical Analysis of Accelerated Life Test Data with a Mix of Failure Modes," *IEEE Trans. on Reliability,* October 1975, pp. 230–237.

69. W. Nelson, "Charts for Confidence Limits and Tests for Failure Rates," *Journal of Quality Technology,* Vol. 4, October 1972, pp. 190–195.

70. W. Nelson and V. C. Thompson, "Weibull Probability Papers," *Journal of Quality Technology,* Vol. 3, No. 2, April 1971, pp. 45–50.

71. W. Nelson, "Hazard Plot Analysis of Incomplete Failure Data," *Annual Symposium on Reliability,* 1969, pp. 391–403.

72. W. Nelson, "Hazard Plotting Methods for Analysis of Life Data with Different Failure Modes," *Journal of Quality Technology,* Vol. 2, No. 3, July 1970, pp. 126–149.

73. W. Nelson, "Theory and Applications of Hazard Plotting for Censored Failure Data," *Technometrics,* Vol. 14, No. 4, November 1972, pp. 945–966.

74. W. Nelson, "Hazard Plotting for Incomplete Failure Data," *Journal of Quality Technology,* Vol. 1, No. 1, January 1969, pp. 27–52.

75. W. Nelson, "Life Data Analysis by Hazard Plotting," *Evaluation Engineering,* Vol. 9, No. 5, 1970, pp. 37–40.

76. E. T. Parascos, "A New Approach to Establishment and Maintenance of Equipment Failure Rate Data Bases," Failure Prevention and Reliability, *The Design Engineering Technical Conference,* Chicago, IL, September 1977. Available from the ASME.

77. S. Pollock, "The Utilization of Integrated Data Collection and Analysis in Reliability/ Maintainability Programs," *Annals of Reliability and Maintainability,* 1967, pp. 149–174.

78. R. F. Powell and H. H. Scott, "Analyzing and Interpreting Field Failure Data," *Annual Symposium on Reliability,* 1970, pp. 94–100.

79. F. Proschan, "Theoretical Explanation of Observed Decreasing Failure Rate," *Technometrics,* Vol. 5, No. 3, August 1963, pp. 375–383.

80. R. P. Riddick, "The Effect of Scheduled Repair Cycles on Marine Equipment Reliability," *Proceedings Annual Symposium on Reliability,* 1967.

81. R. B. Schwartz and S. M. Seltzer, "Failure Distribution Analysis," *Annals of Reliability and Maintainability,* 1965, pp. 817–838.

82. M. L. Shooman and S. Sinkar, "Generation of Reliability and Safety Data by Analysis of Expert Opinion," *Proceedings Annual Reliability and Maintainability Symposium,* 1977, pp. 186–193.

83. M. L. Shooman and S. Tenenbaum, "Hazard Function Monitoring of Airline Components," *Proceedings Annual Reliability and Maintainability Symposium,* 1974, pp. 383–390.

84. R. B. Shurman, "Time Dependent Failure Rates for Jet Aircraft," *Annual Reliability and Maintainability Symposium,* 1978, pp. 198–203.

85. N. Z. Singapurwalla, "Time Series Analysis of Failure Data," *Annual Reliability and Maintainability Symposium Proceedings,* 1978, pp. 107–112.

86. K. G. Sorenson and P. M. Besuner, "A Workable Approach to Extending the Life of

Expensive-Life Limited Components," *The Energy Technology Conference*, Houston, TX, September, 1977. Published by ASME, New York.

87. R. G. Stokes and F. N. Stehle, "Some Life-Cycle Estimates for Electronic Equipment: Methods and Results," *Proceedings of the Annual Symposium on Reliability*, 1968, pp. 27–83.

88. J. R. Taylor and R. H. Lochner, "Statistical Analysis of Field Data," *Annals of Reliability and Maintainability*, 1965, pp. 905–913.

89. M. G. Tomassetti, "Communication Problems of Experience Data Collection and Use," *EOQC-EFNMS, Second Seminar on Reliability Data Banks*, Stockholm, March 1977. See reference 6.

90. J. L. Tomsky, "A Statistical Model for Early Detection of Increasing Failure Rates," *Annual Reliability and Maintainability Symposium Proceedings*, 1978, pp. 498–503.

91. J. L. Tomsky, T. R. Chow, and L. D. Schiller, "System Reliability Estimation from Several Data Sets," *Annual Reliability and Maintainability Symposium Proceedings*, 1976, pp. 18–24.

92. W. E. Vesely, "Failure Data and Risk Analysis," The Energy Technology Conference Proceedings, Houston, TX, September 1977. Published by ASME, New York.

93. M. H. Walker, "The Uncertainty of Reliability Predictions," *Proceedings of the IAS Aerospace Systems Reliability Symposium*, 1962, pp. 70–72. Available from ASME, New York.

94. J. B. Whitney, "A Likelihood Analysis of Some Common Distributions," *Journal of Quality Technology*, Vol. 6, No. 4, October 1974, pp. 182–187.

95. R. L. Williams, "System Reliability Analysis Using the GO Methodology," *The Energy Technology Conference Proceedings*, Houston, TX, September 1977. Published by ASME, New York.

96. P. A. Wodehouse, "The ESA Electronic Components Data Bank Demonstration," *EOQC-EFNMS, Second Seminar on Reliability Data Banks*, Stockholm, March 1977. See reference 6.

97. F. S. Wood, "The Use of Individual Effects and Residuals in Fitting Equations to Data," *Technometrics* Vol. 15, No. 4, November 1973, pp. 677–695.

98. J. Yasuda, "Correlation Between Lab Tests & Field Part Failure Rates," *IEEE Trans. on Reliability*, Vol. R-26, No. 2, 1977, pp. 82–84.

Miscellaneous

99. M. L. Shooman, *Probablistic Reliability: An Engineering Approach*, McGraw-Hill, New York, 1968.

100. N. Mann, R. E. Shafer, and N. D. Singpurwalla, *Methods for Statistical Analysis of Reliability and Life Data*, Wiley, New York, 1974.

101. M. Lloyd and M. Lipow, *Reliability: Management, Methods, and Mathematics*, Prentice-Hall, Englewood Cliffs, NJ, 1961.

102. M. Abramowitz and I. A. Stegun, *Handbook of Mathematical Functions with Formulas, Graphs and Mathematical Tables,*" National Bureau of Standards, June 1964. Available from the Superintendent of Documents, U.S. Government Printing Office, Washington, D.C., 20402.

9
Life Cycle Costing

9.1 INTRODUCTION

Due to the current tight economic situation, the equipment procurement process is experiencing a new trend; that is, a newly purchased item must be supported for its entire life. In many instances the equipment acquisition cost is lower than the equipment field cost (support cost) over the entire life cycle. Generally, the salvage value of a product is low. Therefore, when purchasing a new product, procurement management now looks at the entire life cycle cost of a product rather than just considering just the initial cost of an item.

In many countries government and other large industrial organizations evaluate and outline the entire life cycle cost (LCC) of a product, especially when developing or procuring an expensive item. For example, it would be very difficult not to consider the LCC when purchasing the following products:

1. Air traffic control system
2. Aircraft
3. Military tanks and trucks

For any of the above, a purchasing decision can not be justified without taking into account the equipment life cycle cost. Depending on the product type, the product ownership cost (logistics and operating costs) varies from 10 to 100 times the procurement cost [114].

Wide dissemination of the term *Life Cycle Costing* (LCC) was first given [114] in a document entitled "Life Cycle Costing in Equipment Procurement," published by the Logistics Management Institute, Washington, D.C., in April 1965. The United States government definition of the LCC [7] is

The life cycle of a product is the sum of all costs to the government of procurement and ownership of that product over its entire life span.

This chapter presents some of the main branches of life cycle costing. At the end, an extensive list of references is included for the benefit of readers who wish to delve further in the subject.

9.2 ECONOMIC ANALYSIS

One of the principles of life cycle costing is that the costs over the entire life of the system are to be computed by taking into consideration the time value of money. In other words, the amount of money received or spent today does not have the same buying value if it were received or spent a year later and so on. The commonly used LCC techniques in this section are presented for those persons who are not familiar with engineering economics or discounted cash flow analysis.

9.2.1 Simple Interest

Simple interest is the simplest form of interest. This method is used when the interest is charged on the principal sum or the original amount invested or borrowed. This method does not include an interest charge on the accrued interest. The simple interest, I_s, may be calculated by using the following formula:

$$I_s = p \times i \times m \qquad (9.1)$$

where i is the annual interest rate
  m is the number of years
  p is the principal sum or the original amount borrowed or invested

The total amount of money, *A*, after *m* number of years of investment is

$$A = p + I_s = p + p \times i \times m = p(1 + i \times m) \qquad (9.2)$$

EXAMPLE 1

If a sum of $5000 is invested for 6 years at an interest rate of 8 percent annually, calculate the earned interest, I_s, on the principal sum and the total amount of money, *A*, after 6 years.
  Using Eq. (9.1) we get

$$I_s = \frac{\$5,000 \times 6 \times 8}{100} = \$2400$$

The total amount, A, is calculated from Eq. (9.2), i.e.,

$$A = p + I_s = \$5000 + \$2400 = \$7400$$

9.2.2 Compound Interest (Single-Payment, Compound-Amount Factor Technique).

This method is used quite commonly. The yearly or periodically earned interest is added to the original amount or principal thereon which itself earns interest. The total amount of money, A, after m number of periods or years can be obtained by using

$$A = p(1 + i)^m \qquad (9.3)$$

EXAMPLE 2

If a sum of \$2000 is invested for 5 years at a 5 percent annual interest rate compounded annually, find the total amount of money, A, after 5 years.
   Use Eq. (9.3) for $p = \$2000$, $i = 5/100$ and $m = 5$ to get

$$A = \$2000 \ (1 + 0.05)^5 = \$2,552.60$$

9.2.3 Present Value Method (Single Payment)

The formula for the present value, p, results directly from Eq. (9.3)

$$p = A/(1 + i)^m \qquad (9.4)$$

EXAMPLE 3

A company wishes to buy \$15,000 worth of equipment in the distant future. If the annual interest rate of 5 percent is compounded annually, then calculate the sum of money, p, that has to be deposited in a bank today in order to purchase such equipment after 5 years.
   Utilizing Eq. (9.4) for $A = \$15,000$, $i = 5/100$, and $m = 5$ years, we get

$$p = \$15,000/(1 + 0.05)^5 = \$11,752.90$$

A company has to deposit \$11,752.90 in a bank in order to procure the equipment in 5 years.

9.2.4 Equal-Payment Series Compound-Amount Method

This technique is sometimes known as an annuity. This method is used when a series of equal payments, p, are deposited at the end of each year. Figure 9.1 depicts such a situation. The amount will be zero at the end of the first

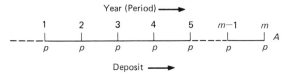

Figure 9.1 A series with equal payments.

year until p amount of money is deposited. The total amount, A, at the expiration of the second year will be

$$A = p(1 + i) \tag{9.5}$$

After p amount of money is added to Eq. (9.5) at the end of the second year, then Eq. (9.5) is rewritten

$$A = p(1 + i) + p \tag{9.6}$$

At the end of the third year, the total sum of money, A, will be

$$\begin{aligned} A &= [p(1 + i) + p](1 + i) + p \\ &= p(1 + i)^2 + p(1 + i) + p \end{aligned} \tag{9.7}$$

One should note here that an additional amount, p, is added at the end of the third year to Eq. (9.7).

In similar fashion, at the end of m years, Eq. (9.7) generalizes to

$$A = p(1 + i)^{m-1} + p(1 + i)^{m-2} + \cdots + p(1 + i) + p \tag{9.8}$$

In the above expression, we have assumed that p amount of money is added annually for $(m - 1)$ years and the p amount is deposited at the end of m years. Equation (9.8) may be rewritten

$$A = p[1 + (1 + i) + (1 + i)^2 + (1 + i)^3 + \cdots + (1 + i)^{m-1}] \tag{9.9}$$

The above expression is a geometric series. To find sum of the series, multiply both sides of Eq. (9.9) by $(1 + i)$

$$A(1 + i) = p[(1 + i) + (1 + i)^2 + (1 + i)^3 + (1 + i)^4 + \cdots + (1 + i)^m] \tag{9.10}$$

Subtract Eq. (9.9) from Eq. (9.10), then rearrange to get

$$A = p\left[\frac{(1 + i)^m - 1}{i}\right] \tag{9.11}$$

EXAMPLE 4

A company invests \$15,000 at the end of each year for 5 years at an annual interest rate of 5 percent compounded annually. Find the total amount of money accumulated, A, at the end of the 5-year period. Utilizing Eq. (9.11) for $p = \$15,000$, $i = 0.05$, and $m = 5$ years, we get

$$A = \$15,000 \left[\frac{(1 + 0.05)^5 - 1}{0.05} \right] = \$82,884.47$$

9.2.5 Sinking Fund Method (Equal Payment Series)

The following formula for p is obtained by rearranging Eq. (9.11)

$$p = A \left[\frac{i}{(1 + i)^m - 1} \right] \tag{9.12}$$

EXAMPLE 5

A reliability manager desires to find out how much money he should deposit at the end of each year over the next 5 years at a compounded annual interest rate of 4 percent in order to purchase equipment worth \$50,000 at the end of the specified period. Furthermore, he desires to make equal contributions at the end of each year.
   Using Eq. (9.12) for $A = \$50,000$, $i = 0.04$ and $m = 5$ years, we get

$$p = \$50,000 \left[\frac{0.04}{(1.04)^5 - 1} \right] = \$9,231.36$$

The reliability manager must deposit \$9,231.36 at the end of each year to have the required sum of \$50,000.

9.2.6 Present-Worth Method (Equal Payment Series)

This method deals with computing the present worth of an amount of money, p, to be paid at the end of each coming m years. The present value of the first deposit or payment (to be made after 1 year) is $p/(1 + i)$. The present value of the second deposit (to be made after 2 years) is $p/(1 + i)^2$. Similarly, one can write the following present value (PV) expression for payments at the end of each coming m years:

$$PV = \frac{p}{1 + i} + \frac{p}{(1 + i)^2} + \cdots + \frac{p}{(1 + i)^m} \tag{9.13}$$

Equation (9.13) is a geometric series. To find the sum of this series multiply both sides of Eq. (9.13) with $1/(1 + i)$ to get

$$\frac{PV}{(1+i)} = \frac{1}{(1+i)} \left[\frac{p}{1+i} + \frac{p}{(1+i)^2} + \cdots + \frac{p}{(1+i)^m} \right] \qquad (9.14)$$

Subtract Eq. (9.13) from Eq. (9.14) to get

$$PV = p \left[\frac{1 - (1+i)^{-m}}{i} \right] \qquad (9.15)$$

EXAMPLE 6

It is expected that a newly installed system will save $20,000 per year for the next 5 years. If the annual interest rate is 15 percent, then find the present value of these savings.

To find the present value, PV, use Eq. (9.15) for $p = \$20,000$; $i = 0.15$, and $m = 5$ years

$$PV = 20,000 \left[\frac{1 - (1+0.15)^{-5}}{0.15} \right]$$

$$= \$67,043.10$$

9.2.7 Capital Recovery Method (Equal Payment Series)

The formula for this technique is obtained by rearranging Eq. (9.15)

$$p = PV \left[\frac{i}{1 - (1+i)^{-m}} \right] \qquad (9.16)$$

EXAMPLE 7

Suppose a small manufacturer has borrowed $500,000 at 10 percent annual interest rate to purchase equipment. This money must be repaid in equal payments at the end of each year over the coming 5 years. Calculate the amount, p, to be repaid at the end of each year.

Using Eq. (9.16) we get

$$p = \$500,000 \left[\frac{0.1}{1 - (1+0.1)^{-5}} \right] = \$131,898.74$$

9.2.8 Economic Analysis Techniques Application to LCC Problems

This section presents an application of some of the economic evaluation techniques presented in the earlier sections.

Suppose the salvage value of a system is SV dollars after m years of field service. At the end of each year, the system maintenance cost is MC dollars. Annual compound interest rate on the money is i percent. If the acquisition cost of the system is AC dollars, then find an expression for the present value of the total life cycle cost to the buyer.

The present value of the maintenance cost, PMC, is given by Eq. (9.15)

$$PMC = MC \left[\frac{1 - (1 + i)^{-m}}{i} \right] \tag{9.17}$$

Using Eq. (9.4), the present worth of the salvage value, PSV, of the system is

$$PSV = \frac{SV}{(1 + i)^m} \tag{9.18}$$

The present worth of the total life cycle cost, PLCC, using Eqs. (9.17) and (9.18) is

$$PLCC = AC + MC \left[\frac{1 - (1 + i)^{-m}}{i} \right] - \frac{SV}{(1 + i)^m} \tag{9.19}$$

where AC is the system acquisition cost.

EXAMPLE 8

Suppose the procurement cost of equipment is $1,000,000. The equipment's useful life as predicted by the manufacturer is 20 years. After that many years of service, its estimated salvage value will be $50,000. The equipment maintenance cost at the end of each year is estimated to be $10,000. In addition, the annual compound interest rate will be approximately 10 percent. Calculate the present value of the equipment total life cycle cost.

To calculate present value of the equipment's total LCC for AC = $1,000,000, MC = $10,000, SV = $50,000, and $i = 0.1$ we use Eq. (9.19) to get

$$PLCC = \$1,000,000 + \$10,000 \left[\frac{1 - (1 + 0.1)^{-20}}{0.1} \right] - \left[\frac{\$50,000}{(1 + 0.1)^{20}} \right]$$

$$= \$1,077,703.50$$

Therefore, the LCC cost of the equipment to the buyer will be $1,077,703.50.

9.2.9 Break-Even Analysis Method

This technique is used where one desires to compare competing alternatives. This method may be better illustrated by solving one simple numerical example on equipment selection.

EXAMPLE 9

Assume a production manager desires to choose one machine out of two machines. The acquisition and unit production costs of machines I and II are

	MACHINE I	MACHINE II
Acquisition cost	$25,000	$20,000
Unit production cost	$2	$4

Determine the most profitable machine by performing a graphical break-even analysis.

The break-even graphical analysis is presented in Fig. 9.2. Both machines' total of acquisition (fixed) and unit production costs are plotted against number of production units separately. As shown in Fig. 9.2, the break-even point is at 2500 units. Machine II is more economical for a production run not to exceed 2500 units. For higher production quantities, machine I will be more profitable. Finally, it may be added that the selection of machines is also dependent on such factors as production schedule, saleability forecast, maintenance, obsolescence, etc.

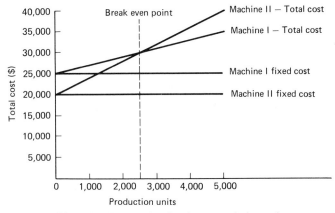

Figure 9.2 Two product break-even analysis graph.

9.3 LIFE CYCLE COST MODELS

This section presents LCC models. LCC models may vary for different equipment and their varying operating environments. An equipment life cycle cost versus reliability curve is shown in Fig. 9.3. This figure shows that as the reliability increases, the equipment procurement cost increases accordingly. However, the increase in the reliability decreases the ownership cost. There are various LCC models used to estimate cost by various manufacturers and users to fulfill their need. Some of them are as follows:

9.3.1 Life Cycle Cost Model I

This model [111] simply breaks down the system LCC into two branches of recurring and nonrecurring costs. Equation (9.20) defines such a relationship

$$\text{LCC} = \alpha + \beta \tag{9.20}$$

where α denotes equipment nonrecurring costs
  β denotes equipment recurring costs

The nonrecurring cost α is

$$\alpha = \sum_{i=1}^{10} \text{NC}_i \tag{9.21}$$

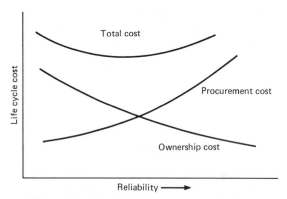

Figure 9.3 LCC plus its subcost versus reliability.

where NC_i is the ith nonrecurring cost: $(i = 1)$ training, $(i = 2)$ support, $(i = 3)$ transportation, $(i = 4)$ acquisition, $(i = 5)$ test equipment, $(i = 6)$ installation, $(i = 7)$ research and development, $(i = 8)$ LCC management, $(i = 9)$ reliability and maintainability improvement, and $(i = 10)$ equipment qualification approval.

Similarly, the equipment recurring cost breakdowns are

$$\beta = \sum_{i=1}^{5} C_i \tag{9.22}$$

where C_i is the ith recurring cost: $(i = 1)$ inventory, $(i = 2)$ labor, $(i = 3)$ maintenance, $(i = 4)$ operating, and $(i = 5)$ support.

Sometimes the model LCC equation is written as

$$\text{LCC} = \text{ownership cost} + \text{procurement cost}$$
$$= \text{OC} + \text{PC} \tag{9.23}$$

9.3.2 Life Cycle Cost Model II

This is another model [2] used to estimate the LCC of a system. The LCC is composed of four major cost components as defined in Eq. (9.24)

$$\text{LCC} = \sum_{i=1}^{4} C_i \tag{9.24}$$

where C_1 represents the research and development cost
C_2 represents the production and construction cost
C_3 represents the operation and support cost
C_4 represents the retirement and disposal cost

Further breakdowns of the above costs are presented in [2].

9.3.3 Life Cycle Cost Model III

This model [122] also has four major cost components. The system life cycle cost expression is

$$\text{LCC} = C_p + C_{dp} + C_{ap} + C_{op} \tag{9.25}$$

where C_{op} denotes costs of the operational phase
$\quad\quad C_{ap}$ denotes costs of the acquisition phase
$\quad\quad C_{dp}$ denotes costs of the definition phase
$\quad\quad C_p$ denotes costs of the conceptual phase

Definition phase and conceptual phase costs are relatively small as compared to the costs of the acquisition and operational phases. C_p and C_{dp} are essentially labor effort costs. Acquisition and operational phases costs represent a major portion of the equipment LCC. Both C_{ap} and C_{op} may be subdivided as follows:

Cost of Acquisition Phase. C_{ap} is defined by

$$C_{ap} = \sum_{i=1}^{4} C_i \tag{9.26}$$

where C_1 is the cost of program management
$\quad\quad C_2$ is the cost of personal acquisition
$\quad\quad C_3$ is the cost of the prime system
$\quad\quad C_4$ is the cost of support equipment

The prime system cost, C_3, is composed of transportation, testing, production, facilities, documentation, installation, design and development, and initial spares and repair component costs.

Advanced training, basic personnel acquisition and manning level costs are the main three elements of the personnel acquisition cost, C_2.

The main three subcomponents of the program management cost are the costs of system management, systems test, and engineering.

Cost of Operational Phase. The operational phase cost expression is

$$C_{op} = \sum_{i=1}^{3} C_{opi} \tag{9.27}$$

where C_{op1} denotes costs of maintenance
$\quad\quad C_{op2}$ denotes the functional operating expense
$\quad\quad C_{op3}$ denotes the operational administrative expense

The operational administrative expense is composed of investment and holding, spares inventory, and administrative and operational program management costs.

Functional operating expense subcomponents are operational consumables and manning costs. The operational manning cost includes labor costs to operate the system under study. Labor costs include direct and indirect expenses.

The following costs are part of the maintenance cost:

1. Equipment downtime costs
2. Costs of personnel replacement
3. Costs of maintenance manning
4. Costs of maintenance consumables
5. Costs of maintenance facilities
6. Costs of repairs and spare parts

9.3.4 Life Cycle Cost Model IV

To estimate a major system LCC, the United States Navy uses a model whose equation is

$$LCC = C_t + C_{as} + C_i + C_{rd} + C_{os} \qquad (9.28)$$

where C_t represents the cost of termination
C_{as} represents the cost of associated systems
C_i represents the cost of investment
C_{rd} represents the cost of research and development
C_{os} represents the cost of support and operating

The investment cost components are acquisition and government investment costs. The government investment costs includes the cost of project management, training, transportation, system test and evaluation, and so on. Other costs components of Eq. (9.28) are self-explanatory. For detailed cost breakdowns, consult reference 7.

9.3.5 Life Cycle Cost Model V

The following LCC expression is used by the United States Army Material Command to estimate life cycle cost of new equipment or systems:

$$LCC = C_{rd} + C_{os} + C_i \qquad (9.29)$$

where C_{rd} is the cost of research and development
C_{os} is the cost of support and operating
C_i is the cost of investment

The other LCC elements are self-explanatory. Those who wish to delve further into the breakdown of the element costs should consult reference 7.

9.3.6 Life Cycle Cost Model VI

This model [7] is used to estimate the LCC of an aircraft turbine engine. The LCC is defined by

$$LCC = \sum_{i=1}^{5} C_i \qquad (9.30)$$

where C_i is ith LCC cost element: ($i = 1$) base maintenance, ($i = 2$) development, ($i = 3$) production quantity, ($i = 4$) depot maintenance, and ($i = 5$) component improvement.

9.3.7 Life Cycle Cost Model VII

This model [100] is used to estimate LCC of switching power supplies. This model was used to compare the cost of existing unit with the cost of new design. The LCC is expressed

$$LCC = C_i + C_f \qquad (9.31)$$

where C_i denotes the initial cost
C_f denotes the cost of failure

The following expression represents the subcomponents of the failure cost:

$$C_f = \lambda t (C_r + C_s) \qquad (9.32)$$

where λ is the unit constant failure rate
C_r is the cost of repair
C_s is the cost of spares
t is the expected life time of system

The spare cost, C_s, is

$$C_s = C_1 + C_2 \qquad (9.33)$$

where C_1 is the spare cost of a unit
C_2 is the fraction of number of spares for each active unit

9.3.8 Life Cycle Cost Model VIII

The main components of this LCC model [54] are initial, operating, and failure costs. The LCC expression of the model is

$$LCC = C_f + C_o + C_i \qquad (9.34)$$

where C_f is the cost of failure
C_o is the cost of operating
C_i is the initial cost

The average cost, C_a, of a failure is

$$C_a = [C_r(MTTR + t_r) + C_{ea}] + C_{rp} + [C_{ls}(MTTR + t_r) + C_{ml}] \qquad (9.35)$$

where t_r is the repaired system response time
MTTR is the mean time to repair (hours)
C_{ea} is the administrative loss cost due to a failure
C_{rp} is the replaced parts cost
C_{ls} is the cost of loss of service (per hour)
C_r is the hourly cost of the repair organization
C_{ml} is the cost of the mission-abort loss which occurs with each cessation of service, and is known as the event-related, one-time cost

The cost of failure, C_f, is expressed as

$$C_f = \frac{C_a l_t}{MTBF} \qquad (9.36)$$

where MTBF is the mean time between failures (hours)
l_t is the system life time (hours)

Substituting Eq. (9.36) in Eq. (9.34) gives

$$LCC = \frac{C_a l_t}{MTBF} + C_o + C_i \qquad (9.37)$$

By replacing $C_a/MTBF$, the failure loss cost rate per hour, with CR in Eq. (9.37), the following equation results:

$$LCC = l_t \, CR + C_o + C_i \qquad (9.38)$$

Equation (9.38) can be used to estimate the LCC of equipment with no built-in redundancy. For built-in equipment redundancy, this expression is modified in reference [54].

9.4 EQUIPMENT COST MODELS USED IN OPERATION

The models presented in this section can be used to estimate various operational phase costs of equipment in a field environment. Some of these models are as follows.

9.4.1 Logistics Support Cost Model

This model [7,59] is used by the United States Air Force Acquisition Logistics Division. The logistics support cost (LSC) is

$$LSC = C_a + C_b + C_c + C_d + C_e + C_f + C_g + C_h + C_i + C_j \quad (9.39)$$

where C_a is the cost of fuel

$\quad C_b$ is the facilities cost

$\quad C_c$ is the spare engines cost

$\quad C_d$ is the support equipment cost

$\quad C_e$ is the technical data and management cost

$\quad C_f$ is the training equipment and training personnel cost

$\quad C_g$ is the cost of initial or replacement spares

$\quad C_h$ is the cost of the on-equipment maintenance

$\quad C_i$ is the cost of the off-equipment maintenance

$\quad C_j$ is the cost of supply management and inventory entry

9.4.2 Support Cost Model

This is another support model [22] developed by the Northrop Corporation. The following are the main eight components of the support cost (SC):

$$SC = \sum_{i=1}^{8} C_i \quad (9.40)$$

where C_1 is the cost of facilities

$\quad C_2$ is the cost of training

$\quad C_3$ is the cost of manpower

$\quad C_4$ is the cost of documentation

$\quad C_5$ is the cost of support equipment

C_6 is the repairable spares repair cost

C_7 is the support equipment spares cost

C_8 is the cost of spares (includes the cost of initial spares plus the 10 years spares cost to support the equipment in question)

9.4.3 Probabilistic Model

This model [2] is used when predicting the probability of availability of a spare part when it is needed. The probability may be calculated by using

$$P_a = \sum_{i=0}^{k} [-q(\ln p)]^i \, p/i! \qquad (9.41)$$

where p is the part reliability ($p = e^{-\lambda t}$)

λ is the part constant failure rate

t is time

k is the number of available spare items in the inventory

q is a particular part type quantity used in equipment or system

P_a denotes probability of availability of a particular spare part type when it is needed

9.5 A DEVELOPMENT COST MODEL

This model [79] is based on the simple reasoning that the cost of equipment development will increase monotonically in order to improve reliability due to the design effort. Therefore, it is reasonable to assume that when the equipment reliability increases, the logistic cost decreases monotonically. The following expression defines the development cost, DC, of an equipment:

$$DC = C_1 + C_2 \left(\frac{\ln R_1}{\ln R_2} \right)^\beta \qquad (9.42)$$

where C_1 is the cost independent of reliability (the basic fixed cost)

C_2 is the cost of facilities, labor, and so on, in order, to produce a reliable system (variable cost)

β is a constant which takes empirical values, for example, when $\beta = 2$, then to double the meantime between failures, the cost increases four times C_2

R_1 is the standard design reliability

R_2 is the improved design reliability

9.6 METHODS FOR COST ESTIMATION

Inputs to the earlier equipment LCC and operational phase models are costs which must be estimated in order to use these models. There are various ways of estimating costs. Cost estimating procedures, in general, are of five types [59]:

1. Using expert opinion
2. Catalog price per unit
3. Cost estimation with specific analogy
4. Estimation of cost-to-cost
5. Estimation of non–cost-to-cost

Expert Opinion Method. This method is used when no other cost data is available or the confidence in the available data is not very high. This procedure uses the opinion of experts who have a number of years of cost related experience on the system, subsystem, or component for which the cost data is required. Sometimes the estimates produced by such a method are used in conjunction with results obtained by using methods (b) to (e).

Catalog Price Per Unit Method. A part for which the price has to be calculated is identified as a specific type. Then the cost estimator simply uses the catalog price. This catalog price is obtained by finding the average of past cost data of numerous items.

Cost Estimation With Specific Analogy. To estimate costs, this method draws an analogy of equipment under study to some earlier similar type of product. Both new and analogous products characteristics used are operating, design, and performance. These characteristics are known to be cost predictive.

Cost-to-Cost Estimation Method. This method concerns itself with estimating the specific product cost as a percentage of some other important equipment cost.

Non–Cost-to-Cost Estimation Method. The performance, size, weight or operating characteristics are known as the noncost parameters of a product. Therefore, the product costs are estimated as function of one or more of these product parameters.

9.7 COST MODEL SELECTION

The cost model selection is one of the most important areas of life cycle costing. One of the main limitations of the model selection is the availability of the cost data. Availability of accurate data at an early stage in the life cycle of a product is limited and generally other cost-related information is also of vague nature. At this stage the cost analyst should try to use a simple model with a small number of input requirements [2]. During this phase he or she should use simple accounting methods and straightforward arithmetic calculations to obtain the end life cycle cost result. An analytical model [2] should include the following factors:

1. All the important factors must be incorporated into the model. In other words, it should be comprehensive.
2. The model design should be simple and straightforward. The design should permit timely implementation whenever it is needed in problem solving.
3. The model should differentiate the associated factors with a problem. The important factors should be highlighted and the less important suppressed (with the knowledge of others).
4. The dynamic characteristics of an item or an equipment must be represented in the model. The model should be able to incorporate any situational changes successfully, without affecting the main objective of the model.
5. The design of the model should allow the analyst to evaluate any item of the equipment independent from other elements when desired.
6. The design of the model should be flexible enough to accommodate easily any expansion or modification to the model.

9.7.1 Benefits of Analytical Models

Some of these benefits [2] are as follows:

1. Identification of high risk and uncertainty areas.
2. Identification of the data requirements and any associated complexity with data.
3. The future events such as reliability parameters, cost estimates, requirements of logistics and so on, can be forecast by using the cost model.
4. Past unexplained situations can be explained by the usage of the analytical model.
5. Alternative solutions to a problem may be compared and the best solution determined by the use of the analytical model.
6. The evaluation of the system as an entity is possible with the model. Furthermore, the model makes it possible to consider all the system variables simultaneously.

9.8 RELIABILITY IMPROVEMENT WARRANTY

The reliability improvement warranty (RIW) [141] is concerned with providing incentive and motivation to manufacturers to produce reliable equipment which will have a small number of failures with low repair costs in the field. RIW-related examples are

1. In one case, the United States Air Force and a manufacturer signed a RIW contract in which the manufacturer was obligated to repair and replace a failed item during operational use at a fixed price.
2. In another case, the Air Force signed a fixed price contract in which it provided incentive to the manufacturer to improve product reliability and maintainability.
3. Sometimes in an RIW contract, the mean time between failures of the equipment in the field is the guarantee imposed. During the specified time intervals, the field mean time between failures (MTBF) is assessed and compared with the guaranteed MTBF. The manufacturer is obligated to take corrective measures at no cost to the user to improve the equipment MTBF if the assessed MTBF is less than the guaranteed MTBF.

   The warranty period is a matter of negotiation between the customer and the user of equipment in question. The duration of the period may be subject to various factors important to both the user and the manufacturer.

Before the equipment reliability improvement warranty [156] commitment, the following analysis must be accomplished:

1. *Product/Program Screening.* This analysis is concerned with identification of programs/projects whose potential for RIW benefits is high.
2. *RIW Proposals Evaluation.* Purpose of these analyses is twofold: (a) verification of reliability realism; and (b) positive and negative incentives balance determination.
3. *Request For Proposal RIW Requirements Assessment.* The objective of this analysis is threefold:
   a. Assuring that the manufacturer/user obligations are realistic.
   b. Assuring a reasonable balance between risks and benefits.
   c. Making sure the RIW terms have continuity with the written technical specifications.
4. *Cost Analysis of RIW/Non-RIW.* This analysis is concerned with establishing RIW cost effectiveness.

9.8.1 Reliability Improvement Warranty Cost Models

RIW pricing models become helpful to contractors or manufacturers when they are confronted with RIW decision-making problems. A manufacturer,

when pricing a RIW, estimates the cost of RIW obligations and associated profit. Two RIW price models are given in the following sections.

RIW Pricing Model. The pricing equation for the first RIW model [146] is

$$P = p + C_1 + Y + C_2 + C_D \qquad (9.43)$$

where $Y =$ $(NR\ t_D\ C_r)/MTBF_a$
    P is the fixed warranty price paid to the manufacturer
    p represents contractor's profit
    C_1 is the manufacturer's warranty associated fixed cost
    C_D represents damages associated with a failure to meet turn-around time requirement
    C_2 is the improvement actions cost to achieve mean time between failures ($MTBF_a$)
    $MTBF_a$ denotes MTBF achieved (RIW period average)
    C_r is the per unit repair cost to the manufacturer
    t_D is the warranty period duration
    R denotes the per calendar time usage rate in operating time
    N denotes the number of items to be purchased by the customer

The right hand side of Eq. (9.41) reflects the manufacturer's risk exposure. Figure 9.4 shows the manufacturer's cost versus MTBF. At a certain MTBF, the sum of repair cost and cost of reliability improvement has a minimum value. At this minimum value, the contractor will have maximum profit.

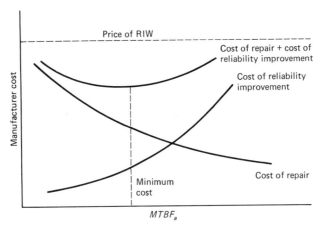

Figure 9.4 Cost versus MTBF.

RIW Pricing Model with MTBF Guarantee. In this pricing model [146] the manufacturer guarantees that by certain specified time the system will meet specified MTBF's. If the required MTBF is higher than the achieved MTBF at a certain mutually agreed point in time, the manufacturer guarantees the consignment spares.

The RIW pricing model expression is

$$P = p + C_1 + Y + C_2 + C_D + C_{cs}K \qquad (9.44)$$

where K is the consignment spares quantity
C_{cs} is each consignment spare cost

Other symbols are the same as the symbols used in Eq. (9.43).
K may be estimated by using the following formula [146]:

$$K = m \left[\frac{\text{MTBF}_1}{\text{MTBF}_a} - 1 \right] \qquad (9.45)$$

where m is the spares target level
MTBF_a is the MTBF measured or achieved
MTBF_1 denotes the MTBF guaranteed

9.9 SUMMARY

This chapter has covered the following areas:

1. Economic analysis
2. Life cycle cost models
3. Equipment cost models in operation
4. A development cost model
5. Methods for cost estimation
6. Cost model selection
7. Reliability improvement warranty

The minute details of cost breakdowns of some of the costs used in LCC and other models are not presented. However, the references given provide these cost breakdowns. Other selected literature on life cycle costing and reliability improvement warranty is listed at the end of the chapter.

Finally, while we may say that there is no shortage of availability of LCC and other cost models, obtaining the cost and other data can be a trying task.

9.10 REFERENCES

Books and Reports

1. R. K. Beach, *Life-Cycle Cost Equations,* Division of Building Research, National Research Council, Ottawa, Canada, September 1979.
2. B. Blanchard, *Design and Manage to Life Cycle Cost,* M/A Press, P.O. Box 92, Forest Grove, OR, 1978.
3. *Department of Defense Guide LCC-1, Life Cycle Costing Procurement Guide,* U.S. Government Printing Office, Washington, D.C. 20402, July 1970.
4. *Department of Defense Guide LCC-2, Casebook-Life Cycle Costing in Equipment Procurement,* U.S. Government Printing Office, Washington, D.C. 20402, July 1970.
5. *Department of Defense Guide LCC-3, Life Cycle Costing Guide for System Acquisition,* U.S. Government Printing Office, Washington, D.C. 20402, January 1973.
6. "Life Cycle Costing" (special issue), *Defense Management Journal,* Vol. 12, No. 1, U.S. Government Printing Office, Washington, D.C. 20402, January 1976.
7. M. Earles, *Factors, Formulas and Structures for Life Cycle Costing,* privately published, 89 Lee Drive, Concord, MA, 01742.
8. C. Gillespie, *Standard and Direct Costing,* Prentice-Hall, Englewood Cliffs, NJ, 1962.
9. J. D. S. Gibson, *Supplemental Life Cycle Costing Program Management Guidance,* 2nd ed., Joint AFSC/AFLC Commander's Working Group on Life Cycle Cost, ASD/ACL Wright Patterson AFB, OH, 45433, January 1976.
10. J. W. Griffith, *Life Cycle Cost Workbook,* U.S. Department of Commerce, December 1979. Available from the Superintendent of Documents, U.S. Government Printing Office, Washington, D.C.
11. D. P. Haworth, *The Principles of Life Cycle Costing,* Industrialization Form., Vol. 6, No. 3–4., Dept. of Architecture, Graduate School of Design, Harvard University, Cambridge, MA, 02138, 1975.
12. F. C. Jelen (ed.), *Cost and Optimization Engineering,* McGraw-Hill, New York, 1970.
13. J. E. Kerman and L. J. Meker, *Life Cycle Cost Procurement Guide,* Joint AFSC/AFLC Commander's Working Group on Life Cycle Cost, ASD/ACL, Wright Patterson AFB, OH, 45433, July 1976.
14. R. J. Kapsch, *Life Cycle Costing Techniques Applicable to Law Enforcement Facilities,* U.S. Department of Justice, October 1974. Superintendent of Documents, U.S. Government Printing Office, Washington, D.C. U.S.A.
15. *Logistics Spectrum,* The Society of Logistics Engineers, 3322 South Memorial Parkway, Suite 2, Huntsville, AL, 35801.
16. L. J. Menker, *Life Cycle Cost Analysis Guide,* Joint AFSC/AFLC Commander's Working Group On Life Cycle Cost, ASD/ACL, Wright Patterson AFB, OH, 45433, November 1975.
17. MIL-STD-499A (USAF), Engineering Management, Headquarter's, Air Force Systems Command, Attention SDDE, Andrews AFB, MD, 20331.
18. P. F. Ostwald, *Cost Estimating for Engineering and Management,* Prentice-Hall, Englewood Cliffs, NJ, 1974.
19. J. Popper, *Modern Cost Engineering Techniques,* McGraw-Hill, New York, 1970.
20. M. R. Seldon, *Life Cycle Costing: A better method of Government Procurement,* Westview Press, 5500 Central Avenue, Boulder, CO, 80301, 1978.
21. R. T. Ruegg, J. S. McConnaughey, G. Thomas Sav, and K. A. Hockenbery, *Life Cycle Costing,* Department of Energy, September 1978. Superintendent of Documents, U.S. Government Printing Office, Washington, D.C.

Articles

22. M. E. Alarcon, and L. M. Donaldson, "Support Cost Comparison Methodology," *Annual Reliability and Maintainability Symposium,* 1979, pp. 254–261. Available from the IEEE.

23. AMCP 706–133 ADA 026 006, *Cost Factors and Analysis,* article 7, Available from the NTIS, Springfield, VA, 22161.

24. S. Banerjee and L. Waverman, "Life Cycle Costs of Coal and Nuclear Generating Stations," Royal Commission on Electric Power Planning, Ontario Government, July 1978.

25. R. K. Barasia, "Development of a Life Cycle Management Cost Model," *Annual Reliability and Maintainability Symposium,* 1978, pp. 254–259.

26. I. Bazovsky, "Life Cycle Profits Instead of Life Cycle Costs," *Annual Reliability and Maintainability Symposium,* 1974, pp. 377–382.

27. R. L. Baglow, "The Reliability Parameter and its Importance For Life Cycle Management," *Microelectronics and Reliability,* Vol. 14, 1975, pp. 91–105.

28. R. G. Bertchy, "Effect of Reliability Programs on Life Cycle Cost—A Case History," *Microelectronics and Reliability,* Vol. 17, 1978, pp. 9–14.

29. R. E. Biedenbender, "Cost-to-Produce," *Annual Reliability and Maintainability Symposium,* 1973, pp. 481–483.

30. B. S. Blanchard, "Cost Effectiveness, System Effectiveness, Integrated Logistics Support and Maintainability," *IEEE Trans. on Reliability,* Vol. R-16, 1967, pp. 117–125.

31. S. J. Blewitt, "Cost and Operational Effectiveness of Reliability and Maintainability Improvements," *Annual Reliability and Maintainability Symposium,* 1974, pp. 417–421. Available from the IEEE.

32. W. H. Bolden, "The AN/ARC-164 Radio: Life Cycle-Cost Savings," *Microelectronics and Reliability,* Vol. 17, 1978, pp. 15–23.

33. C. E. Booth, "Computer Simulations of Life Cycle Cost Elements," *Annual Technical Conference Transactions,* ASQC, 1969, pp. 731–740.

34. C. R. Borchers and R. E. Churchill, "System Design Trade-Offs and Economic Planning," *Proc. 1966 Annual Symposium on Reliability,* 1966, pp. 276–282. Available from the IEEE.

35. A. Boerckel, "The Formula For Survival—Optimum Quality At Optimum Cost," *Annual Reliability and Maintainability Symposium,* 1973, pp. 373–378. Available from the IEEE.

36. A. J. Boness and A. N. Schwartz, "A Cost Benefit Analysis of Military Aircraft Replacement Policies," *NRLQ,* vol. 16, 1969, pp. 237–257.

37. C. Brook and R. Barasia, "A Support System Life Cycle Cost Model," *Annual Reliability and Maintainability Symposium,* 1977, pp. 297–302. Available from the IEEE.

38. H. A. Brode, "Cost of Ownership—An Overview Life Cycle Costs: Evaluation of Avionic System Reliability Improvements," *Annual Reliability and Maintainability Symposium,* 1975, pp. 212–216. Available from the IEEE.

39. E. R. Brussell, K. Pope, and K. M. Tasugi, "Cost of Ownership—Industry Viewpoint—Parametric Analysis of Operating and Support," *Annual Reliability and Maintainability Symposium,* 1975, pp. 217–220. Available from the IEEE.

40. J. R. Canada and H. M. Wodworth, "Methods for Quantifying Risk in Economic Analyses of Capital Projects," *The Journal of Industrial Engineering,* Vol. XIX, January 1968, pp. 32–37.

41. T. V. Caver, "New Maintenance Manuals Help to Reduce Life-Cycle Costs," *Defense Management Journal,* Vol. 14, No. 5, 1978, pp. 26–31.

42. T. V. Caver, "Life-Cycle Cost: Attitude and Latitudes," *Defense Management Journal,* Vol. 15, No. 4, 1979, pp. 12–17.

43. F. R. Carey, "System Reliability Cost Tradeoff Methodology," *Tenth Annual West Coast Reliability Symposium,* 1969, pp. 91–104.

44. G. D. C. Chiang, "Mock Trial: Goody & et al vs Big George Motor Company," *Annual Reliability and Maintainability Symposium,* 1979, pp. 83–84. Available from the IEEE.

45. T. D. Cox and B. L. Morrison, "An Approach For Determining Optimum RAM Requirements," *Annual Reliability and Maintainability Symposium,* 1977, pp. 378–384. Available from the IEEE.

46. A. Coppola, "Reliability As a Capital Investment, *Annual Reliability and Maintainability Symposium,* 1974, pp. 349–357. Available from the IEEE.

47. G. B. Cohen, "Reliability Influence on Life Cycle Costs," *Eight Annual Reliability and Maintainability Conference,* 1969, pp. 452–459. Available from the IEEE.

48. M. B. Darch, "LCC—The Implications on Internal Data Collection," *Microelectronics and Reliability,* Vol. 17, 1978, pp. 25–29.

49. A. M. Daine, "Determining Total Equipment Cost," *Second National Convention on Military Electronics,* Washington, D.C., July 1958.

50. T. A. Daly and P. H. Oskesman, "Commercial Reliability Programs—A Good Investment," *11th National Symposium on Reliability and Quality Control,* 1965, pp. 385–391.

51. W. L. Davidson and B. J. Landstra, "A Logistics Analysis and Ranking Model (ALARM)," *Reliability and Maintainability Symposium,* 1975, pp. 1–14. Available from the IEEE.

52. E. F. Dertinger, "Status of Reliability Requirements in Government Contracts," *Reliability and Quality Control Symposium,* 1965, pp. 125–135. Available from the IEEE.

53. R. DeSantis, I. Nathan, and R. Reddy, "Xerox Servicing Cost Model," *Annual Reliability and Maintainability Conference,* 1969, pp. 285–293. Available from the IEEE.

54. D. B. Dickinson and L. Sessen, "Life Cycle Cost Procurement of Systems and Spares," *Annual Reliability and Maintainability Symposium,* 1975, pp. 282–286. Available from the IEEE.

55. T. E. Dixon and R. M. Anderson, "Implementation of the Design-to-Cost Concept," *Proceedings Annual Reliability and Maintainability Symposium,* 1975, pp. 224–229. Available from the IEEE.

56. A. Dushman, "Effect of Reliability On Life Cycle Inventory Cost," *Annual Symposium on Reliability,* 1975, pp. 549–561. Available from the IEEE.

57. D. P. Dunbar, "Reliability of System Design and Testing," *Defense Management Journal,* Vol. 14, No. 6, 1978, pp. 26–30.

58. S. Duham and J. Catlin, "Total Life Cost and the 'Ilities,' " *Annual Reliability and Maintainability Symposium,* 1973, pp. 491–495.

59. D. R. Earles, "LCC—Commerical Application Ten Years of Life Cycle Costing," *Annual Reliability and Maintainability Symposium,* 1975, pp. 74–85. Available from the IEEE.

60. D. R. Earles, "Design To Operation and Support Costs," *Reliability and Maintainability Symposium,* 1974, pp. 149–154. Available from the IEEE.

61. M. Ebenfelt and S. Ogren, "Some Experiences From the Use of An LCC Approach," *Reliability and Maintainability Symposium,* 1974, pp. 142–146. Available from the IEEE.

62. M. Ebenfelt and O. Waak, "LCC—Defense Application, Some Comments on LCC as a Decision Making Tool," *Annual Reliability and Maintainability Symposium,* 1975, pp. 63–73. Available from the IEEE.

63. M. A. Eimstad, "Reliability and Maintainability Technical and Cost Relationships," *Annual Symposium on Reliability,* 1968, pp. 331–337. Available from the IEEE.

64. L. F. Eliel, "Logistics Effect Model (LEM) Applications," *Annual Reliability and Maintainability Symposium,* 1978, pp. 260–266. Available from the IEEE.

65. G. E. Eusties, "Reduced Support Costs For Shipboard Electronic Systems," *Annual Reliability and Maintainability Symposium,* 1977, pp. 316–319. Available from the IEEE.

66. T. L. Fagan, "A Computerized Cost Model For Space Systems," *8th Reliability and Maintainability Conference,* 1969, pp. 85–89. Available from Gordon and Breach Science Publishers, New York.

67. C. M. Fishman and M. J. Slovin, "Life Cycle Cost Impact On High Reliability Systems," *Reliability and Maintainability Symposium,* 1974, pp. 358–362. Available from the IEEE.

68. T. J. Fody, "The Procurement of Window Air Conditioners, Using Life Cycle Costing, *Annual Reliability and Maintainability Symposium,* 1977, pp. 81–88. Available from the IEEE.

69. B. Fox, "Total Annual Cost, a Reliability Criterion," *Proceedings of the 10th National Symposium on Reliability and Quality Control,* 1964.

70. B. Freunkin, P. W. Hodge, and M. J. Sabra, "An Econometric Approach to Redundancy Selection," *GIDEP,* Report No. 347-40.00.00-K4-27, 1972, pp. 1–8.

71. J. S. Gansler, "Application of Life Cycle Costing to the DOD System Acquisition Decision Process," *Reliability and Maintainability Symposium,* 1974, pp. 147–148. Available from the IEEE.

72. A. S. Goldman, "Problems in Life Cycle Support Cost Estimation," *Naval Research Logistics Quarterly,* Vol. 16, 1969, pp. 111–120.

73. J. S. Greenberg, "The Economic Implications of Unreliability," *Annual Reliability and Maintainability Symposium,* 1976, pp. 293–299. Available from the IEEE.

74. G. R. Greenfield, "Cost Analysis of Reliability Programs," *Seventh Annual West Coast Reliability Symposium,* 1966, pp. 93–107.

75. J. F. Grosson, "Designing to a Cost: An Update on NAVY Initiatives," *Defense Management Journal,* Vol. 15, No. 1, 1979, pp. 7–10.

76. G. W. Hart, "DOD Should Share its Experience in Life Cycle Costing," *Defense Management Journal,* Vol. 14, No. 1, 1978, pp. 16–18.

77. G. Harvey, "Life Cycle Costing—A Review of the Technique," *Management Accounting,* June 1976.

78. J. T. Henderson, "A Computerized LCC/ORLA Methodology," *Annual Reliability and Maintainability Symposium,* 1979, pp. 51–55. Available from the IEEE.

79. A. M. Hevesh, "Cost of Reliability Improvement," *Annual Symposium on Reliability,* 1969, pp. 54–61. Available from the IEEE.

80. G. P. Huber, "Multiplicative Utility Models in Cost Effectiveness Analysis," *The Journal of Industrial Engineering,* Vol. XIX, 1968, pp. 17–19.

81. M. Huetinck, M. Lipow, and N. R. Garner, "Cost Effective Tradeoffs for Sequentially Deployed Systems," *Tenth Annual West Cost Reliability Symposium,* 1969, pp. 1–20.

82. W. S. Jones, "Is Reliability the Key to Reduced Support Costs?" *Defense Management Journal,* Vol. 14, No. 3, 1978, pp. 30–35.

83. R. F. Johnson and L. Siegler, "Cost Effective Spares Provisioning Models for Airline Operations," *Tenth Annual West Coast Reliability Symposium,* 1969, pp. 165–216.

84. V. C. Jordan, "Life Cycle Costing—Introduction," *10th Reliability and Maintainability Conference,* 1971, pp. 224–228. Available from the IEEE.

85. I. Katz and R. E. Cavender, "Weapon-System Life Cycle Costing," *10th Reliability and Maintainability Conference,* 1971, pp. 225–230. Available from the IEEE.

86. R. J. Kaufman, "Life Cycle Costing: A Decision Making Tool for Capital Equipment Acquisition," *Cost and Management Accountant (Canada),* March–April 1970, pp. 21–28.

87. T. G. Kiang, "Life Cycle Costing—A New Dimension For Reliability Engineering Challenge," *Proc. 30th Annual Technical Conference,* ASQC, Toronto, 1976.

88. J. Kogan, "Life-Cycle Cost of Lifting—Applicances," *Annual Reliability and Maintainability Symposium,* 1979, pp. 28–33. Available from the IEEE.

89. A. W. Kratzke and J. Ganger, "Cost Optimization Subject to Availability Constraints," *Annual Reliability and Maintainability Symposium,* 1979, pp. 73–78. Available from the IEEE.

90. C. H. Langwost, "Life Cycle Costing Applied In Evaluating Alternative Short Take-Off

and Landing (STOL) Aircraft and Tracked Air Cushion Vehicles (TACU) As Modes of Transportation, *10th Annual Reliability and Maintainability Conference,* 1971, pp. 231–235. Available from the IEEE.

91. R. Laskin and W. L. Smithhisler, "The Economics of Standard Electronic Packaging," *Annual Reliability and Maintainability Symposium,* 1979, pp. 67–72. Available from the IEEE.

92. H. L. Leve and M. M. Platte, "Effects of Reliability Variations on System Cost, *Annual Symposium on Reliability,* 1966, pp. 283–295. Available from the IEEE.

93. M. O. Locks, "Maintainability and Life-Cycle Costing," *Annual Reliability and Maintainability Symposium,* 1978, pp. 251–253. Available from the IEEE.

94. J. E. Lott, "Reliability Predictions and System Support Costs," *IEEE Trans. on Reliability,* Vol. 16, 1967, pp. 126–132. Available from the IEEE.

95. J. L. Madden, "Cost-Effective Maximum Times For Aircraft Engines," *Annual Symposium on Reliability,* 1966, pp. 118–124. Available from the IEEE.

96. G. R. McNichols, "Treatment of Uncertainty in Life Cycle Costing," *Annual Reliability and Maintainability Symposium,* 1979, pp. 56–61. Available from the IEEE.

97. R. L. McLaughlin and H. D. Voegtlen, "Ground Electronic Equipment Support Cost vs Reliability and Maintainability," *5th National Reliability and Quality Control Symposium,* 1959, pp. 36–42. Available from the IEEE.

98. W. C. Messecar, "Role of Life Cycle Costing in Fleet Planning Decisions," *10th Reliability and Maintainability Conference,* 1971, pp. 236–240.

99. E. B. Morrison, "Will Life Cycle Costing Rule Out the Low Bidder," *Technical Conference Transactions,* ASQL, 1969, pp. 343–347.

100. D. Monteith and B. Shaw, "Improved R.M. and LCC for Switching Power Supplies," *Annual Reliability and Maintainability Symposium,* 1979, pp. 262–265. Available from the IEEE.

101. A. Munoz, "Minimize Downstream Costs Through Applied Maintainability," *Annual Reliability and Maintainability Symposium,* 1979, pp. 266–270. Available from the IEEE.

102. V. O. Muglia, A. S. Cici, R. H. Waln, "Optimum Life-Cycle Costing," *Annual Reliability and Maintainability Symposium,* 1974, pp. 369–376. Available from the IEEE.

103. V. O. Muglia, "Reliability and Maintainability Management Using Risk Analysis," *Annual Reliability and Maintainability Symposium,* 1976, pp. 230–233. Available from the IEEE.

104. I. Nathan, "A Reliability Cost-Benefit Analysis Study," *Annual Reliability and Maintainability Symposium,* 1972, pp. 308–318. Available from the IEEE.

105. I. Nathan, "Management Decision Utilizing Cost-Effectiveness Modeling," *IEEE Transactions on Reliability,* Vol. R-18, 1969, pp. 54–63.

106. E. T. Parascos, "Reliability Engineering and Underground Equipment Failure, Cost and Manufacturer's Analysis," *3rd Annual Reliability Engineering Conference for the Electric Power Industry,* 1976, pp. 25–30.

107. J. R. Peronnet, "The Logistics of Life Cycle Cost," *Microelectronics and Reliability,* Vol. 19, 1979, pp. 23–30.

108. G. S. Peratino, "Air Force Approach to Life Cycle Costing," *Annual Symposium On Reliability,* 1968, pp. 154–187. Available from the IEEE.

109. P. C. Pedrick, "Survey of Life Cycle Costing Practices of Non-Defense Industry," *Annual Symposium On Reliability,* 1968, pp. 188–192. Available from the IEEE.

110. G. A. Raymond, "Reliability vs the Cost of Failure," *4th National Symposium on Reliability and Quality Control,* 1957, pp. 187–188. Available from the IEEE.

111. H. Reiche, "Life Cycle Cost," *Reliability and Maintainability of Electronic Systems,"* 1980, Computer Science Press, Potomac, MD, pp. 3–23.

112. S. I. Rosenthal, "Failure Free Warranty (Reliability Improvement) Seller Viewpoints,"

Annual Reliability and Maintainability Symposium, 1975, pp. 80–86. Available from the IEEE.

113. H. Rosenberg and J. H. Witt, "Effects on LCC of Test Equipment Standardization," *Annual Reliability and Maintainability Symposium,* 1976, pp. 287–292. Available from the IEEE.

114. W. J. Ryan, "Procurement Views of Life Cycle Costing," *Annual Symposium on Reliability,* 1968, pp. 164–168. Available from the IEEE.

115. J. C. Harty and L. Siegler, "A Practical Life Cycle/Cost of Ownership Type Procurement VIA Long Term/Multi Year 'Failure Free Warranty' (FFW) Showing Trial Procurement Results," *Tenth Reliability and Maintainability Conference,* 1971, pp. 241–251. Available from the IEEE.

116. L. S. Schlosser, "Practical Innovation Can Mean Lower Life-Cycle Costs," Defense Management Journal,Vol. 14, 1978, pp. 32–35.

117. M. J. Shumaker and J. C. DuBuisson, "Trade-Off of Thermal Cycling vs Life-Cycle Costs," *Annual Reliability and Maintainability Symposium,* 1976, pp. 300–490. Available from the IEEE.

118. B. A. Schmidt, "Preparation For LCC Proposals and Contracts," *Annual Reliability and Maintainability Symposium,* 1979, pp. 62–66. Available from the IEEE.

119. J. P. Solomond and G. A. Marreglia, "Cost Optimizing System to Evaluate Reliability (Coster)," *Annual Reliability and Maintainability Symposium,* 1977, pp. 385–390. Available from the IEEE.

120. D. Sternlight, "The Fast Deployment Logistic Ship Project: Economic Design and Decision Technique," *NRLQ,* Vol. 17, 1970, pp. 373–387.

121. F. Stehle, P. R. Oyerly, and A. Scarfile, "The Development and Application of A Cost-Effectiveness Assessment Procedure For Shipboard Electronic Equipments," *5th Reliability and Maintainability Conference,* Vol. 5, 1966, pp. 61–74.

122. N. C. Stordahl and J. L. Short, "The Impact and Structure of Life Cycle Costing," *Annual Symposium on Reliability,* 1968, pp. 509–515. Available from the IEEE.

123. R. G. Stokes and F. N. Stehle, "Some Life-Cycle Cost Estimates for Electronic Equipment: Methods and Results," *Annual Symposium on Reliability,* 1968, pp. 169–183. Available from the IEEE.

124. A. Stratton, "The Principles and Objectives of Cost-Effectiveness Analysis," *The Aeronautical Journal of the Royal Aeronautical Society,* Vol. 72, 1968, pp. 43–53.

125. L. Session, "Life Cycle Cost Testing," *Evaluation Engineering.* January/February 1974, pp. 15–20.

126. D. Steinlight, "Life Cycle Costing as an Economic Design and Decision Tool," *Naval Research Logistics Quarterly,* Vol. 18, No. 14, 1971.

127. T. G. Terborgh, "Business Investment Management," *The Machinery and Applied Products Institute,* Washington, D.C., 1967.

128. C. D. Weiner, "Contractor Initiatives For Reliability and Maintainability," *Annual Reliability and Maintainability Symposium,* 1978, pp. 243–250. Available from the IEEE.

129. R. B. Werden, "A Practical Method of Optimizing System Life Cycle Costs vs Availability," *Microelectronics and Reliability,* Vol. 17, 1978, pp. 1–7.

130. G. F. Walters and J. A. McCall, "Software Quality Metrics For Life Cycle Cost-Reduction," *IEEE Transactions on Reliability,* Vol. R-28, 1979, pp. 212–220.

131. E. S. Winland, "Cost-Effectiveness Analysis For Optimal Reliability and Maintainability," *11th National Symposium on Reliability and Quality Control,* 1965, pp. 107–114. Available from the IEEE.

132. H. W. Wynholds and J. P. Skratt, "Weapon System Parametric Life Cycle Cost Analysis," *Annual Reliability and Maintainability Symposium,* 1977, pp. 303–309. Available from the IEEE.

Reliability Improvement Warranty (RIW)

133. J. R. Anderson, "Warranties—The Easy Way Out," *Proceedings Annual Reliability and Maintainability Symposium,* 1979, pp. 413–415.

134. H. Bayer, "Long-Term Commercial Warranty," *Proceedings Annual Reliability and Maintainability Symposium,* 1978, pp. 50–54. Available from the IEEE.

135. H. Balaban and B. Retterer, "The Use of Warranties for Defense Avionics Procurement," *Proceedings Annual Reliability and Maintainability Symposium,* 1974, pp. 363–368. Available from IEEE.

136. H. Balaban, D. Cuppett, and G. Harrison, "The F-16 RIW Program," *Proceedings Annual Reliability and Maintainability Symposium,* 1979, pp. 79–82. Available from IEEE.

137. W. R. Blischke, "Calculation of the Cost of Warranty Policies as a Function of Estimated Life Distribution," *Naval Research Logistics Quarterly,* Vol. 22, 1975, pp. 681–696.

138. W. J. Bonner, "A Contractor View of Warranty Contracting," *Annual Reliability and Maintainability Symposium,* 1976, pp. 351–356. Available from the IEEE.

139. T. N. Bowers, "Reliability Guarantee vs. Cost of Capital," *Tenth Annual West Coast Reliability Symposium,* 1969, pp. 157–164. Available from the IEEE.

140. T. A. Budue, "Manufacturings Contribution to Reliability," *Seventh National Symposium on Reliability and Quality Control,* 1961, pp. 144–148. Available from the IEEE.

141. P. O. Chelson, "Can We Expect ECPs Under RIW?" *Proceedings Annual Reliability and Maintainability Symposium,* 1978, pp. 204–209.

142. R. C. Day and L. E. McIntyre, "RIW Data Collection and Reporting Method," *Proceedings Annual Reliability and Maintainability Symposium,* 1978, pp. 66–72. Available from the IEEE.

143. W. W. Flottman and M. R. Worstell, "Mutual Development, Application and Control of Supplies Warranties of American Airlines," *Proceedings of Annual Reliability and Maintainability Symposium,* 1977, pp. 213–221. Available from the IEEE.

144. R. M. Genet, "Government Depot Maintenance Warranties," *Proceedings of the Annual Reliability and Maintainability Symposium,* 1976, pp. 363–365. Available from the IEEE.

145. P. Gottfried and R. L. Madison, "Warranty Analysis, Industrial and Commercial Product Reliability," *Annual Technical Conference Transactions, American Society of Quality Control,* 1965, pp. 300–305.

146. R. K. Gates, R. S. Bicknell, and J. E. Bortz, "Quantitative Models Used in the RIW Decision Process," *Proceedings Annual Reliability and Maintainability Symposium,* 1977, pp. 229–236. Available from the IEEE.

147. A. F. Grimm, "A Method for Predicting Warranty Costs," *Technical Conference Transactions, American Society of Quality Control,* 1976, pp. 205–210.

148. A. J. Hanter and C. W. Strempke, "TACAN RIW Program," *Proceedings Annual Reliability and Maintainability Symposium,* 1978, pp. 62–65. Available from the IEEE.

149. C. A. Hardy and R. J. Allen, "Reliability Improvement Warranty Techniques and Applications," *Proceedings Annual Reliability and Maintainability Symposium,* 1977, pp. 222–228. Available from the IEEE.

150. R. Kowalski and R. White, "Reliability Improvement Warranty (RIW) and the Army Lightweight Doppler Navigation System (LDNS)," *Proceedings Annual Reliability and Maintainability Symposium,* 1977, pp. 237–241. Available from the IEEE.

151. F. J. Kreuze, "Reliability Aspects of 'Full Life' Product Warranty," *Proceedings of the Annual Reliability and Maintainability Symposium,* 1979, pp. 73–78. Available from the IEEE.

152. O. Markowitz, "Aviation Supply Office FFW/RIW Case History #2, ABEX Pump," *Proceedings Annual Reliability and Maintainability Symposium,* 1976, pp. 357–362. Available from the IEEE.

153. O. Markowitz, "Failure Free Warranty/Reliability Improvement Warranty, Buyer Viewpoints," *Annual Technical Conference Transactions, American Society of Quality Control,* 1975, pp. 87–97.

154. O. Markowitz, "A New Approach Long Range Fixed Price Warranty Within Operational Environments—For Buyer/User," *Proceedings Tenth Reliability and Maintainability Conference,* 1971, pp. 252–257. Available from the IEEE.

155. P. O. Nerber, "Warranty Cost Estimates for Avionic Sub-Systems," *Proceedings Annual Symposium on Reliability,* 1969, pp. 270–275. Available from the IEEE.

156. D. G. Newman, "USAF Experience with RIW," *Proceedings Annual Reliability and Maintainability Symposium,* 1978, pp. 55–61. Available from the IEEE.

157. J. Noah and H. L. Eskew, "Cost-of-Ownership-Pitfalls," *Proceedings Annual Reliability and Maintainability Symposium,* 1975, pp. 221–224. Available from the IEEE.

158. E. A. Polgar and J. H. Yueh, "A Model for Determination of Incentive Fee," *Proceedings Eleventh National Symposium on Reliability and Quality Control,* 1965, pp. 115–127. Available from the IEEE.

159. B. L. Retterer, "Consideration for Effective Warranty Application," *Proceedings Annual Reliability and Maintainability Symposium,* 1976. pp. 346–350. Available from the IEEE.

160. J. Rose and E. Phelps, "Cost of Ownership Application to Airplane Design," *Proceedings Annual Reliability and Maintainability Symposium,* 1976, pp. 287–292. Available from the IEEE.

161. R. R. Shorey, "Factors in Balancing Government and Contractor Risk With Warranties," *Proceedings Annual Reliability and Maintainability Symposium,* 1976, pp. 366–368.

162. R. M. Springer, "RIW With Cost Sharing," *Proceedings Annual Reliability and Maintainability Symposium,* 1977, pp. 391–395. Available from the IEEE.

163. C. H. Karr and G. L. Wagner, "Reliability and Maintainability—Today's Warranty Application," *Proceedings Reliability and Maintainability Symposium,* 1976, pp. 491–495. Available from the IEEE.

164. C. D. Weimer, "Contractor Incentives for Reliability and Maintainability," *Proceedings Reliability and Maintainability Symposium,* pp. 243–250, 1978. Available from the IEEE.

10
Maintenance Engineering

10.1 INTRODUCTION

Maintenance engineering interfaces with product design and product support. Today this is a distinct discipline, although it has some overlaps with reliability and maintainability engineering. Therefore, a reliability and maintainability engineer should have some knowledge of maintenance engineering. Taking this view into consideration, this chapter is devoted to maintenance engineering. This discipline is equal in importance to other disciplines of engineering. A newly manufactured engineering product must be maintained properly in the field in order to obtain its effective performance.

Maintenance is defined as all actions essential in order to retain a product in or restoring it to a satisfactory operational condition [228].

Maintenance and maintenance engineering with well-defined interfaces are two different areas [41]. Maintenance engineering is a planning and analysis function whereas the maintenance is a function that carries out maintenance physically.

Some of the contributing objectives [41] of maintenance engineering are:

1. Maintenance operations improvement
2. Maintenance organization improvement
3. Minimizing the requirement of maintenance skills
4. Reducing the complexity effect
5. Maintenance amount and frequency minimization
6. Minimizing the supply support
7. Preventive maintenance frequency optimization
8. Optimum utilization of maintenance facilities
9. Improving maintenance training and associated publications

References 54 and 110 list a number of published articles on the availability of maintained systems and the grouping of preventive maintenance policies, respectively. At the end of this chapter, an extensive list of books and articles on maintenance and maintainability engineering is presented. References 1 to 159, and 228 and 229 are on maintenance engineering and references 159 to 228 are on maintainability engineering. Some of the references listed in both areas have noticeable overlaps. The closer ties between both disciplines is one of the main reasons for including references on maintainability engineering. In addition, one section of this chapter is devoted to maintainability engineering.

10.2 DEFINITIONS USED IN MAINTENANCE ENGINEERING

This section presents some of the commonly used definition in maintenance engineering [228].

Corrective Maintenance. The actions that must occur in order to repair a product that has failed to a satisfactory operational condition.

Preventive Maintenance. The maintenance done to retain a product in satisfactory operational condition, by providing inspection, detection, and correction of initial stage failures.

Down Time. An element of time in which a system or equipment is unable to carry out its intended functions satisfactorily.

Up Time. An element of time in which a system or equipment is either operating or in an alert or reacting state.

10.3 MAINTENANCE MANAGEMENT METHODS

Machines, plant, buildings, and so on deteriorate with time and use; therefore, they require maintenance [30, 229]. Furthermore, another reason to control maintenance is that the modern machinery is very expensive and sophisticated and thus requires better management and maintenance methods. It costs money to maintain a product in the field. Effective maintenance can reduce the maintenance and production costs by increasing the useful life of an item.

The maintenance management function is to retain systems and equipment running at the minimum cost to the company. The theory of maintenance management methods described in this section is the most suitable theory extracted from a review of several books and publications [229]. The following main areas of maintenance management methods are discussed briefly:

1. Maintenance organization
2. Preventive maintenance
3. Maintenance planning
4. Maintenance scheduling
5. Work orders
6. Work measurement
7. Maintenance cost budget
8. Maintenance training
9. Stores and spares control

10.3.1 Maintenance Organization

No one maintenance organization can be most useful in all situations. The organization should be tailored according to need. This need may be defined by technical, personnel, or geographical situations. The items concerned with the maintenance organization are:

1. *Maintenance Management Function.* This is divided into primary and secondary functions:
   Primary Functions: (a) Existing equipment maintenance, (b) inspection and lubrication of equipment, (c) existing equipment modification, and (d) new equipment installation.
   Secondary Functions: (a) Storage, (b) salvage, (c) waste disposal, (d) noise and pollution control, and (e) insurance administration.
2. *Maintenance Role in the Plant.* The role of maintenance department may be affected by the following four important aspects of an organization as a whole:

   a. The plant type
   b. The equipment type
   c. The type of services
   d. The skill types

   These factors determine the maintenance part in the company as a whole.
3. *The Maintenance Management Role.* The following are associated with the role of management:

   a. *Vertical Authority:* As short a line of vertical authority as possible is recommended, because as the line of vertical authority extends, the communication of orders and information can become confusing. The information and orders are transmitted clearly and quickly with a short gap between the top and bottom of organizational hierarchy.

   b. *Span of Control:* At the higher level of maintenance management hierarchy between three to six persons should report to an individual. At the lower level 18 to 25 persons are recommended per individual.

   c. *Enunciation of Functions:* Each maintenance staff member's job and responsibility must be clearly defined on paper or communicated verbally to avoid misunderstanding.

   d. *Maintenance and Production Department Heads:* To avoid confusion both these people's status must be equal in the hierarchy of management. For example, if the status of production head is higher than his or her counterpart, the following may result:

     i. Equipment/system misuse by the operators

     ii. Untimely equipment release for servicing and repair by the production department

     iii. Equipment purchase without considering the maintenance aspect

10.3.2 Preventive Maintenance

This is performed on equipment and systems to prevent the breakdown of equipment. When it is necessary, a preventive maintenance program on equipment must be established. The preventive maintenance schedule should take the production demand into consideration. Two main constituents of preventive maintenance are inspection and lubrication.

1. *Inspection*

   The objective of inspection is that (i) unexpected breakdowns are prevented, (ii) accuracy of the equipment is maintained, and (iii) unnecessary maintenance is prevented. The following key areas are associated with inspection:

   a. *Equipment Selection for Inspection:* Critical equipment and facilities are the prime candidates for inspection. Equipment requiring inspection should be looked at with regard to condition of equipment, plant type, and its importance to the activities of the organization. Furthermore, when establishing a criterion for equipment inspection, an item should be included for inspection, if (i) its malfunctioning will hold up production, (ii) its failure will endanger the health and safety of an employee or employees, (iii) a processed material is being wasted, or (iv) it may cause damage to other items.

   b. *Inspection Check List:* A list should be established for each item. The list should contain the list of items to be inspected during the inspection period.

c. *Inspection Frequency:* A time interval should be established between equipment inspections. The equipment analysis is the basis for establishing the time interval. In addition, the manufacturer's recommendations are also considered.

d. *Manning of Inspection:* The inspection should be performed by a senior craftperson or a responsible person. To make sure the inspection is performed properly, it should be double checked on sampling basis by a maintenance management person.

e. *Reasons for Unnecessary Inspection:* Possible reasons are (i) a standby unit is available, (ii) inspection and planning costs are higher than the inspection gains, (iii) the equipment useful life is more than the required service life.

2. *Lubrication*

An effective equipment lubrication management system performs lubrication in the right quantity at the right time by following the right method. The following are some of the advantages of lubrication: (a) it extends the failure-free equipment life, (b) it reduces the wear and tear, and (c) the solid friction is converted to fluid friction.

A proper lubrication planning system should be developed to carry out the lubrication function effectively.

10.3.3 Maintenance Planning

There are three types of planning: long-range, short-range, and day-to-day planning.

1. *Long-Range Maintenance Planning:* This type of planning is performed by senior maintenance management such as the director, manager, etc. It projects the need of maintenance say 5 years or 8 years ahead. The long range planning requires a forecast of the following two input factors:

   a. The maintenance facility and equipment requirement changes
   b. The production machinery changes

2. *Short-Range Maintenance Planning:* This type of planning is normally carried out by medium level maintenance management. The short-range planning forecasts the needs over a year.

3. *Day-to-Day Maintenance Planning:* This type of planning is normally performed by the lead people and technicians. The day-to-day planning is concerned with daily spares and materials need, labor allocation, tool needs, reporting and recording data for cost allocation and analysis, and so on.

10.3.4 Maintenance Job Scheduling

This is carried out in consultation with the production department. Maintenance jobs scheduling should be prepared in two phases: long and short periods. The long period scheduling involves 6 to 8 weeks, in order to get the approval of departments concerned, staffing, etc. The short period for scheduling is between 1 to 2 weeks. The maintenance job scheduling should be flexible enough to incorporate emergency and priority jobs. Usually 25 percent of the available labor is reserved for emergency and minor jobs. The remaining 75 percent is scheduled. A scheduler normally fulfills the task of scheduling. The critical path method can be used. This is discussed in detail in Section 10.5. Nowadays this technique, when computerized, becomes a very powerful tool in maintenance scheduling.

10.3.5 Work Orders

A work order is used to authorize and direct an individual or a department to perform a certain task. The work order tells the maintenance craftperson about the scheduled starting and finishing time and date of the job, and the method and facilities required. Usually, a work order incorporates

1. Work order number
2. Priority of the job
3. Allocation of personnel
4. Job location
5. Job description
6. Reason for maintenance
7. Starting and finishing time and date of the job
8. Labor and material, estimated and actual costs, and so on.

Work Request. This is used, usually, by the production and other departments to inform the maintenance department about repair or other work to be performed on an equipment. The work request should not be used in the place of a work order by the maintenance department. Emergency jobs may be performed without a work request, although one must be filled soon afterward.

10.3.6 Work Measurement

Work measurement is concerned with determining the term of time and amount of work involved in a job. The following are the two objectives of work measurement:

1. To measure the time spent by machines or people for each activity in order to complete their work
2. To establish a standard time for a job.

Benefits from using work measurement in maintenance work are (1) equipment down time reduction, (2) preventive maintenance improvement, (3) maintenance delays reduction, and (4) decrease in cost due to increase in performance.

The following are some of the techniques used for the determination of time standards:

1. Work sampling
2. Guess estimates
3. Universal maintenance standards (UMS)
4. Predetermined motion time systems (PMTS)
5. Time studies
6. Statistical analysis of past data

10.3.7 Maintenance Cost Budgeting

The maintenance manager is usually responsible for developing a maintenance cost budget for each fiscal year. Each year's budget should reflect the thinking of the company, in order for it to become a living and vital tool. The following four types of cost budgets are used in maintenance:

1. Budget of the service department
2. Budget of the materials
3. The repair budget
4. Budget of the maintenance facility

10.3.8 Maintenance Training

In order to perform effective maintenance in a factory, maintenance personnel training is necessary. The training need of the personnel is determined by the maintenance manager or engineer. There are basically two types of training:

1. Training of apprentices: Apprentices are trained in a company to ensure continuing availability of trained craftpeople.
2. Training of staff: This includes technical and administrative personnel such as maintenance planners, equipment engineers, job inspectors, schedulers, time keepers, cost estimators, and so on. Training for these people is provided in the form of special courses according to their need.

10.3.9 Stores and Spares Control

This is one of the most important areas of maintenance engineering. The timely availability of spares, standby equipment, and other items are vital to the effectiveness of maintenance organization. On the other hand, one should note that overstocking of spares will be costly to the organization because capital is tied up. Furthermore, inventory will be subject to deterioration, obsolescence, and so on. To have an effective stores and spares control, among other things, the spare parts required for each equipment must be listed and inventory control models should be developed.

10.4 AN INVENTORY CONTROL MODEL

The beginning of inventory control management was probably in 1915, when the first economic lot size expression was developed [230]. Further work on this idea was published in 1931 [231]. This model deals with a situation when an inventory of a certain spare part is maintained in order to fulfill a requirement from the maintenance department. When the stock of such items falls to a predetermined point, then steps are taken to purchase a replenishment quantity of items in question. One of the main objectives of this mathematical model is to determine the optimum value of the procurement quantity.

Figure 10.1 represents the model idealized situation diagram. According to Fig. 10.1, the quantity α of a certain spare part is ordered at one time. The α number of spare parts are ordered at the item reorder level and received at the exact time when the stock level is fully depleted. At this point, the stock is replenished with α spare parts, and the cycle repeats as shown in Fig. 10.1.

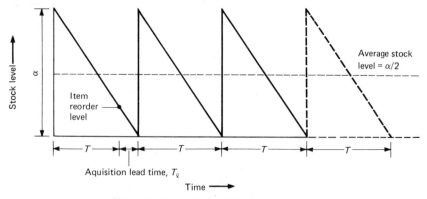

Figure 10.1. Inventory control model diagram.

The total increment cost, TC, associated with model is

$$TC = C_h + C_p \qquad (10.1)$$

where C_h is the total annual incremental holding cost of spare parts in inventory

C_p is the total annual incremental preparation cost to order spare parts quantity

$$C_h = \frac{C_u \alpha}{2} \qquad (10.2)$$

C_u is the spare part holding cost per item per year
$\alpha/2$ is the average stock level of spares

$$C_p = \frac{C_o Y}{\alpha} \qquad (10.3)$$

C_o is the single order placement cost
Y is the annual requirement of spares quantity

By substituting Eqs. (10.2) and (10.3) in Eq. (10.1) we get

$$TC = \frac{C_u \alpha}{2} + \frac{C_o Y}{\alpha} \qquad (10.4)$$

In order to find optimum value of α, differentiate Eq. (10.4) with respect to α and set the derivative equal to zero

$$\frac{dTC}{d\alpha} = \frac{C_u}{2} - \frac{C_o Y}{\alpha^2} = 0 \qquad (10.5)$$

Rearranging Eq. (10.5), we get the optimal lot size

$$\alpha_o = \left(\frac{2 Y C_o}{C_u} \right)^{1/2} \qquad (10.6)$$

Thus, the optimum number of orders, n, per year is

$$n = \frac{Y}{\alpha_o} \qquad (10.7)$$

The optimum time, T_o, between orders is from Eq. (10.7)

$$T_o = \frac{1}{n} = \frac{\alpha_o}{Y} \qquad (10.8)$$

EXAMPLE 1

Suppose $Y = 1000$ spares per year, $C_u = \$10$ per spare part per year, and $C_o = \$50$ per order, calculate the optimum value of α_o, n, and T_o using Eqs. (10.6), (10.7), and (10.8), respectively.
By using Eq. (10.6)

$$\alpha_o = \left(\frac{2 \times 1000 \times 50}{10}\right)^{1/2} = 100 \text{ spares}$$

Similarly, using Eqs. (10.7) and (10.8) we get

$$n = \frac{1000}{100} = 10 \text{ orders}$$

and

$$T_o = \frac{100}{1000} = 0.1 \text{ years}$$

There are several other inventory control models which may be found in references 234 and 235.

10.5 MAINTENANCE PLANNING AND CONTROL TECHNIQUES

There are basically two techniques which are used in maintenance planning and control. These two methods are known as the critical path method (CPM) and the Program Evaluation and Review Technique (PERT) [232]. The critical path method was developed in the fifties by Walker and Kelley of DuPont and Remington Rand, respectively, to control maintenance at the DuPont facilities. The development of the PERT is associated with controlling and scheduling the Polaris missile project of the United States Navy. Both techniques were released to the public in the late fifties.

The basic theory and symbols for both methods are the same. CPM is used where the estimates of activity duration are reasonably predictable, for example, in construction industry. PERT has application in research and development projects where the activities' duration estimates are much more uncertain. The PERT activity expected duration time, T, is calculated from the following formula:

$$T = (t_o + 4t_m + t_p)/6 \qquad (10.9)$$

where t_o denotes the optimistic estimate of an activity duration; in other words, the shortest time estimated to complete an activity

t_p denotes the pessimistic estimate of an activity duration; in other words, the longest time estimated to complete an activity

t_m denotes the most likely estimate of an activity duration; in other words, the time in between optimistic and pessimistic duration times.

EXAMPLE 2

A PERT activity has an earliest completion time, $t_o = 10$ hours, a latest completion time, $t_p = 20$ hours, and a most likely completion time, $t_m = 14$ hours. Find the expected duration, T, time of the activity. By using Eq. (10.9)

$$T = (10 + 4 \times 14 + 20)/6 = 14.33 \text{ hours}$$

10.5.1 CPM and PERT Symbols and Definitions

The following symbols are used to develop CPM and PERT networks:

A circle denotes an event or a node. A point in time is represented by an event. An event can be a start or an end of an activity or activities. Usually each event is denoted with a number.

An activity is represented by a continuous arrow. It denotes that work has to be accomplished in order to progress to another event. Money, materials, time, and labor are consumed to accomplish an activity.

A dotted arrow represents a dummy activity. The symbol denotes a restraint. This is an imaginary activity which is accomplished in zero time without the use of any resources. To show a proper relationship between

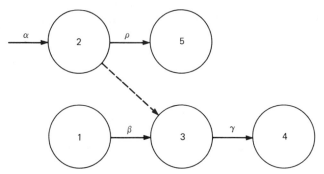

Figure 10.2. A CPM network with a dummy activity.

activities of a network, a dummy activity is used. Figure 10.2 shows the usage of a dummy activity. Before starting activity γ, activities α and β must be accomplished. However, activity ρ can start after the completion of only activity α.

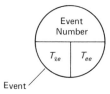

This is the commonly used symbol to represent an event. The top half of the circle is used to hold the event number. The bottom half of the circle is divided into two parts, where the left half is used for latest event time, T_{le}, and the right half is used for earliest event time, T_{ee}.

The following are the frequently used definitions in CPM and PERT networks analysis:

Latest event time (T_{le}).   This is the time which indicates that without delaying the completion of the project, an event may be reached or an activity may be accomplished.

Earliest event time (T_{ee}).   This is the time which indicates that the earliest time an activity can be accomplished or an event can be reached.

Critical path.   This is the longest path through the CPM or PERT network. The largest sum of activity duration times of all the paths of a network is the total time of the critical path. Any delay in the completion of activities on the critical path will delay the entire project completion date.

EXAMPLE 3

The following data is given for a certain project.

IDENTIFICATION OF AN ACTIVITY	IMMEDIATE PREDECESSORS ACTIVITY OR ACTIVITIES	ACTIVITY DURATION TIME (TIME TO ACCOMPLISH ACTIVITY IN HOURS)
A	—	20
B	A	10
C	B	30
D	C,A	40
E	D,B	5

Find the critical path of the network. The network shown in Fig. 10.3 has the following paths:

1. *A-D-E* (20 + 40 + 5 = 65 hours)
2. *A-B-E* (20 + 10 + 5 = 35 hours)
3. *A-B-C-D-E* (20 + 10 + 30 + 40 + 5 = 105 hours)

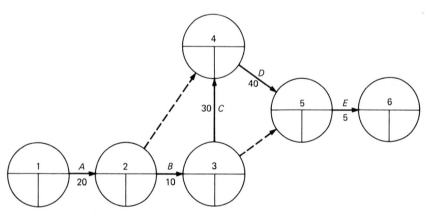

Figure 10.3. A CPM network.

The critical path is represented by activities *A-B-C-D-E*. This is the longest path of the network of a project whose duration will be 105 hours.

Formulas for Network Analysis. A network with two events is given in Fig. 10.4. The meaning of the symbols in Fig. 10.4 are as follows:

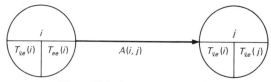

Figure 10.4. A two-event network.

i denotes the ith event
j denotes the jth event
$A(i,j)$ denotes the duration of the activity between events i and j
$T_{le}(i)$ denotes the latest event time of the event i
$T_{ee}(j)$ denotes the earliest event time of the event j
$T_{ee}(i)$ denotes the earliest event time of the event i
$T_{le}(j)$ denotes the latest event time of the event j.

For activity (i,j):

$$\text{Latest finish time} = T_{le}(j) \tag{10.10}$$

$$\text{Earliest start time} = T_{ee}(i) \tag{10.11}$$

$$\text{Earliest finish time} = T_{ee}(i) + A(i,j) \tag{10.12}$$

$$\text{Latest start time} = T_{le}(j) - A(i,j) \tag{10.13}$$

$$\text{Interfering float} = T_{le}(j) - T_{ee}(j) \tag{10.14}$$

$$\text{Free float} = T_{ee}(j) - T_{ee}(i) - A(i,j) \tag{10.15}$$

$$\text{Total float} = t_{le}(j) - T_{ee}(i) - A(i,j) \tag{10.16}$$

Steps to Determine the Critical Path of a Network

1. Compute T_{le} for each event by following a *backward* pass of the network in question. To calculate the latest event time of event i, use the following formula:

$$T_{le}(i) = \text{Minimum for all succeeding event } j \text{ of } [T_{le}(j) - A(i,j)] \tag{10.17}$$

The network's last event T_{le} and T_{ee} are equal.

2. Compute T_{ee} for each event by following a *forward* pass of the network in question. To calculate the earliest event time of event j, use the following formula:

$$T_{ee}(i) = \text{Maximum for all preceding event } i \text{ of } [T_{ee}(i) + A(i,j)] \qquad (10.18)$$

The network's first event earliest event time is equal to zero.

3. Find the critical path of the network. Procedures to obtain the critical path are given below:

   a. Choose those network events whose $T_{le} = T_{ee}$. If these events are along a single path from the beginning to the end, then this is the critical path of the network. However, if there is more than one path, then one should use the method described in (b).

   b. For each path obtained in (a), compute each activity total float. The critical path is the one which has the *minimum* sum of the total floats.

EXAMPLE 4

For the network shown in Fig. 10.3, calculate each event's earliest and latest times using Eqs. (10.18) and (10.17), respectively. Figure 10.5 shows the resulting network with T_{le}'s and T_{ee}'s.

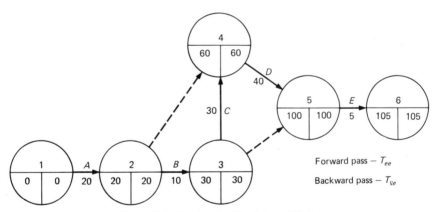

Figure 10.5. A network with T_{le}'s and T_{ee}'s.

10.6 MAINTENANCE MEASURING INDICES

This section presents indices which are useful to measure maintenance performance. Numerous indices should be used in an effort to reduce the maintenance cost. The indices should be used in combination to fulfill the objective of management because no one index is adequate to fulfill this need. One of the main objectives of these indices is that by using the past performance as a point of reference, they indicate the trends of the ratios in question. The following selective indices [115, 229] are presented in this section:

Broad Indicators. The following three indices fall into this category:

1. The maintenance cost ratio (MCR) is

$$MCR = \frac{\alpha}{\beta} \qquad (10.19)$$

where α is the total cost of maintenance
β is the total sales

For all industries, the average value of MCR is 5 percent. There is wide variation among different industries. For the chemical and steel industries the average MCR's are 6.8 and 12.8 percent, respectively.

2. The maintenance cost to value of facilities ratio is

$$MCVR = \frac{\alpha}{\gamma} \qquad (10.20)$$

where MCVR denotes the maintenance cost to value of facility ratio
γ denotes the total cost of investment in plant and equipment

For example, in the chemical and steel industries, the average MCVR's are 3.8 and 8.6 percent, respectively.

3. The total maintenance cost to total output ratio is

$$MCOR = \frac{\alpha}{\theta} \qquad (10.21)$$

where θ is the total output.

The common factor between all these indicators is the cost of maintenance which will indicate the performance of the organization in relation to the total output, sales, and so on.

The need for the short-term prediction is not necessary because these ratios are sensitive to the short-term period. A 1-year interval should be considered when computing the values of these indicators.

Preventive Maintenance Indicators.

1. Downtime hours
2. The cost of preventive maintenance to cost of repair which is given by

$$CPMR = \frac{\alpha_1}{\rho} \tag{10.22}$$

where α_1 is the cost of preventive maintenance
ρ is the cost of repair

3. The following index relates the preventive maintenance man-hours to maintenance man-hours:

$$PMMR = \frac{\mu_{pm}}{\mu_m} \tag{10.23}$$

where μ_{pm} denotes the total preventive maintenance man-hours
μ_m denotes the total maintenance man-hours

The ratio should be between 20 and 40 percent.

Work Order System Indicators.   The following two ratios should be used in order to have an effective work order system:

1. The ratio of the planned jobs completed by due dates to the number of planned jobs index is

$$PJTPJR = \frac{PJ}{TPJ} \tag{10.24}$$

where **PJ**   is the number of planned jobs completed by due dates
TPJ is the total number of planned jobs

2. This ratio is concerned with obtaining the percentage of planned jobs accomplished within estimated costs. This index is known as the planned jobs cost control indicator.

An Indicator for Material Control.   This indicator is

$$JAMPJR = \frac{PJAM}{TPJ} \times 100 \qquad (10.25)$$

where PJAM is the number of planned jobs waiting for material.

An Indicator for Backlog Control.   One of the most useful indicators for the backlog control is

$$MHPMHB = \frac{PMH}{TMB} \qquad (10.26)$$

where PMH is the number of planned man-hours
      TMB is the number of backlog man-hours

Job Planning Indicator.   This indicator is concerned with the percentage of total maintenance jobs accomplished through planning.

Emergency Work Indicator.   This index is

$$MEJTMR = \frac{MHEJ}{TMMH} \qquad (10.27)$$

where MHEJ   is the total man-hours spent on emergency jobs
      TMMH is the total maintenance man-hours

The lower the value of this indicator, the better the planning and preventive maintenance operations are.

An Indicator for Forecasting Effectiveness.   This ratio is

$$MHMHFR = \frac{MRS}{MRF} \qquad (10.28)$$

where MRF is the total forecasted maintenance hours
      MRS is the total maintenance hours spent on maintenance

Productivity Indicator.   This indicator is associated with measuring productivity. The ratio is

$$\text{MLCMC} = \frac{\text{MLC}}{\text{MMC}} \tag{10.29}$$

where MMC is the total cost of the maintenance materials

MLC is the total cost of the maintenance labor

The time intervals used to compute values of ratios in Eqs. (10.21) to (10.29) are not specifically defined. However, it is suggested that a 1-month period should be considered. Otherwise, the period should be chosen according to the specific need.

10.7 MATHEMATICAL MODELS USED IN MAINTENANCE

This section presents two mathematical models which can be used for maintenance decision making. These are

10.7.1 Equipment Replacement Model

This model [233] is used to find the optimum replacement interval of an item where the increasing trend in the item maintenance cost is predictable. For the rising maintenance cost of an item, the expected yearly total cost, C_t, of that item is

$$C_t = C_0 + [x(\beta - 1)/2] + \text{IC}/\beta \tag{10.30}$$

where β denotes the number of years of the item life

IC is the acquisition cost of the item

x is the yearly increase in the maintenance cost

C_0 is the yearly nonvarying maintenance plus operating costs

In Eq. (10.30), the interest rate is neglected.

To find the optimum value of β differentiate Eq. (10.30) with respect to β, then set the resulting expression equal to zero.

$$\frac{dC_t}{d\beta} = \frac{x}{2} - \text{IC}/\beta^2 = 0 \tag{10.31}$$

Rearranging Eq. (10.31), we get

$$\beta_0 = \left(\frac{2\text{IC}}{x}\right)^{1/2} \tag{10.32}$$

where β_0 is the optimum replacement period of the item.

EXAMPLE 5

Find the optimum value for β in Eq. (10.31) if the equipment IC = \$1000 and x = \$200 per year.

$$\beta_0 = \sqrt{\frac{2 \times 1000}{200}} = 3.162 \text{ years}$$

10.7.2 A Model for Equipment Parts Replacement

This model represents an equipment [19] which is inspected after a long period by statutory requirement. For example, this period could be 1 year. The following were assumed to develop this model:

1. Between statutory inspections the equipment operating cost increases. This increase in the cost could be due to the deterioration of some equipment parts.
2. In order to reduce the operating cost of the equipment, some of these deteriorating components are replaced at a certain interval in between the inspection periods.
3. Replacement intervals in between the inspection intervals are equally spaced.

Therefore, the main objective of this model is to find the part replacement period, x_p, with the minimum sum of equipment operating and part replacement costs.

The following notation is associated with this model:

K_1 is the single replacement cost
$K(x)$ is the equipment operating cost at time x (Note that after replacement, this cost is in per unit time.)
m is the number of replacements
x_p is the period between parts replacement
$[0, X]$ is the interval between statutory inspections

The total cost, $K(x_p)$, of the model whose diagram is shown in Fig. 10.6, is

$$K(x_p) = \alpha + \beta \tag{10.33}$$

where α is the equipment operating cost between inspections
    β is the equipment part replacement cost between inspections

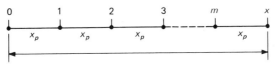

Period between statutory inspections

Figure 10.6. Equipment parts replacement model.

$$\beta = mK_1 \tag{10.34}$$

$$\alpha = (m + 1)\left[\int_0^{x_p} K(x)\, dx\right] \tag{10.35}$$

where the quantity in brackets is the operating cost for each time interval x_p. By substituting Eqs. (10.35) and (10.34) in Eq. (10.33) we get

$$K(x_p) = (m + 1)\left(\int_0^{x_p} K(x)\, dx\right) + mK_1 \tag{10.36}$$

From Fig. 10.6

$$X = (m + 1)x_p \tag{10.37}$$

Rewriting Eq. (10.37) in terms of m gives the following results:

$$m = (X - x_p)/x_p \tag{10.38}$$

By substituting Eq. (10.38) in Eq. (10.36), the resulting expression is

$$K(x_p) = \frac{X}{x_p}\left[\int_0^{x_p} K(x)\, dx\right] + K_1(X - x_p)/x_p \tag{10.39}$$

To find the optimal value for x_p, substitute a particular cost function for $K(x)$, integrate this function over the time interval $[0, x_p]$, differentiate Eq. (10.39) with respect to x_p, set the resulting expression equal to zero, and rearrange the equation in terms of x_p.

10.8 MAINTAINABILITY

This is a characteristic of product design and installation. Maintainability may be defined [228] as the probability that a system/equipment/subsystem/component will be kept in or restored to a satisfactory condition within a specified period, after performing maintenance according to prescribed procedures and resources.

Maintainability as an engineering discipline is in its third decade. The attention on certain maintainability features when designing systems was first focused on in the late fifties. The result of this effort, in 1959, was the publication of MIL-M-26512 (USAF) [159]. Ever since the late fifties, several publications on maintainability have appeared [159–227].

10.8.1 Maintainability Measures

The following are the measures of maintainability:

Mean Time to Repair (MTTR). The system MTTR is given by the following expression:

$$\text{MTTR} = A/B \tag{10.40}$$

where $A \equiv \sum\limits_{j=1}^{k} t_j \lambda_j$

λ_j denotes the system's jth repairable component constant failure rate

t_j denotes that due to jth component failure, the time required to repair the system/equipment

k represents the number of repairable components

B denotes the summation of λ_j from $j = 1, 2, 3, \ldots, k$.

Equipment Repair Time (ERT). This is a median of the distribution of equipment repair times. For various distributions, the formulas for ERT are

Normal distribution

$$\text{Equipment repair time} = \text{MTTR} \tag{10.41}$$

This is due to the normal tribution's symmetry in which case the mean is equal to the median.

Exponential
For this distribution

$$\text{Equipment repair time} = \frac{\text{MTTR} \times 69}{100} \tag{10.42}$$

Log normal
For this distribution

$$\text{Equipment repair time} = \text{MTTR} \ (e^{\sigma^2/2})^{-1} \qquad (10.43)$$

where σ denotes the logarithmic standard deviation.

Mean Corrective Maintenance Time (MCMT). This includes only the active maintenance time. No idle time is included. It is

$$\text{MCMT} = A(B)^{-1} \qquad (10.44)$$

where $A \equiv \Sigma \ \lambda_j \ m_{cj}$
- λ_j is the jth component constant failure rate
- m_{cj} denotes the system's active maintenance time due to the failure of jth component
- B   denotes the summation of λ_j.

Mean Preventive Maintenance Time (MPMT). This is an equipment's active preventive maintenance times arithmetic mean. Equipment mean preventive maintenance time is

$$\text{MPMT} = C(D)^{-1} \qquad (10.45)$$

where $C \equiv \Sigma \ f_j \ m_{pj}$
- f_j denotes the frequency of the jth preventive maintenance task
- m_{pj} denotes the jth preventive maintenance task's equipment active maintenance time
- D   denotes the summation of f_j.

Mean Active Maintenance Time (MAMT). This is a mean of the distribution of all maintenance action times. This includes both corrective and preventive maintenance actions. It is defined

$$\text{MAMT} = \frac{A + C}{B + D} \qquad (10.46)$$

where $A \equiv \Sigma \ \lambda_j \ m_{cj}$
$ B \equiv \Sigma \ \lambda_j$
$ C \equiv \Sigma \ f_j \ m_{pj}$
$ D \equiv \Sigma \ f_j$

Geometric Mean Time to Repair (MTTR$_g$). This is associated with the log normal distribution. It is given by Eq. (10.43). Furthermore, the MTTR$_g$ is identical to the ERT of the log normal distribution.

10.8.2 Maintainability Functions

When a repair begins on an item at $t = 0$ it is useful to predict the probability that the repair will be accomplished in a time, t. To do this, the maintainability functions are used. The maintainability function, $m(t)$, for any distribution can be obtained from the following expression by substituting the probability density function of the distribution in question:

$$m(t) = \int_0^t f(t) \, dt \tag{10.47}$$

$f(t)$ is the repair time probability density function
t is time

The probability density functions of distributions whose maintainability functions are presented below are given in Chapter 2.

For the three repair time distributions, the maintainability equations are as follows.

Exponential. This distribution is applied to corrective maintenance. The maintainability function, $m(t)$, is found by substituting the exponential probability density function in Eq. (10.47) and integrating

$$m(t) = 1 - e^{-\mu t} \tag{10.48}$$

where $\mu = 1/\text{MTTR}$, where the MTTR denotes the mean time to repair. The μ is the constant repair rate.
t is the repair time variable
$m(t)$ denotes the probability that repair will be accomplished in time t, given that it was started at $t = 0$

Weibull. This distribution is not that popular in the maintainability work. By substituting the Weibull probability density function in Eq. (10.47) and integrating, the following maintainability expression results

$$m(t) = 1 - e^{-(t/\alpha)^\beta} \tag{10.49}$$

where α is the scale parameter
   β is the shape parameter

The mean maintenance time, t_m, is

$$t_m = \alpha \Gamma \left(\frac{\beta + 1}{\beta} \right) \tag{10.50}$$

Gamma. This is a quite commonly used distribution in maintainability engineering. By substituting the gamma probability density function in Eq. (10.47) and integrating, the following maintainability expression is obtained:

$$m(t) = \frac{\alpha^\beta}{\Gamma(\beta)} \int_0^t x^{\beta-1} e^{-\alpha x} dx \tag{10.51}$$

where β is the shape parameter
   α is the scale parameter

The mean, m_g, of the gamma distribution is

$$m_g = \frac{\beta}{\alpha} \tag{10.52}$$

Maintainability functions for other distributions are given in reference [159].

10.9 SUMMARY

This chapter presents the following areas of maintenance engineering:

1. Maintenance engineering definitions
2. Maintenance management methods
3. Inventory control modeling
4. Maintenance planning and control techniques
5. Maintenance measuring indices
6. Mathematical models used in maintenance
7. Maintainability

Maintainability is discussed briefly in this chapter because of its interrelationship with the maintenance engineering. The author has tried to focus more on the structure of the maintenance concepts than the minute details. An extensive list of selected references is presented on maintenance (1–158) and maintainability engineering (159–228).

10.10 REFERENCES

Maintenance

Books

1. R. E. Barlow and F. Proschan, *Statistical Theory of Reliability and Life Testing,* Holt, Reinhart and Winston, New York, 1975.
2. I. Bazovsky, *Reliability Theory and Practice,* Prentice-Hall, Englewood Cliffs, NJ, 1961.
3. B. S. Blanchard and E. E. Lowery, *Maintainability: Principles and Practices,* McGraw-Hill, New York, 1969.
4. E. S. Buffa, *Modern Production Management,* Wiley, New York, 1969, pp. 598–616.
5. R. Clements and D. Parkes, *Manual of Maintenance,* Business Publications Limited, London, 1966.
6. R. H. Clifton, *Principles of Planned Maintenance,* Arnold, London, 1974.
7. W. Cooling, *Low Cost Maintenance Control,* AMA Company, New York, 1978.
8. A. S. Corder, *Maintenance Management Techniques,* McGraw-Hill (UK) Ltd., 1976.
9. C. C. Corder, *Organizing Maintenance,* British Inst. of Management Publication, U.K., 1963.
10. I. B. Gertsbakh, *Models of Preventive Maintenance,* North-Holland, New York, 1977.
11. E. B. Feldman, *Industrial Housekeeping, Its Technology and Technique,* Macmillan, New York, 1963.
12. F. Gradon, *Maintenance Engineering: Organization and Management,* Wiley, New York, 1973.
13. H. Grothus, *Maintenance Activities Time Standards,* Plant Eng. Institute, Dorsten, 1969.
14. A. Hald, *Statistical Theory with Engineering Applications,* Wiley, New York, 1952.
15. J. E. Heintzelman, *The Complete Handbook of Maintenance Management,* Prentice-Hall, Englewood Cliffs, NJ, 1976.
16. T. G. Hicks, *Successful Engineering Management,* McGraw-Hill, New York, 1966, pp. 233–247.
17. T. M. Husband, *Maintenance Management and Terotechnology,* Saxon House, Farnborough, U.K., 1976.
18. W. G. Ireson, *Reliability Handbook,* McGraw-Hill, New York, 1966.
19. A. K. S. Jardine, *Maintenance, Replacement and Reliability,* Wiley, New York, 1973.
20. A. K. S. Jardine, *Operational Research in Maintenance,* Manchester University Press, Barnes and Nobles, New York, 1970.
21. S. M. Johnson, *Deterioration, Maintenance and Repair of Structures,* McGraw-Hill, New York, 1965.
22. D. W. Jorgenson, J. J. McCall, and R. Radner, *Optimal Replacement Policy,* North-Holland, Amsterdam, 1967.
23. A. Kelly and H. Harris, *Management of Industrial Maintenance,* Newnes, Butterworth, London, 1978.
24. B. T. Lewis and W. W. Pearson, *Maintenance Management,* J. F. Rider, New York, 1963.
25. B. T. Lewis and L. M. Tow, *Readings in Maintenance Management,* Cahners Books, Boston, 1973.
26. J. L. McKinley and R. D. Bent, *Maintenance and Repair of Aerospace Vehicles,* McGraw-Hill, New York, 1967.
27. E. Molloy (ed.), *Maintenance Engineer's Pocket Book,* Newnes London, 1953.
28. L. C. Morrow, *Maintenance Engineering Handbook,* McGraw-Hill, New York, 1966.

29. P. M. Morse, *Queues, Inventories and Maintenance,* Wiley, New York, 1958.
30. E. T. Newbrough, *Effective Maintenance Management: Organization Motivation and Control in Industrial Maintenance,* McGraw-Hill, New York, 1967.
31. A. A. Niazi, *Management of Maintenance,* Asia Publishing House, London, 1967.
32. *Organization of Maintenance Conference Proceedings 1969.* Available from the Iron and Steel Institute, London, U.K.
33. V. Z. Priel, *Systematic Maintenance Organization,* McDonald & Evans, London, 1974.
34. R. Reed, *Plant Location, Layout, and Maintenance,* Irwin, Homewood, IL, 1967.
35. *Report by the Working Party on Maintenance Engineering,* April 1970, Ministry of Technology, (HM 80), London, U.K.
36. T. F. Sack, *A Complete Guide to Building and Plant Maintenance,* Prentice-Hall, Englewood Cliffs, NJ, 1971.
37. H. Stewart, *Guide to Efficient Maintenance Management,* Business Publications, London, 1963.
38. *South-Eastern Plant Engineering and Maintenance Conference Proceedings,* Available from American Institute of Plant Engineers, Cincinnati, U.S.A.
39. *Techniques of Plant Engineering and Maintenance Proceedings, Vol. 1–24,* Clapp & Poliak, New York.
40. *Terotechnology in Iron & Steel Works, Conference Proceedings,* 1972, Available from the Iron and Steel Institute, London, U.K.
41. U.S. Military Handbook, *Maintenance Engineering Techniques,* available from NTIS, Springfield, Virginia, 22161, USA.
42. *Western Plant Eng. and Maintenance Conference Proceedings,* available from the American Institute of Plant Engineers, Cincinnati, OH, U.S.A.
43. E. N. White, *Maintenance Planning, Control & Documentation,* Gower Press, Epping, 1973.

Articles

44. G. H. Almy, "Allocation, Assessment, and Demonstration of System MTTR," *Annals of Reliability and Maintainability,* 1971, pp. 172–177.
45. W. Armitage, and A. K. S. Jardine, "Maintenance Performance—A Decision Problem," *International Journal of Production Research,* Vol. 7, 1968, p. 15.
46. H. J. Bajaria, "Maintainability Considerations in Transportation Industry," *Annual Reliability and Maintainability Symposium,* 1978, pp. 45–49.
47. R. E. Barlow and L. C. Hunter, "Optimum Preventive Maintenance Policies," *Operations Research,* Vol. 8, No. 1, 1960, pp. 90–100.
48. R. A. Barnard and T. D. Matteson, "Military Aircraft Maintenance," *Proceedings Annual Reliability and Maintainability Symposium,* 1975, pp. 596–600.
49. K. G. Barnes, "Premature Performance of Scheduled Maintenance," *Annals of Reliability and Maintainability Conference,* 1971, pp. 318–327.
50. Ye. Yu. Barzilovich, "Maintenance of Complex Technical Systems II Survey," *Engineering Cybernetics,* No. 1, 1967, pp. 63–75.
51. B. A. Basker and T. M. Husband, "Determination of Optimal Overhaul Intervals and Inspection Frequencies—A Case Study," *Microelectronics and Reliability,* Vol. 17, 1978, pp. 313–315.
52. B. A. Basker and T. Husband, "Maintenance Engineering, The Current State of Art," *Production Engineering,* February 1976.
53. R. B. Beattie, W. W. Harris, and A. P. Olbrich, "Contract Maintenance—How Far to Go?", *Chemical Engineering,* April 1961, p. 170.
54. M. Berg and B. Epstein, "Grouping of Preventive Maintenance Policies," *Proceedings of*

the NATO Conference on Reliability Testing and Reliability Evaluation, North-Holland, Amsterdam, September 1972.

55. B. Bergman, "Some Graphical Methods for Maintenance Planning," *Proceedings Annual Reliability and Maintainability Symposium,* 1977, pp. 467–471.

56. H. M. Blegen and B. Nylehn, "Organizing the Maintenance Function," *International Journal of Production Research,* Vol. 7, 1968, p. 23.

57. W. Bleikardt, "Digital Transmission Network Maintenance Aspects," *Annual Reliability and Maintainability Symposium,* 1978, pp. 460–471.

58. R. L. Boviard, "Characteristics of Optimal Maintenance Policies," *Management Science,* Vol. 7, No. 3, 1961, pp. 238–254.

59. R. Bratt, "Field Maintenance Organization Analysis Models," *Proceedings Annual Reliability and Maintainability Symposium,* 1980, pp. 24–28.

60. D. B. Brown and H. F. Martz, "Simulation for the Maintenance of a Deteriorating Component System," *IEEE Trans. on Reliability,* February 1971, pp. 28–32.

61. G. J. H. Buss, "Maintenance Incentives—Part II, Organization of Manufacture," *Works Engineering and Factory Services,* May 1971, pp. 35–38.

62. G. J. H. Buss, "Maintenance Incentives—Part I," *Works Engineering and Factory Services,* April 1971, pp. 14–18.

63. R. C. Cavendish, "The Use of Predetermined Time Systems in Planning and Controlling Maintenance Activities," *The MTM Journal of Methods-Time Measurement,* Vol. 18, No. 2, 1973, pp. 14–35.

64. P. K. W. Chan, and T. Downs, "Optimization of Maintained Systems," *IEEE Trans. on Reliability,* Vol. 29, April 1980, pp. 42–44.

65. H. H. Cho, "On the Proper Preventive Maintenance," *National Symposium on Reliability and Quality Control,* 1963, pp. 431–438.

66. S. Crabbe and K. Ahlborg, "On Condition Maintenance Programs," *Proceedings Annual Symposium on Reliability,* 1970, pp. 300–307.

67. J. Creson, "Exponential Distribution Analysis of Corrective Maintenance Times," *ASQC Journal,* Vol. 2, No. 4, June 1964.

68. C. Derman, G. J. Leibermann, S. M. Koss, "A Renewal Decision Problem," *Management Science,* Vol. 24, No. 5, January 1978.

69. D. I. Dhavale, and G. L. Otterson, "Maintenance by Priority," *Industrial Engineering,* Vol. 12, No. 2, February 1980, pp. 24–27.

70. D. K. Donnelly, "Computers for Plant and Equipment Maintenance," *The Canadian Mining and Metallurgical Bulletin,* February, 1976.

71. W. R. Downs, "Predicting Maintenance Time Distribution of a Complex System," *10th Annual West Coast Reliability Symposium,* February 1969, pp. 37–47.

72. W. R. Downs, "System maintainability Verification—The Paired-Time-Comparison (PTC) Method," *Annual Reliability and Maintainability Symposium,* 1979, pp. 280–284.

73. J. D. Duckworth and R. P. Quintana, "Computer Support in Air Force Maintenance," *Proceedings Annual Reliability and Maintainability Symposium,* 1978, pp. 11–16.

74. W. R. Downs and C. Kasparian, "A Method of Synthesizing Repair Times," *Annals of Reliability and Maintainability Conference,* 1971, pp. 178–180.

75. R. P. Fischer, "The Effectiveness of On-Board Maintenance," *IEEE Trans. on Reliability,* March 1964, pp. 41–47.

76. B. J. Flehinger, "A General Model for the Reliability of Systems Under Various Preventive Maintenance Policies," *Ann. Math. Statist.,* Vol. 33, No. 1 (1962), pp. 137–156.

77. W. F. Ford and J. W. Foster, "Generalized Procedure for Evaluation of Maintenance Aids," *Annals of Reliability and Maintainability Conference,* 1971, pp. 291–300.

78. J. W. Foster and R. S. Morris, "Maintenance Engineering Analysis Techniques," *10th Reliability and Maintainability Conference,* 1971.

79. O. N. Gabrielson, "Frequency of Maintenance," *Annals of Reliability and Maintainability Conference,* July 1966, pp. 428–433.

80. I. B. Gertsbakh, "Optimum Rule for Maintenance of a System with Many States," *Engineering Cybernetics,* No. 5, 1964, pp. 39–45.

81. W. P. Gillespie and C. H. Wysowski, "Teamwork Maintenance," *Industrial Engineering,* August 1974, pp. 26–29.

82. D. Gregor, D. Harmon, and P. Pates, "Maintainability Estimating Relationships," *Annual Reliability and Maintainability Symposium Proceedings,* 1975, pp. 18–25.

83. P. D. Guy, "Measurement of Maintenance," *Works Management,* Feb. 1973, pp. 5–9.

84. P. D. Guy, "How Maintenance can be Measured," *Maintenance Engineering,* Sept. 1972, pp. 48–50.

85. P. D. Guy, "Measurement of Maintenance," *Works Management,* February 1973, pp. 5–9.

86. R. C. Hall, R. C. Schneider, and C. A. Pell, "Analysis of the Incidence of Unscheduled Maintenance," *Annual Reliability and Maintainability Conference,* 1979, pp. 275–279.

87. M. J. Harris and A. Kelly, "The Maintenance Spares Inventory and its Control," *The Plant Engineer,* Vol. 19, No. 10, Oct. 1975, pp. 22–25.

88. D. J. Harrahy and M. L. Hinkle, "Inventory Control Models for Logistics Planning and Operational Readiness with Cost Constraints," *Annals of Reliability and Maintainability Conference,* 1967, pp. 436–448.

89. N. A. J. Hastings, "The Repair Limit Replacement Method," *Operational Research Quarterly,* Vol. 20, No. 3, pp. 337–349, 1969.

90. H. F. Heap and S. F. Nowlan, "Reliability Centered Maintenance," *Annual Reliability and Maintainability Symposium,* 1978, pp. 38–44.

91. K. L. Helland, "Motivating Management on Maintainability," *Annual Reliability and Maintainability Symposium,* 1978, pp. 32–37.

92. L. Hunter, "Optimum Checking Procedures," in *Statistical Theory of Reliability,* M. Zelen (ed.), Univ. of Wisconsin Press, Madison WI, 1963, pp. 96–108.

93. T. M. Husband, "Research Objectives in the Organization of Maintenance," *Management Education and Development,* Vol. 6, Part I, April 1975, pp. 3–11.

94. A. K. S. Jardine and A. J. L. Kirk, "Maintenance Policy for Sugar Refinery Centrifuges," *Proceedings of the Institution of Mechanical Engineers,* Vol. 187, 1973, London, pp. 679–686.

95. A. K. S. Jardine, T. J. Goldrick, and J. Stender, "The Annual Maintenance Cost Limit Approach for Vehicle Fleet Replacement," *Proceedings of the Institution of Mechanical Engineers,* Vol. 190, 1976, pp. 71–80, London.

96. A. K. S. Jardine, "Determination of Optimal Maintenance Times," *The Plant Engineer,* Vol. 13, No. 6, June 1969, pp. 109–114.

97. A. K. S. Jardine, "Solving Industrial Replacement Problems," *Proceedings Annual Reliability and Maintainability Symposium,* 1979, pp. 136–14.

98. J. J. Johnston and W. E. Sharp, "Maintenance Costs: Some Trends," *Factory,* August 1973, pp. 38–40.

99. M. D. Jones and R. Mielec, "Logistics Supportability Testing," *Proceedings Annual Reliability and Maintainability Symposium,* 1978, pp. 17–21.

100. H. P. Jost, "Down With Maintenance," *Journal of I. of Mech. E.,* April 1972, London.

101. J. F. Kalabach, "Effect of Preventive Maintenance on Reliability," *Sixth National Symposium on Reliability and Quality Control,* 1958, pp. 484–488.

102. I. Katz, "A New Concept of Planned Inspections," *Annals of Reliability and Maintainability,* July 1966, pp. 416–420.

103. E. Kay, "The Effectiveness of Preventive Maintenance," *Int. Jr. Prod. Res.,* Vol. 14, No. 3, 1976, pp. 329–344.

104. A. Kelly and M. J. Harris, "Simulation—An Aid to Maintenance Decisions," *The Plant Engineer,* November 1971, pp. 43–46.
105. A. Kelly, "Work Measurement for Maintenance Managers," Part 1, 2, and 3, *Factory,* September, October, and November 1972, pp. 32, 33, 35 and 45; 32, 33, 35 and 37; 53.
106. M. Klein, "Inspection-Maintenance-Replacment Under Markovian Deterioration," *Management Science,* 9, 1963, pp. 25–32.
107. P. Kolesar, "Minimum Cost Replacement Under Markovian Deterioration," *Manag. Sci.,* Vol. 12, No. 96, 1966, pp. 694–706.
108. P. B. Lawler and R. J. McNichols, "Determination of Optimum Replacement Time for a System Composed on N Independently Failing Subsystems Given the Subsystem's Ages," *Annals of Reliability and Maintainability,* 1971, pp. 313–317.
109. L. G. Leffler, "Maintenance Reserve for Large Systems," *Proceedings Annual Reliability and Maintainability Symposium,* 1975, pp. 444–448.
110. C. H. Lie, H. L. Hwang, and F. A. Tillman, "Availability of Maintained Systems—A State-of-the-Art Survey," *AIIE Transactions,* September 1977, pp. 247–259.
111. H. E. Lynch, R. S. Morris, R. J. McNichols, and D. R. Shreve, "Techniques for Determining the Number of Spares with a Prechosen Probability Level," *Annals of Reliability and Maintainability,* 1971, pp. 181–187.
112. R. L. Madison, "An Analysis of the Effects of Maintenance on Parts Replacement," *Third National Symposium on Reliability and Quality Control,* 1957, pp. 19–29.
113. H. Mahlooji, L. L. George, and Po-Wen-Hu, "Optimal Replacement and Build Policies," *Proceedings Annual Reliability and Maintainability Symposium,* 1974, pp. 147–152.
114. B. H. Mahon, "Some Practical Aspects of Data Collection and Analysis in Reliability Studies," *Proceedings of the NATO Conference on Reliability Testing and Reliability Evaluation,* North-Holland, Amsterdam, 1972.
115. "Maintenance Managers' Guide," prepared by the Energy Research and Development Administration, Division of Construction Planning and Support, December 1976, Washington, D.C.
116. "Maintenance Problems," Reader's Views, *Journal of I. of Mech. E.,* Feb. 1972, p. 69, London.
117. L. Mann, and E. R. Coates, "Evaluating a Computer for Computer Management," *Industrial Engineering,* Vol. 12, No. 2, Feb. 1980, pp. 28–32.
118. J. J. McCall, "Maintenance Policies for Stochastically Failing Equipment," *Management Science,* 11, 5, 1965, pp. 493–524.
119. D. G. McCullough, "Yardstick for Maintenance Performance," *Chemical Engineering,* Oct. 29, 1973, pp. 120–124.
120. R. J. McNichols and T. R. Rogers, "An Opportunistic Preventive Replacement Policy for a System with Independently Failing Subsystems," *Logistics Spectrum,* Vol. 3, No. 4, 1969, pp. 29–37.
121. R. I. Moeller, "Coordination of Maintenance Planning During Equipment Development," *Sixth National Symposium on Reliability and Quality Control,* 1958, pp. 330–334.
122. A. G. Mona, "Some Economic Aspects of Maintenance Versus Redundancy for Manned Space Stations," *Proceedings Annual Symposium on Reliability,* 1970, pp. 404–409.
123. A. Munoz, "Minimize Downstream Costs Through Applied Maintainability," *Annual Reliability and Maintainability Symposium,* 1979, pp. 266–270.
124. R. Myers and R. S. Dick, "Some Considerations of Scheduled Maintenance," *Proceedings Annual Reliability and Maintainability Symposium,* 1963, pp. 343–356.
125. J. R. Nelson, "Maintainability Pay-Offs During Weapon-System Test: The Value of Appropriate Testing," *Proceedings Annual Reliability and Maintainability Symposium,* 1975, pp. 26–29.
126. F. S. Nowlan, "The Relationship Between Reliability and the Periodicity of Scheduled

Maintenance," *Fifth Annual West Coast Reliability Symposium on Reliability in American Industry,* 1964, pp. 1–28. Available from ASME.

127. M. R. Nusbaum, "Maintenance Replacement Sequence Selection," *10th Reliability and Maintainability Conference,* 1971, pp. 287–290.

128. K. S. Park, "Gamma Approximation for Preventive Maintenance Scheduling," *AIEE Transactions,* Vol. 7, No. 4, 1975, pp. 393–397.

129. D. Parkes, "The Problems of the Maintenance Engineer," *Journal of I. Mech. E.,* May 1971, London.

130. V. B. Parr, "A Systems Unavailability Trade-Off Program," *Reliability and Maintainability Conference,* 1970, pp. 319–325.

131. L. F. Pau, "Specification of an Automatic Test System vs. an Optimum Maintenance Policy and Equipment Reliability," *Proceedings Annual Reliability and Maintainability Symposium* 1979, pp. 142–146.

132. E. L. Peterson, "Maintainability Derivations Using the Analytical Maintenance Model," *IEEE Trans. on Reliability,* June 1968, pp. 111–114.

133. E. L. Peterson and H. B. Loo, "Maintainability Risk Analysis Using the Analytical Maintenance Model," *Annals of Reliability and Maintainability Conference,* 1967, pp. 498–504.

134. L. J. Phaller, "Operational Influences on Maintainability," *Annual Reliability and Maintainability Symposium,* 1978, pp. 218–221.

135. R. P. Phillips, "Preventive Maintenance Simplified," *Iron and Steel Engineer,* Feb. 1970, pp. 64–69.

136. W. P. Pierskalla and J. A. Vockler, "A Survey of Maintenance Models: The Control and Surveillance of Deteriorating Systems," *Naval Research Logistics Quarterly,* Vol. 23, 1976, pp. 353–358.

137. L. Pilborough, "Is Planned Maintenance the Answer to Unscheduled Shutdowns?," *Process Engineering,* August 1972, pp. 54–56.

138. A. O. Plait, "The Navy's New Maintenance Data Collection Subsystem," *Annals of Reliability and Maintainability Conference,* 1970, pp. 319–335.

139. P. Rozenblit, "Estimating Optimal Preventive Maintenance Strategies," *Engineering Cybernetics,* No. 5, 1973, pp. 758–761.

140. J. W. Ruszkiewicz, "Maintenance Goals and Management Control," *Industrial Engineering,* Vol. 12, No. 2, February 1980, pp. 22–23.

141. H. E. Schock, "Simulating Multi-skill Maintenance: A Case Study," *Annual Reliability and Maintainability Symposium,* 1979, pp. 7–12.

142. B. F. Shelly, "Maintenance Manhour Distributions—A Case Study," *Annual Symposium on Reliability,* 1966, pp. 704–711.

143. H. C. Slawson, "Management Aspects of Maintainability and Maintenance Analysis," *Eighth Reliability and Maintainability Conference,* 1969, pp. 460–465.

144. J. R. Smith, "Reliability and Maintenance Design Considerations," *Third National Symposium on Reliability and Quality Control,* 1956, pp. 264–269.

145. "Thoughts on Terotechnology," Readers' Views, *Journal of I. of Mech. E.,* July 1972, London.

146. E. Turban, "The Use of Mathematical Models in Plant Maintenance," *Management Science,* Vol. 13, No. 6, Feb. 1967, pp. 342–358.

147. R. C. Vergin and M. Scriabin, "Maintenance Scheduling for Multicomponent Equipment," *AIIE Transactions,* Vol. 9, No. 3, 1977, pp. 297–305.

148. E. P. Virene, "Waiting Line Queuing Effects on Availability," *Proceedings Annual Symposium on Reliability,* 1969, pp. 162–167.

149. B. J. Voosen, "Planet II—Planned Logistics and Evaluation Technique," *Annals of Reliability and Maintainability,* 1971, pp. 301–312.

150. Y. T. Wang and C. L. Proctor, "Optimal Maintenance Policy for Systems That Experience State Degradations," *Microelectronics and Reliability,* Vol. 14, 1975, pp. 199–202.
151. L. A. Washburn, "Maintenance Replication Rate Prediction and Measurement," IEEE Trans. on Reliability, May 1967, pp. 49–52.
152. K. Weir and B. Tiger, "Analysis of Maintenance Man Loading via Simulation," *IEEE Trans. on Reliability,* August 1971, pp. 164–169.
153. E. L. Welker, "Relationship Between Equipment Reliability, Preventive Maintenance Policy, and Operating Costs," *5th National Symposium On Reliability and Quality Control,* 1959, pp. 170–180.
154. B. B. White, "Program Standard Help Software Maintainability," *Annual Reliability and Maintainability Program,* 1978, pp. 94–98.
155. M. A. Wilson, "The Learning Curve in Maintenance Analysis," *Annals of Reliability and Maintainability,* July 1966, pp. 434–443.
156. M. A. Wilson, "Effects of Maintenance and Support Factors on the Availability of Systems," *Annals of Reliability and Maintainability,* 1967, pp. 562–572.
157. R. C. Woodman, "Replacement Policies for Components that Deteriorate," *Operational Research Quarterly,* Vol. 18, No. 2, 1967.
158. N. L. Wu, "Scheduling Techniques Even Out Routine Maintenance Work Load," *Industrial Engineering,* July 1971, pp. 17–21.

Maintainability

Books

159. AMCP 706–133, ADA 026 006, Maintainability Engineering Theory and Practice, January 1976. Available from the NTIS, Springfield, VA 22161.
160. B. S. Blanchard and E. E. Lowery, *Maintainability: Principles and Practices,"* McGraw-Hill, New York, 1969.
161. W. Cox and C. E. Cunningham, *Applied Maintainability Engineering,* Wiley, New York, 1972.
162. E. B. Feldman, *Building Design for Maintainability,* McGraw-Hill, New York, 1975.
163. A. S. Goldman and T. B. Slattery, *Maintainability: A Major Element of System Effectiveness,* Wiley, New York, 1964.
164. *Maintainability Handbook for Electronic Equipment Design,* 1960, Available from the US Naval Training Device Center, New York.
165. D. J. Smith and A. H. Babb, *Maintainability Engineering,* Wiley, New York, 1973.

Articles

166. *Aerospace Reliability and Maintainability Conference Proceedings,* Available from the Society of Automotive Engineers, New York, 1964.
167. E. J. Althaus and H. D. Voegtlen, "A Practical Reliability and Maintainability Model and Its Applications," *Proceedings Eleventh National Symposium on Reliability and Quality Control,* pp. 41–48.
168. *Automotive Support System Symposium for Advanced Maintainability Proceedings,* San Diego, CA, 1974. Available from the IEEE.
169. H. S. Balaban, "Reliability Demonstration: Purposes, Practices and Value," *Proceedings 1975 Annual Reliability and Maintainability Symposium,* pp. 246–248.
170. C. F. Barber, "Expanding Maintainability Concepts and Techniques," *IEEE Trans. on Reliability,* Vol. R-16, May 1967, pp. 5–9.

171. B. H. Batchelor, "A Discussion on Demonstrating Maintainability," *Proceedings Aerospace Reliability and Maintainability Conference,* 1963, pp. 344–352.

172. G. T. Bird, "MIL-STD-471, Maintainability Demonstration," *Journal of Quality Technology,* Vol. 1, No. 2, April 1969, pp. 134–148.

173. R. M. Burton and G. T. Howard, "Optimal Design for System Reliability and Maintainability," *IEEE Trans. on Reliability,* Vol. R-20, No. 2, May 1971, pp. 56–60.

174. S. R. Calabro and P. Sgouros, "Maintainability Prediction in Department of Defense Development Programs," *IEEE Trans. on Reliability,* May 1967.

175. G. B. Cohen, "Survey of Reliability and Maintainability Demonstration Techniques," *Seventh Reliability and Maintainability Conference,* 1968, pp. 75–81.

176. *Conference on Maintainability of Electronic Equipment Proceedings,* Electronic Industries Association, New York, 1958.

177. J. de Corlieu, "Maintainability Diagnosis Techniques," *Proceedings Annual Symposium on Reliability,* 1966, pp. 133–142.

178. J. E. Daveau, "Suggested Improvements for Maintainability Demonstration," *Proceedings Annual Symposium on Reliability,* 1969, pp. 572–579.

179. H. S. Dorick, "Maintainability Indices for Equipment Designers," *Reliability and Maintainability Conference,* May 1963, pp. 513–517.

180. W. R. Downs, "Maintainability Analysis versus Maintenance Analysis—Interfaces and Discrimination," *Proceedings Annual Reliability and Maintainability Symposium,* 1976, pp. 476–481.

181. *Engineering Foundation Conference on Application of Signature Analysis to Machinery Reliability and Performance, Proceedings,* New England College, Henniker, NH, 1975. Available from the Institute of Environmental Sciences, U.S.A.

182. E. R. Fallon, "Requirements for Air Force Weapon System Maintainability," *Proceedings Aerospace Reliability and Maintainability Conference,* 1963, pp. 326–331. Available from the ASME.

183. A. L. Goel, "Panel Discussion on Some Recent Developments in Reliability Demonstration, Introductory Remarks," *Proceedings Annual Reliability and Maintainability Symposium,* 1975, pp. 244–245.

184. T. I. Goss, "New Maintainability Demonstration Tests," *Proceedings Annual Symposium on Reliability and Maintainability,* 1975, pp. 231–238.

185. D. D. Gregor, "Maintainability: F-5/7–38 Design Decisions," *Annals of Reliability and Maintainability,* Vol. 4, 1965, pp. 509–525.

186. G. H. Griswold, D. A. Topmiller, "Maintainability Prediction for Avionic Equipment," *Reliability and Maintainability Conference,* 1965, pp. 49–58.

187. R. L. Harrington, "Assessment of Marine Systems' Reliability and Maintainability," *Annals of Reliability and Maintainability,* 1970, pp. 298–318.

188. D. Heimann, "Availability—Concepts and Definitions," *Proceedings Annual Reliability and Maintainability Symposium,* 1976, pp. 482–490.

189. J. L. Hesse, "Maintainability Analyses and Prototype Operations," *Proceedings Annual Reliability and Maintainability Symposium,* 1975, pp. 194–199.

190. A. H. Hevesh, "Maintainability of Phased Array Radar Systems," *IEEE Trans. on Reliability,* May 1967.

191. R. F. Hynes, "Techniques in Optimizing Spare-Parts Provisioning," *Proceedings Annual Symposium on Reliability,* 1967, pp. 511–526.

192. *International Automatic Support System Symposium for Advanced Maintainability,* Arlington, TX, 1973, Available from IEEE.

193. L. T. Jones, "Maintainability Parameters Using the Consensus Method," *Proceedings Annual Reliability and Maintainability Symposium,* 1978, pp. 492–497.

194. R. Kamm, "Designing for Maintainability," *Proceedings Aerospace Reliability and Maintainability Conference,* 1963, pp. 353–355.

195. E. D. Karmiol, W. T. Weir, and J. S. Youtcheff, "The Reliability/Maintainability Relationship in Aerospace Programs," *Annals of Reliability and Maintainability,* Vol. 4, 1965, pp. 261–271.

196. J. Klion and K. C. Klotzkin, "Analysis Techniques for Maintainability Specification," *Annual Symposium on Reliability Proceedings,* 1967, pp. 501–510.

197. R. A. Kowalski and W. O. Stevenson, "An Evaluation of Design Sensitive Maintainability-Prediction Techniques," *Proceedings Annual Reliability and Maintainability Symposium,* 1975, pp. 200–206.

198. H. R. Leuba, "System Improvement Through Maintainability Demonstration," *Annals of Reliability and Maintainability,* 1970, pp. 174–177.

199. D. W. Lowry, "Maintainability Demonstration Test Performed on a Computer System," *Proceedings Annual Symposium on Reliability,* 1970, pp. 308–318.

200. N. Maroulis, "On the Meaning of Quantified Reliability," *IEEE Trans. on Reliability,* June 1963, pp. 39–48.

201. F. D. Mazzola, "Maintainability Demonstration," *IEEE Trans. on Reliability,* Vol. R-16, May 1967, pp. 37–42.

202. F. D. Mazzola, "Maintainability Demonstration," *Annals of Reliability and Maintainability,* July 1966, pp. 463–468.

203. O. A. Meykar, "Maintainability Terminology Supports the Effectiveness Concept," *IEEE Trans. on Reliability,* Vol. R-16, May 1967, pp. 10–15.

204. W. Nelson, "Producer Risk Tables for Method and Maintainability Demonstration Plan of MIL-STD-471," *Annals of Reliability and Maintainability,* 1971, pp. 188–194.

205. J. J. Naresky, "Reliability and Maintainability Research in the United States Air Force," *Annals of Reliability and Maintainability,* 1966, pp. 769–787.

206. E. P. O'Connell, "Maintainability Prediction Assessment: A New Concept," *Annual Symposium on Reliability,* 1967, pp. 527–533.

207. D. Oksoy, "An Exact Analysis of the Method-one Maintainability Demonstration Plan in MIL-STD-471," *IEEE Trans. on Reliability,* Vol. R-21, 1972, pp. 207–211.

208. E. L. Peterson, "Maintainability Design Requirements for Future Space Systems," *IEEE Trans. on Reliability,* Vol. R-15, May 1966, pp. 17–21.

209. E. L. Peterson, "Maintainability Application to System Effectiveness Quantification," *IEEE Trans. on Reliability,* Feb. 1971, pp. 3–7.

210. O. T. Perry, "A Statistical Approach to Maintainability Allocation," *Annals of Reliability and Maintainability,* Vol. 10, 1971, pp. 164–171.

211. T. L. Regulinski, "Systems Maintainability Modeling," *Proceedings Annual Symposium on Reliability,* 1970, pp. 449–457.

212. *Reliability and Maintainability Conference Proceedings,* available from Gordon and Breach, New York, 1965–1971.

213. *Reliability and Maintainability Symposium Proceedings,* 1972–1980, available from the IEEE.

214. B. L. Rettere and H. D. Voegtlen, "Advanced Maintainability Techniques for Aircraft Systems," *Proceedings Aerospace Reliability and Maintainability Conference,* 1963, pp. 429–433.

215. L. V. Rigby, "Results of Eleven Separate Maintainability Demonstrations," *IEEE Trans. on Reliability,* Vol R-16, May 1967.

216. C. M. Ryerson, "Physics of Maintainability," *Annals of Reliability and Maintainability,* Vol. 6, 1967, pp. 389–391.

217. J. V. Sanderson, "Design Disclosure Format for Design Review and Maintainability," *IEEE Trans. on Reliability,* Vol. R-16, May 1967.

218. J. Schmee, "Application of Sequential t-test to Maintainability Demonstration," *Proceedings Annual Reliability and Maintainability Symposium,* 1975, pp. 239–243.

219. R. J. Seigel, "Maintainability Design," *Annals of Reliability and Maintainability,* 1971, pp. 162–163.

220. E. P. Simshauser, "Maintainability Prediction by Function," *Annals of Reliability and Maintainability,* 1966, pp. 421–427.

221. T. B. Slattery, "Derivation of Maintainability Requirements," *Proceedings Aerospace Reliability and Maintainability Conference,* 1963, pp. 326–331, available from the ASME.

222. R. R. Stanton, "Maintainability Program Requirements, Military Standard 470," *IEEE Trans. on Reliability,* Vol. R-16, May 1967.

223. L. A. Washburn, "Maintenance Replication Rate Prediction and Measurement," *IEEE Trans. on Reliability,* Vol. R-16, May 1967.

224. R. A. Westland, D. T. Hanifan, "A Reliability-Maintainability Trade-Off Procedure," *Reliability and Maintainability Conference,* 1964, pp. 600–611.

225. W. H. Widawsky, "Reliability and Maintainability Parameters Evaluated with Simulation," *IEEE Trans. on Reliability,* August 1971, pp. 158–163.

226. M. A. Wilson and T. L. Fagan, "Maintainability in Space—A Survey," *IEEE Trans. on Reliability,* Vol. R-16, May 1967.

227. T. O. Wright, "Maintainability, The Measure of Availability," *Eighth Reliability and Maintainability Conference,* July 1969, pp. 466–469.

Miscellaneous:

228. MI1-STD-721B, Definitions of Effectiveness Terms for Reliability, Maintainability, Human Factors, and Safety, August 1966, Department of Defense, Washington, D.C., 20301.

229. B. Singh (Dhillon), A Study of Maintenance Management Methods, M.Sc. Thesis, September 1973. Available from the Department of Mechanical Engineering, University of Wales, Swansea, U.K.

230. F. Harris, *Operations and Cost (Factory Management Series),* A. W. Shaw, Chicago, IL, 1915, pp. 48–52.

231. F. E. Raymond, *Quantity and Economy in Manufacture,* D. Van Nostrand, Princeton, NJ, 1931.

232. E. S. Buffa, *Operations Management: Problems and Models,* Wiley, New York, 1972.

233. W. J. Fabrycky, P. M. Ghare, and P. E. Torgersen, *Industrial Operations Research,* Prentice-Hall, Englewood Cliffs, NJ, 1972.

11
Medical Equipment Reliability

11.1 INTRODUCTION

A few years ago industrialists and engineers considered health care a secluded area, "for doctors only." However, in many cases, engineers collaborated to improve the design of equipment* such as intensive care units and heart-lung machines. Although these contributions were important to health care, the main body of the medical systems was left unexplored. Nowadays, there are encouraging signs that both engineers and industrialists are breaking away from their traditional outlook. This may be due to the increasing cost of medical care, stringent government requirements on medical devices, the public outcry for the better quality of health care services, etc. Some of these factors have triggered a new interest in improved health care equipment reliability. Following the success of reliability and maintainability engineering in the aerospace and military fields, many of the advances made over the last three decades are now beginning to be applied to medical equipment.

The first serious thinking of medical equipment reliability occurred in the late sixties, while the seventies witnessed further emphasis on the field. This has resulted in a large number of publications.

Some of the early published literature on medical equipment electrical hazards is presented in references 2, 4, 15, 52, 54. However, serious thinking on medical equipment or instrumentation reliability was not initiated until the second part of the sixties [3, 5, 7, 8, 10, 13, 16, 17, 19, 20, 23, 24, 26, 28, 29, 35, 36, 37, 41, 43, 46, 49, 50, 53]. Since 1969, some equipment reliability improvement ideas borrowed from the military and the aerospace fields are

---

* Readers should note that the term "equipment" as used in this chapter is equivalent to a device or component, and vice versa.

beginning to be applied in the medical field [33, 12]. An extensive list of references are presented at the end of the chapter for those readers who wish to delve further in the subject.

11.2 UNITED STATES HEALTH CARE DATA

Some of the findings for the year 1974, which reference [11] quotes from the 1975 edition of the *American Hospital Association Guide to the Health Care Field* are that

1. There were 7174 American Hospital Association (AHA) registered hospitals with total assets worth 52 billion dollars.
2. For the year 1974, 35,506,000 patients were admitted to these hospitals for an average stay of seven days.
3. There were a total of 1,513,000 beds in all of these 7174 AHA hospitals. The average bed occupancy rate was 77.2 percent for the year 1974.
4. These hospitals, with the exception of resident physicians, interns, and students, employed three million persons.

The above four items lead us to conclude that (1) everyone of us is a potential consumer of hospital or hospital-related health care services; and (2) the health care system is a big organization.

11.3 MEDICAL DEVICES

A modern hospital uses over 5000 different devices. These devices range from a simple tongue depressor to a complex pacemaker. In the early seventies [25], these devices plus dental equipment and materials represented a $2 billion dollar market yearly. This market, at that time, was growing at an annual rate between 10 to 15 percent. This leads one to believe that the amount spent to manufacture medical-related equipment, devices, and components is fairly large. Therefore, the improvement in reliability of such items will lead to potential monetary savings and will also reduce the chances of device failure.

A medical device or equipment can be subject to the following situations:

1. Misuse
2. Poor maintenance
3. Users unfamiliar with the device or not aware of the capability of a device in question

Medical equipment may be categorized [7, 38] into the following three classifications:

Classification I. Under this category, equipment which can be directly or indirectly responsible for a patient's death or serious injury is included. The following are some of these equipment:

1. Respirators
2. Cardiac pacemakers
3. Cardiac defibrillators
4. Electrocardiographic monitors
5. Spectrophotometers
6. X-ray machines (in some cases)

The reliability requirements for such equipment must be very rigid.

Classification II. Most of the routine or semiemergency diagnostic or therapeutic electronic equipment used on patients fall into this category. This classification also includes most of the equipment used in the laboratory. Some of these equipment are

1. Electronyography equipment
2. Electroencephalograph recorders and monitors
3. Electrocardiograph recorders and monitors
4. Colorimeters
5. Gas analyzers
6. Hemoglobinometers
7. Ultrasound devices
8. Diathermy devices
9. Electrocautery devices
10. X-ray equipment

If we compare the equipment of category II to those in category I, the main difference is the lack of real-time use. By no means is the equipment listed in this category less important than the equipment in category I. What it really meant is that the failure of the category II equipment does not create the same emergency for patient as does the classification I equipment. This implies that for the second category devices there is time for repair or replacement. The reliability requirements for such devices are less rigid than for the devices in the first classification.

Classification III. The equipment included under this category are those whose malfunctioning is not critical to life or welfare of a patient. These items are mainly used for the patient's convenience. Some of the equipment included under this category are

1. Wheelchairs
2. Electric beds
3. Television receivers

The reliability requirements for this category of equipment are not very rigid. Finally, note that there is a considerable overlap among the equipment used in classifications I and II. Therefore, care must be taken when designing equipment which could be used in both categories. For example, the electro-cardiograph monitor is essentially a category II equipment, but could be used in a category I situation. Great care must be taken from the reliability point of view when designing such equipment.

11.4 UNITED STATES GOVERNMENT REGULATIONS ON MEDICAL EQUIPMENT

Today, medical devices [45] are more under United States Government control due to the "Medical Device Amendments of 1976" which became Public Law 94–295 in May 1976. To control medical devices, the Food and Drug Administration (FDA) may use one or more of the following three (devices are categorized into three classes*) procedures:

1. *General controls* are applied to control class I devices. These controls are good manufacturing practices, restricted device requirements, inspections, etc.
2. *Performance standards* are used to control class II devices. Medical device safety and effectiveness are subject to regulatory standards.
3. *Premarket approval requirement* is used to control class III devices. This requirement deals with obtaining an approval prior to device marketing.

Class I and II device controls are also included in the class III device controls. Furthermore, the class I controls are also included in the controls of class II. The following are some forecasts of events that will occur due to the Food and Drug Administration regulations impact [45] on medical device safety, effectiveness, and manufacturers:

1. Increase in product liability losses
2. Decrease in device productivity

---

* The three categories of devices presented in this section do not correspond to the three categories of devices defined in section 11.3.

3. Increase in medical device cost
4. Faster removal of "quack" devices from the market
5. Marginal increase in medical device safety and/or effectiveness
6. Proportionate decrease in the number of small manufacturers in the device manufacturing business
7. Foreign manufacturer advantage over the domestic manufacturers because of their less expensive product which is due to less rigid home government regulations

11.5 MEDICAL DEVICE RELIABILITY IMPROVEMENT PROCEDURES AND TECHNIQUES

These are presented in the following three sections:

11.5.1 Medical Equipment Design Specification

The device design specification is one of the main factors in the device reliability improvement. If the design specification does not clearly outline the device reliability, safety, and other requirements clearly and effectively, then a manufactured device may not be able to perform its required functions in the field reliably, safely, and effectively. Design specifications are written in many shapes and formats; however, an equipment specification should cover the following six [31] areas:

1. Scope statement
2. Related references to specifications
3. Material, construction, and functional requirements details
4. Requirements for quality assurance
5. Marking and packaging requirements details
6. Notes, comments, and other relevant information

11.5.2 Bio-Optronics Approach to Produce Reliable Medical Devices

A thirteen-step approach is presented in this section to manufacture reliable medical devices. This procedure [32] was developed by Bio-Optronics. The product development plan is as follows:

1. Perform analysis on existing specific medical problems.
2. To solve a specific medical problem, develop a product concept.
3. Determine the medical device operational environments.
4. Evaluate who will be operating the proposed product.

5. Construct an engineering prototype device.
6. Perform tests on the prototype device in the laboratory.
7. Test the prototype in real-life field environments.
8. Go back and change the product design accordingly to fulfill the real-life field requirements.
9. Repeat the laboratory and real-life (field) tests on the modified product.
10. Construct pilot devices to perform tests.
11. Have pilot units tested in the field environment by impartial specialists.
12. Release the device for production.
13. Follow-up on the device field performance and support with device maintenance.

This approach practiced by Bio-Optronics is a good example of how to produce reliable and safe medical devices.

11.5.3 Reliability Evaluation Techniques

The two basic approaches to calculate equipment reliability measures are simulation and analytical methods. Both of these procedures employ mathematical models of the system. Simulation proceeds by performing sampling experiments on the mathematical model of the system. Simulation experiments are virtually the same as ordinary statistical experiments except that they are performed on a mathematical model rather than the actual system. Analytical methods, on the other hand, proceed directly by the solution of the equations describing the system model. The various commonly used methods for calculating reliability measures are identified in Fig. 11.1. The analytical

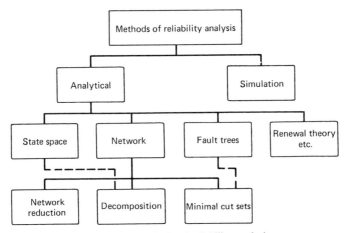

Figure 11.1. Methods of reliability analysis.

methods are discussed in Chapters 3 and 4 of this book. These techniques can be used to predict reliability of medical devices or systems.

11.6 RELIABILITY MODELS

In this section we present reliability evaluation models of a single unit, parallel, and standby systems. The reliability techniques of Fig. 11.1 can be used to compute reliability of these models.

11.6.1 Single Unit Reliability Prediction

The following single unit reliability expression is taken from Chapter 3:

$$R(t) = e^{-\int_0^t \lambda(t)\, dt} \tag{11.1}$$

where $R(t)$ is the single unit reliability at time t

$\lambda(t)$ is the single unit instantaneous failure rate or hazard rate

For the constant unit failure rate

$$\lambda(t) = \lambda \tag{11.2}$$

Substituting Eq. (11.2) in Eq. (11.1), we get

$$R(t) = e^{-\lambda t} \tag{11.3}$$

To obtain the mean time to failure (MTTF) integrate the system or component reliability function over the time interval $[0, \infty]$ to get

$$\text{MTTF} = \int_0^\infty R(t)\, dt \tag{11.4}$$

EXAMPLE 1

Supposing a pacemaker has a constant failure rate, $\lambda = 0.001$ failure/year, calculate the pacemaker reliability for a 10-year mission and mean time to failure. To compute pacemaker reliability for a 10-year mission, substitute the given data for λ and t in Eq. (11.3) to get

$$R(10) = e^{-(0.001)(10)} = 0.99$$

Pacemaker reliability is equal to 0.99 for the 10-year mission.

To calculate pacemaker mean time to failure (MTTF) substitute given data for λ in Eq. (11.3) and then substitute the resulting expression in Eq. (11.4) to get

$$\text{MTTF} = \int_0^\infty e^{-(0.001)t}\,dt$$

$$= \frac{1}{0.001} = 1000 \text{ years}$$

11.6.2 Parallel System

A two active unit parallel system is shown in Fig. 11.2.

For example, Fig. 11.2 might represent a mathematical model of two active, redundant electrocardiographic monitors. This system will only fail if both electrocardiographic monitors fail. This type of redundancy can be used to increase system reliability.

Using one of the reliability evaluation techniques presented in Chapter 3, the Fig. 11.2 network reliability, R_p, is

$$R_p = 1 - \prod_{i=1}^{2} (1 - R_i) \tag{11.5}$$

where R_i is the ith electrocardiograph monitor reliability.

In the case of constant failure rates, λ_1 and λ_2 of monitors 1 and 2, respectively, substituting Eq. (11.3) in Eq. (11.5), we get

$$R_p(t) = 1 - \prod_{i=1}^{2} (1 - e^{-\lambda_i t}) \tag{11.6}$$

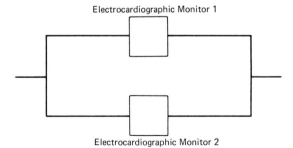

Electrocardiographic Monitor 1

Electrocardiographic Monitor 2

Figure 11.2. A two active unit parallel system.

For identical electrocardiographic monitors, Eq. (11.6) reduces to

$$R_p(t) = 1 - (1 - e^{-\lambda t})^2 \qquad (11.7)$$

EXAMPLE 2

Suppose two identical electrocardiographic monitors are forming an active parallel redundant system as shown in Fig. 11.2. The electrocardiographic constant failure rate is $\lambda = 0.001$ failure/hour. Calculate the system reliability for a 1000-hour mission and the mean time to failure. Assume both electrocardiographic monitors start operating at time $t = 0$ and their failures are statistically independent.

To compute system reliability, substitute the given data for $\lambda = 0.0001$ failure/hour and $t = 1000$ hours in Eq. (11.7) to get

$$R_p(1000) = 1 - (1 - e^{-(0.0001)(1000)})^2$$
$$= 0.991$$

Thus, redundant electrocardiographic monitors reliability is equal to 0.991.

To calculate the mean time to failure (MTTF), substitute Eq. (11.7) in Eq. (11.4) to get

$$\text{MTTF} = \int_0^\infty [1 - (1 - e^{-\lambda t})^2] \, dt$$
$$= \frac{3}{2} \frac{1}{\lambda} \qquad (11.8)$$

Substitute for $\lambda = 0.0001$ failure/hour in Eq. (11.8) to obtain

$$\text{MTTF} = \frac{3}{2} \frac{1}{0.0001} = 15{,}000 \text{ hours}$$

11.6.3 Standby Redundancy

A two-unit standby, redundant system is shown in Fig. 11.3. Figure 11.3 represents a situation with one electrocardiographic monitor operating and the other one on standby status until the operating monitor fails. The reliability, $R_{st}(t)$, for the two-unit standby system with constant unit failure rate is

$$R_{st}(t) = (1 + \lambda t)e^{-\lambda t} \qquad (11.9)$$

where λ is the electrocardiographic constant failure rate
$\quad\; t$ is time

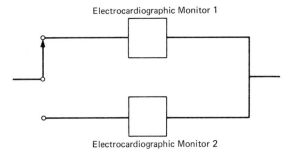

Figure 11.3. A two-unit standby redundant system.

A generalized formula for an n = unit standby system reliability is given in Chapter 3 of this book.

Equation (11.9) is subject to the following three assumptions:

1. The standby unit is as good as new.
2. Unit failures are statistically independent.
3. The switching arrangement is perfect.

EXAMPLE 3

For the identical unit configuration shown in Fig. 11.3, calculate the system reliability and the mean time to failure (MTTF) for given $\lambda = 0.0001$ failure/hour and $t = 1000$ hours.

Using Eq. (11.9) for $\lambda = 0.0001$ failure/hour and $t = 1000$ hours, we get

$$R_{st}(1000) = [1 + (0.0001)(1000)]e^{-(0.0001)(1000)}$$

$$R_{st}(1000) = (1 + 0.1)(0.9048)$$

$$= 0.9953$$

Thus, the two identical unit electrocardiographic monitor standby system reliability is equal to 0.9953. To obtain an expression MTTF, substitute Eq. (11.9) in Eq. (11.4) and then integrate to obtain

$$\text{MTTF} = \int_{0}^{\infty} (1 + \lambda t)e^{-\lambda t}\, dt = \frac{2}{\lambda} \tag{11.10}$$

By substituting the numerical value for λ in (11.10) we get the electrocardiographic monitor standby system MTTF.

$$\text{MTTF} = \frac{2}{0.0001} = 20,000 \text{ hours}$$

11.7 SUGGESTIONS FOR HEALTH CARE PROFESSIONALS AND RELIABILITY ENGINEERS

Here, we present recommendations to both health care professionals and associated reliability engineers who wish to enhance reliability of the health care systems [44]. The health care professionals include hospital administrators, doctors, and manufacturers.

11.7.1 Suggestions for the Health Professionals

1. The responsibility for equipment reliability rests with the manufacturer during the design, development, and manufacturing phases. In the field, it is the user who must use the equipment or device within the specified design conditions. Both the involved parties must accept their responsibility.
2. Recognize that the poor medical device or equipment reliability is due to failures. A device reliability can be improved by positive thinking and actions.
3. Compare the human body and the medical device failures. Both require positive actions by reliability engineers and doctors to improve the reliability of the device and prolong the life of the human, respectively.
4. Realize that probably the biggest single expense in any business organization is the cost of failures. These failures may be related to business systems, people, equipment, etc. Reduction in such failures will decrease business costs.
5. Realize that application of reliability techniques have successfully improved the reliability of aerospace systems. Payoffs obtained from the applications of these techniques to the health care system will be of similar nature.

11.7.2 Suggestions for the Reliability Engineer

1. Not all device failures are equal in importance. Direct your attention to the critical failures of a device in question.
2. Try to forget, for the time being, about the sophisticated reliability methods used to improve aerospace systems reliability. For the time being, consider yourself as an expert on failure correction, failure mode and effect analysis, and simple fault tree procedures only.
3. Realize that the responsibility of equipment reliability rests with the manufacturer during the design, development, and manufacturing phases, but in the field it is the user's.
4. Develop your thinking to be a cost conscious reliability engineer. Carry out cost versus reliability trade-off analyses. Remember that the manage-

ment will only approve those reliability decisions whose benefits are greater than their costs. Some reliability improvement decisions require very little or no additional expense.

5. Keep in mind that a simple failure reporting system in a hospital will be more beneficial in terms of improving medical device reliability than having an extensive inventory of spares or standby units.

6. To obtain immediate results, use design review, failure mode and effect analysis, parts review, and qualitative fault trees.

11.8 AEROSPACE AND MEDICAL EQUIPMENT RELIABILITY COMPARISON

The following are some of the comparisons for the aerospace and medical equipment reliability:

AEROSPACE EQUIPMENT RELIABILITY	MEDICAL EQUIPMENT RELIABILITY
1. A well-established field for reliability	1. This is a relatively new area for reliability
2. Expensive equipment	2. Relatively less costly devices
3. Reliability professionals with past aerospace product reliability experience apply sophisticated reliability techniques	3. Reliability professionals with less past medical device reliability experience apply relatively simple reliability methods
4. Well being of humans directly or indirectly involved	4. Patients lives are involved
5. Large size manufacturing organizations with well established reliability departments	5. Relatively smaller size manufacturers with relatively recent interest with the reliability field

11.9 SUMMARY

This section summarizes the main themes of the chapter:

1. There are over 5000 different types of medical devices in a modern hospital.
2. Reliability awareness in the medical field is relatively new with its beginning in the late sixties.
3. Many of the reliability techniques are directly borrowed from the aerospace and military fields.
4. Some of the problems arising in the medical reliability field are poorly written equipment design specifications, lack of field failure data, varying operational environments, users with poor equipment knowledge, etc.
5. Awareness of reliability engineering in the medical field is due to complying with new legislation, realizing lower equipment life cycle costs, avoiding law suits, providing better and safer health care, and so on.

6. General reliability techniques can be directly applied to improve medical equipment reliability.

An extensive list of references on the subject is presented for readers who wish to explore this topic further.

11.10 REFERENCES

1. T. P. Adams and M. D. Fasono, "The Microcircuit Pacemaker Space Age Spin-Off to Achieve Reliability and Long Life," *Proceedings Annual Reliability and Maintainability Symposium*, 1975, pp. 360–365. Available from the IEEE.
2. G. A. Bousvaros, C. Don, and J. A. Hopps, "An Electrical Hazard of Selective Angiocardiography," *Can. Med. Assoc. J.* 87, 1962, pp. 286–288.
3. J. M. R. Bruner, "Hazards of Electrical Apparatus," *Anesthesiology*, 28 (2), 1967, pp. 396–425.
4. H. B. Burchell, "Electrocution Hazards in the Hospital or Laboratory," *Circulation*, 1963, pp. 1015–1017.
5. R. C. Camishion, "Electrical Hazards in the Research Laboratory," *J. Surgical Res.*, 6; 1966, pp. 221–227.
6. J. W. Cook, "Practical Reliability Monitoring of Medical Electronics Instrumentation," Association for the Advancement of Medical Instrumentation: Annual Workshop, 1972. (Abstract: *Medical Instrumentation*, Vol. 6, No. 2, March 1972, p. 201).
7. J. F. Crump, Safety and Reliability in Medical Electronics," *Annual Symposium on Reliability*, 1969, pp. 320–322. Available from the IEEE.
8. G. B. Devey, "Toward Automated Health Services," *Proceedings of the IEEE*, Vol. 57, No. 11, 1969, p. 1828.
9. C. G. Drury, S. C. Schiro, S. J. Czaja, and R. E. Barnes, "Human Reliability in Emergency Medical Response," *Proceedings Annual Reliability and Maintainability Symposium*, 1977, pp. 38–42. Available from the IEEE.
10. R. O. Egeberg, "Engineers and the Medical Crisis," *Proceedings of the IEEE*, Vol. 57, No. 11, 1969, pp. 1807–1808.
11. G. A. Fairhurst and K. L. Murphy, "Help Wanted," *Proceedings Annual Reliability and Maintainability Symposium*, 1976, pp. 103–106. Available from the IEEE.
12. T. A. Finger, "The Alpha System," *Proceedings Annual Reliability and Maintainability Symposium*, 1976, pp. 92–96. Available from the IEEE.
13. R. Gechman, "Tiny Flaws in Medical Design Can Kill," *Hosp. Top.* 46, 1968, pp. 23–24.
14. W. Greatbatch, "Designing for High Reliability in Medical Electronic Equipment," *Medical Engineering*, Charles Ray (ed.), Yearbook Medical Publishers, Chicago, 1972, p. 1003.
15. J. A. Hopps and O. Z. Roy, "Electrical Hazards in Cardiac Diagnosis and Treatment," *A Med. Electron. Biol. Eng.* 1, 1963, pp. 133–134.
16. J. A. Hopps, "Electrical Shock Hazards—The Engineer's Viewpoint," *Proc. Symp. on New Electrical Hazards in Our Hospitals*, sponsored by Canadian Medical and Biological Engineering Society, Ottawa, Canada, September 1967.
17. J. A. Hopps, "Electrical Hazards in Hospital Instrumentation," *Annual Symposium on Reliability*, 1969, pp. 303–307. Available from the IEEE.
18. L. H. Isaacson and E. F. Taylor, "Specification x — 1414: A Reliability Milestone," *Proceedings Annual Symposium on Reliability*, 1970. Available from the IEEE.
19. J. P. Johnson, "Reliability of ECG Instrumentation in a Hospital," *Annual Symposium on Reliability*, 1967, pp. 314–318. Available from the IEEE.

20. R. J. Johns, "What is Blunting the Impact of Engineering on Hospitals?," *Proceedings of the IEEE,* Vol. 57, No. 11, 1969, pp. 1823–1827.

21. K. S. Kagey, "Reliability in Hospital Instrumentation," *Proceedings Annual Reliability and Maintainability Symposium,* 1973, pp. 85–88. Available from the IEEE.

22. G. B. Krishnamurty, "Community Health Education Programs," *Proceedings Annual Reliability and Maintainability Symposium,* 1976, pp. 97–102. Available from the IEEE.

23. J. R. Landoll and C. A. Caceres, "Automation of Data Acquisition In Patient Testing," *Proceedings of the IEEE,* Vol. 57, No. 11, 1969, p. 100.

24. R. S. Ledley, "Practical Problems in the Use of Computers in Medical Diagnosis," *Proceedings of the IEEE,* Vol. 57, No. 11, 1969, p. 1941.

25. D. M. Link, "Current Regulatory Aspects of Medical Devices," *Proceedings Annual Reliability and Maintainability Symposium,* 1972, pp. 249–250. Available from the IEEE.

26. J. L. Meyer, "Some Instrument Induced Errors in the Electrocardiogram," *J. Amer. Med. Assoc.,* 201, 1967, pp. 351–358.

27. L. A. Micco, "Motivation for the Biomedical Instrument Manufacturer," *Proceedings Annual Reliability and Maintainability Symposium,* 1972, pp. 242–244. Available from the IEEE.

28. J. G. Nevland, "Electrical Shock and Reliability Considerations in Clinical Instruments," *Annual Symposium on Reliability,* 1969, pp. 308–313. Available from the IEEE.

29. J. C. Norman and L. Goodman, "Acquaintance with and Maintenance of Biomedical Instrumentation," *J. Assoc. Advan. Med. Inst.,* Vol. 1, September 1966, p. 8.

30. "Reliability Technology for Cardiac Pacemaker," NBS/FDA Workshop, June 1974, NBS Publication 400–28, U.S. Department of Commerce, National Bureau of Standards, Washington, D.C.

31. J. J. Riordan, "Making Sure that Medical Devices Work," *J. Assoc. Advan. Med. Inst.,* Vol. 6, No. 2, March 1972.

32. H. B. Rose, "A Small Instrument Manufacturer's Experience with Medical Instrument Reliability," *Proceedings Annual Reliability and Maintainability Symposium,* 1972, pp. 251–254. Available from the IEEE.

33. D. F. Simonaitis, R. T. Anderson, and M. P. Kaye, "Reliability Evaluation of a Heart Assistance System," *Proceedings Annual Reliability and Maintainability Symposium,* 1972, pp. 233–241. Available from the IEEE.

34. J. E. Sigdell, "Properties and Limitations of Electronic Instrumentation," *Medical Engineering,* Charles Ray (ed.), Yearbook Publishers, 1972, Chicago, p. 974.

35. I. P. Smirnov and M. A. Shneps-Shneppe, "Medical System Engineering," *Proceedings of the IEEE,* Vol. 57, No. 11, 1969, pp. 1869–1879.

36. Special Issue on Technology and Health Services, *Proceedings of the IEEE,* Vol. 57, No. 11, 1969, pp. 1799–2042.

37. P. E. Stanley, "Monitors That Save Lives Can Also Kill," *Modern Hospital,* Vol. 108, No. 3, 1967, pp. 119–121.

38. B. Stein, "Reliability and Performance Criteria for Electro-Medical Apparatus," *Proceedings Annual Reliability and Maintainability Symposium,* 1973, p. 89. Available from the IEEE.

39. J. U. Stuart, R. Parke, and J. G. Webster, "Patient Monitoring by Radio Telemetry on a Medical Ward," *J. Assoc. Advan. Med. Inst.,* Vol. 6, No. 3, June 1972, pp. 240–245.

40. B. Steindel, "Quality Control in the Practice of Medicine," *11th Annual West Coast Reliability Symposium,* 1970, pp. 197–202. Available from the IEEE.

41. E. M. Swartz, "Product Liability, Manufacturer Responsibility for Defective or Negligently Designed Medical and Surgical Instruments," *De Paul Law Review,* Vol. 18, 1969, pp. 348–407.

42. E. F. Taylor, "Reliability, Risk and Reason in Medical Equipment," *Association for the Advancement of Medical Instrumentation, 5th Annual Meeting,* March 23, 1970.

43. E. F. Taylor, "Reliability: What Happens If . . . ?," *Proceedings of the Annual Symposium on Reliability,* 1969. Available from the IEEE.

44. E. F. Taylor, "The Reliability Engineer in the Health Care System," *Proceedings Reliability and Maintainability Symposium,* 1972, pp. 245–248. Available from the IEEE.

45. E. F. Taylor," The Impact of FDA Regulations on Medical Devices," *Proceedings Annual Reliability and Maintainability Symposium,* 1980, pp. 8–10. Available from the IEEE.

46. E. F. Taylor, "The Effect of Medical Test Instrument Reliability On Patient Risks," *Annual Symposium on Reliability,* 1969, pp. 328–330.

47. C. N. W. Thompson, "Model of Human Performance Reliability in Health Care System," *Annual Reliability and Maintainability Symposium,* 1974, pp. 335–339. Available from the IEEE.

48. E. L. Thomas, "Systems Engineering and Operation Research," *Medical Engineering,* Charles Ray (ed.), Yearbook Publishers, Chicago, 1972, p. 100.

49. J. G. Truxal, "Technology and Health Services," *Proceedings of the IEEE,* Vol. 57, No. 11, 1969, p. 1802.

50. W. Waits, "Planned Maintenance," *Med. Res. Eng.,* **7**, 12, (1968).

51. P. E. Weill, "From Toothbrush to Pacemaker: The FDA is Moving In," *Medical Care Review,* Vol. 36, 1979, pp. 119–121.

52. D. I. Weinberg, J. A. Artley, R. E. Whalen, and M. T. McIntosh, "Electrical Shock Hazards in Cardiac Catheterization," *Circ. Research,* 11, 1962, pp. 1004–1011.

53. J. O. Wear, "Maintenance of Medical Equipment in the Veterans' Administration," *Assoc. for the Advan. of Med. Inst.,* Third Annual Meeting, Houston, TX, July 15, 1968.

54. R. E. Whalen, C. F. Starmer, and H. D. McIntosh, "Electrical Hazards Associated with Cardiac Pacemaking," *N. Y. Acad. Science,* 111, 1964, pp. 922–931.

55. J. Winger, T. Bray, and P. Halter, "Lower Health Costs from High Reliability," *Proceedings Annual Reliability and Maintainability Symposium,* 1979, pp. 203–210. Available from the IEEE.

56. M. Zane, "Patient Care Appraisal," *Proceedings Annual Reliability and Maintainability Symposium,* 1976, pp. 84–91. Available from the IEEE.

12
Power Equipment Reliability

12.1 INTRODUCTION

Power equipment reliability is another important area of reliability theory application. The researchers in the area of power system reliability have developed their own definitions, concepts, and techniques. There is a certain commonality of concepts with other reliability engineering areas but noticeable diversity in definitions and models. For example the term "unavailability of a unit" is known as the "forced outage rate of a unit" in the power system reliability area.

Application of probability methods to generating capacity is evident in the publications of thirties. However, significant contributions to the field did not appear until the second half of the forties [11, 3, 1]. All of these published papers are listed in reference 11.

Most of the published literature on power system reliability up to 1977 is listed in references 11, 7, and 41. Selective books dealing with power system reliability are listed in references 1 to 5. Power equipment reliability evaluation and other mathematical models are presented in the following sections of this chapter.

12.2 POWER SYSTEM RELIABILITY DEFINITIONS AND INDICES

This section presents selective power system reliability definitions and indices [16].

Forced Outage. This situation occurs when an equipment or a unit has to be taken out of service due to a component failure or damage.

Scheduled Outage.   This represents a situation in which an equipment outage is planned in advance.

Forced Derating.   An equipment or a unit operating at a forced derated or lowered capacity due to a component failure or damage.

Forced Outage Hours.   These represent the total number of hours a unit or an equipment spends in the forced outage state.

Service Hours.   These denote the total number of actual operation hours of the equipment or unit.

Forced Outage Rate.   The forced outage rate of an item is defined by the following formula:

$$\text{Forced outage rate} = \left(\frac{\alpha}{\beta + \alpha}\right) \times 100 \qquad (12.1)$$

where α is the number of forced outage hours
   β is the number of service hours

One should note that in the basic reliability theory the "forced outage rate" of a unit is known as the "unavailability" of a unit.

Forced Outage Ratio.   The forced outage ratio, FR, is

$$FR = \frac{FOH}{TUH} \times 100 \qquad (12.2)$$

where FOH is the unit or equipment forced outage hours
   TUH is the unit or equipment total available hours

Mean Forced Outage Duration.   This term is analogous to mean time to repair (MTTR). Mean forced outage duration is obtained from the following relationship:

$$\text{Mean forced outage duration} = \frac{SFOH}{n} \qquad (12.3)$$

where n represents the number of forced outages
   SFOH denotes sum of forced outage hours

Mean Time to Forced Outage. This term is analogous to mean time to failure (MTTF). Mean time to forced outage (MTTF$_{ps}$) is

$$\text{MTTF}_{ps} = \frac{\text{SSH}}{n} \qquad (12.4)$$

where SSH represents the summation of service hours.

This section has presented some of the commonly used power system reliability definitions and indices. However, for the comprehensive list one should consult reference [16].

12.3 EQUIPMENT RELIABILITY MODELS

These models can be used to predict equipment availability. The Markov and other models presented in this section are associated with power equipment. For all of these models it is assumed that failure and repair rates are constant. In addition the unit or system failures are statistically independent. The Markov and other techniques are discussed in Chapter 3. Markovian and non-Markovian models are presented in the following sections:

12.3.1 Single Generator Unit Model

This Markovian model can also be used to represent a transformer, pulverizer, etc. We assume that the generator unit whose availability is being evaluated is a two-state repairable equipment. In other words, it can only be in either the operational or failed state. Other assumptions are as follows:

1. Failures are statistically independent.
2. Failure and repair rates are constant.
3. The repaired system is as good as new.

The transition diagram of the equipment is shown in Fig. 12.1. The following notations are used to develop equations for the model:

i represents the ith state of the generator unit: $i = 0$ (operating), $i = 1$ (failed)

$P_i(t)$ denotes the probability that the generator unit is in state i at time t

λ is the constant generator unit failure rate

μ is the constant generator unit repair rate

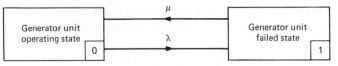

Figure 12.1. Generator unit transition diagram.

The associated system of differential equations with Fig. 12.1 is

$$P_0'(t) + \lambda P_0(t) - \mu P_1(t) = 0 \tag{12.5}$$

$$P_1'(t) + \mu P_1(t) - \lambda P_0(t) = 0 \tag{12.6}$$

At $t = 0$, $P_0(0) = 1$ and $P_1(0) = 0$. The prime denotes differentiation with respect to time t. Solving Eqs. (12.5) and (12.6) by using the Laplace transforms we get

$$P_0(t) = \frac{1}{A} (\mu + \lambda e^{-At}) \tag{12.7}$$

$$P_1(t) = \frac{\lambda}{A} (1 - e^{-At}) \tag{12.8}$$

$$A = \lambda + \mu \tag{12.9}$$

For large t, Eq. (12.7) reduces to

$$A_a = \frac{\mu}{A} \tag{12.10}$$

where A_a is generator unit steady-state availability.

Similarly, the unit steady-state unavailability, A_u is obtained from Eq. (12.8)

$$A_u = \frac{\lambda}{A} \tag{12.11}$$

The above expression corresponds to the unit forced outage rate expression of Eq. (12.1).

EXAMPLE 1

A generator unit constant failure and repair rates data are as follows:

$$\lambda = 0.001 \text{ failure/hour}$$
$$\lambda = 0.01 \text{ repair/hour}$$

Calculate the steady-state unavailability of the generator unit. Utilizing Eq. (12.11) for the given data, the generator unavailability is

$$A_u = \frac{\lambda}{A} = \frac{\lambda}{\lambda + \mu} = \frac{0.001}{0.001 + 0.01} = 0.0909$$

12.3.2 Single Generator Unit Model with a Derated State

Here we present a three-state system Markovian model to represent a power generating unit with a derated state [4]. For example, a generating unit at a coal fired power generating station may be operating at a reduced capacity (say 300 megawatts instead of 500 megawatts) due to one or more pulverizer failures. This mathematical model can represent such a situation. The repairable, three-state generator unit can be either in the operating, derated, or failed state. Such a situation is represented in the state-space diagram shown in Fig. 12.2. The following are assumed in developing this model:

1. Failures are statistically independent.
2. The repaired system is as good as new.
3. All failure and repair rates are constant.

Associated notations with the repairable system model are as follows:

i denotes the generator unit ith state: $i = 0$ (operating), $i = 1$ (derated), $i = 2$ (failed)

$P_i(t)$ denotes the generator unit ith state probability at time t

λ_j denotes jth constant failure rate of the generator unit: $j = 1$ (operating to derated), $j = 2$ (operating to failed), $j = 3$ (derated to failed)

μ_j denotes jth constant repair rate of the generator unit: $j = 1$ (derated to operating), $j = 2$ (failed to operating), $j = 3$ (failed to derated)

s is the Laplace transform variable

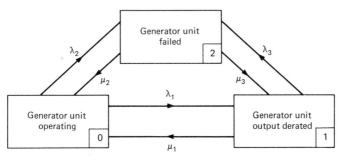

Figure 12.2. A generator unit transition diagram.

Figure 12.2 yields the differential equations

$$P_0'(t) + (\lambda_1 + \lambda_2)P_0(t) = \mu_1 P_1(t) + \mu_2 P_2(t) \tag{12.12}$$

$$P_1'(t) + (\mu_1 + \lambda_3)P_1(t) = \mu_3 P_2(t) + P_0(t)\lambda_1 \tag{12.13}$$

$$P_2'(t) + (\mu_1 + \mu_3)P_2(t) = \lambda_3 P_1(t) + \lambda_2 P_0(t) \tag{12.14}$$

At $t = 0$, $P_0(0) = 1$, $P_1(0) = 0$, $P_2(0) = 0$. The prime denotes differentiation with respect to time t. The Laplace transform of Eqs. (12.12) to (12.14) can be written

$$(s + \lambda_1 + \lambda_2)P_0(s) - \mu_1 P_1(s) - \mu_2 P_2(s) = 1 \tag{12.15}$$

$$-\lambda_1 P_0(s) + (s + \mu_1 + \lambda_3)P_1(s) - \mu_3 P_2(s) = 0 \tag{12.16}$$

$$-\lambda_2 P_0(s) - \lambda_3 P_1(s) + (s + \mu_2 + \mu_3)P_2(s) = 0 \tag{12.17}$$

The coefficients of the above equations are

$$
\begin{array}{ccc|c}
(s + \lambda_1 + \lambda_2) & -\mu_1 & -\mu_2 & 1 \\
-\lambda_1 & (s + \mu_1 + \lambda_3) & -\mu_3 & 0 \\
-\lambda_2 & -\lambda_3 & (s + \mu_2 + \mu_3) & 0
\end{array}
$$

Solution via Cramer's rule yields

$$P_0(s) = \frac{s^2 + s(\mu_1 + \mu_2 + \mu_3 + \lambda_3) + \mu_1\mu_2 + \lambda_3\mu_2 + \mu_1\mu_3}{s^3 + s^2(\mu_1 + \mu_2 + \mu_3 + \lambda_1 + \lambda_2 + \lambda_3) + s(\mu_1\mu_2 + \lambda_3\mu_2 + \mu_1\mu_3 + \mu_2\lambda_1 + \lambda_1\mu_3 + \lambda_1\lambda_3 + \mu_1\lambda_2 + \lambda_2\mu_3 + \lambda_2\lambda_3)} \tag{12.18}$$

$$P_1(s) = \frac{s\lambda_1 + \lambda_1\mu_2 + \lambda_1\mu_3 + \mu_3\lambda_2}{s^3 + s^2(\mu_1 + \mu_2 + \mu_3 + \lambda_1 + \lambda_2 + \lambda_3) + s(\mu_1\mu_2 + \lambda_3\mu_2 + \mu_1\mu_3 + \mu_2\lambda_1 + \lambda_1\mu_3 + \lambda_1\lambda_3 + \mu_1\lambda_2 + \lambda_2\mu_3 + \lambda_2\lambda_3)} \tag{12.19}$$

$$P_2(s) = \frac{s\lambda_2 + \lambda_1\lambda_3 + \mu_1\lambda_2 + \lambda_2\lambda_3}{s^3 + s^2(\mu_1 + \mu_2 + \mu_3 + \lambda_1 + \lambda_2 + \lambda_3) + s(\mu_1\mu_2 + \lambda_3\mu_2 + \mu_1\mu_3 + \mu_2\lambda_1 + \lambda_1\mu_3 + \lambda_1\lambda_3 + \mu_1\lambda_2 + \lambda_2\mu_3 + \lambda_2\lambda_3)} \tag{12.20}$$

The roots of the denominators of Eqs. (12.18) to (12.20) become:

$$c_1, c_2 = \frac{-(\mu_1 + \mu_2 + \mu_3 + \lambda_1 + \lambda_2 + \lambda_3) \pm \sqrt{(\mu_1 + \mu_2 + \mu_3 + \lambda_1 + \lambda_2 + \lambda_3)^2 - 4(\mu_1\mu_2 + \lambda_3\mu_2 + \mu_1\mu_3 + \mu_2\lambda_1 + \lambda_1\mu_3 + \lambda_1\lambda_3 + \mu_1\lambda_2 + \lambda_2\mu_3 + \lambda_2\lambda_3)}}{2} \tag{12.21}$$

Now, Eqs. (12.18) to (12.21) in time domain become

$$P_0(t) = \left(\frac{\mu_1\mu_2 + \lambda_3\mu_2 + \mu_1\mu_3}{c_1c_2}\right)$$
$$+ \left[\frac{\mu_1c_1 + \mu_2c_1 + \mu_3c_1 + c_1\lambda_3 + c_1^2 + \mu_1\mu_2 + \lambda_3\mu_2 + \mu_1\mu_3}{c_1(c_1 - c_2)}\right] e^{c_1 t}$$
$$+ \left\{1 - \left(\frac{\mu_1\mu_2 + \lambda_3\mu_2 + \mu_1\mu_3}{c_1c_2}\right)\right.$$
$$\left. - \left[\frac{\mu_1c_1 + \mu_2c_1 + \mu_3c_1 + c_1\lambda_3 + c_1^2 + \mu_1\mu_2 + \lambda_3\mu_2 + \mu_1\mu_3}{c_1(c_1 - c_2)}\right]\right\} e^{c_2 t}$$

(12.22)

$$P_1(t) = \left(\frac{\lambda_1\mu_2 + \lambda_1\mu_3 + \lambda_2\mu_3}{c_1c_2}\right) + \left[\frac{c_1\lambda_1 + \lambda_1\mu_2 + \lambda_1\mu_3 + \lambda_2\mu_3}{c_1(c_1 - c_2)}\right] e^{c_1 t}$$
$$- \left[\frac{\lambda_1\mu_2 + \lambda_1\mu_3 + \lambda_2\mu_3}{c_1c_2} + \frac{c_1\lambda_1 + \lambda_1\mu_2 + \lambda_1\mu_3 + \lambda_2\mu_3}{c_1(c_1 - c_2)}\right] e^{c_2 t}$$

(12.23)

$$P_2(t) = \left(\frac{\lambda_1\lambda_3 + \mu_1\lambda_2 + \lambda_2\lambda_3}{c_1c_2}\right) + \left[\frac{c_1\lambda_2 + \lambda_1\lambda_3 + \lambda_2\mu_1 + \lambda_2\lambda_3}{c_1(c_1 - c_2)}\right] e^{c_1 t}$$
$$- \left[\frac{\lambda_1\lambda_3 + \mu_1\lambda_2 + \lambda_2\lambda_3}{c_1c_2} + \frac{\lambda_2c_1 + \lambda_1\lambda_3 + \mu_1\lambda_2 + \lambda_2\lambda_3}{c_1(c_1 - c_2)}\right] e^{c_2 t}$$

(12.24)

System operational availability $= P_0(t) + P_1(t)$

$$= \left(\frac{\mu_1\mu_2 + \lambda_3\mu_2 + \mu_1\mu_3 + \lambda_1\mu_2 + \lambda_1\mu_3 + \lambda_2\mu_3}{c_1c_2}\right)$$
$$+ \left[\frac{\mu_1c_1 + \mu_2c_1 + \mu_3c_1 + c_1\lambda_3 + c_1^2 + \mu_1\mu_2 + \lambda_3\mu_2 + \mu_1\mu_3 + c_1\lambda_1 + \lambda_1\mu_2 + \lambda_1\mu_3 + \lambda_2\mu_3}{c(c_1 - c_2)}\right] e^{c_1 t}$$
$$+ \left\{1 - \left(\frac{\mu_1\mu_2 + \lambda_3\mu_2 + \mu_1\mu_3 + \lambda_1\mu_2 + \lambda_1\mu_3 + \lambda_2\mu_3}{c_1c_2}\right)\right.$$
$$\left. - \left[\frac{\mu_1c_1 + \mu_2c_1 + \mu_3c_1 + c_1\lambda_3 + c_1^2 + \mu_1\mu_2 + \lambda_3\mu_2 + \mu_1\mu_3 + c_1\lambda_1 + \lambda_1\mu_2 + \lambda_1\mu_3 + \lambda_2\mu_3}{c_1(c_1 - c_2)}\right]\right\} e^{c_2 t}$$

(12.25)

The availability expression is valid if and only if c_1 and c_2 are negative. As t becomes very large, the system steady-state operational availability equation is expressed as

$$\lim_{t \to \infty} (P_0(t) + P_1(t)) = \left(\frac{\mu_1\mu_2 + \lambda_3\mu_2 + \mu_1\mu_3 + \lambda_1\mu_2 + \lambda_1\mu_3 + \lambda_2\mu_3}{c_1 c_2} \right) \quad (12.26)$$

12.3.3 A Generator Unit Model with Preventive Maintenance

This section presents another repairable, three-state Markov model [3] with two mutually exclusive absorbing states. This model represents a situation when the preventive maintenance is performed on the generator and it is repaired upon failure. The state-space diagram shown in Fig. 12.3 represents such a situation. This model can also represent a repairable, three-state device such as a fluid flow valve [3].

The following assumptions are associated with the model:

1. Failures are statistically independent.
2. All failure and repair rates are constant.
3. The repaired system is as good as new.

The following notation is associated with the model:

i denotes the ith state of the generator unit: $i = 0$ (operating), $i = p$ (preventive maintenance), $i = f$ (failed)
$P_i(t)$ denotes the probability that the generator unit is in state i at time t
λ_p denotes the generator unit preventive maintenance constant rate
λ_f denotes the generator unit constant failure rate
μ_p denotes the generator unit constant repair rate due to preventive maintenance
μ_f denotes the generator unit constant repair rate

The system of differential equations associated with Fig. 12.3 are

$$P_0'(t) + (\lambda_p + \lambda_f)P_0(t) - \mu_p P_p(t) - \mu_f P_f(t) = 0 \quad (12.27)$$

$$P_p'(t) + \mu_p P_p(t) - P_0(t)\lambda_p = 0 \quad (12.28)$$

$$P_f'(t) + \mu_f P_f(t) - P_0(t)\lambda_f = 0 \quad (12.29)$$

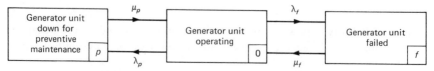

Figure 12.3. A generator unit state-space diagram.

At $t = 0$, $P_0(0) = 1$, $P_1(0) = P_2(0) = 0$. The prime denotes differentiation with respect to time t. Solving Eqs. (12.27) to (12.29) by using Laplace transforms, we get

$$P_0(t) = \frac{\mu_p \mu_f}{k_1 k_2} + \left[\frac{(k_1 + \mu_p)(k_1 + \mu_f)}{k_1(k_1 - k_2)} \right] e^{k_1 t}$$
$$- \left[\frac{(k_2 + \mu_p)(k_2 + \mu_f)}{k_2(k_1 - k_2)} \right] e^{k_2 t} \qquad (12.30)$$

$$P_p(t) = \frac{\lambda_p \mu_f}{k_1 k_2} + \left[\frac{\lambda_p k_1 + \lambda_p \mu_f}{k_1(k_1 - k_2)} \right] e^{k_1 t} - \left[\frac{(\mu_f + k_2)\lambda_p}{k_2(k_1 - k_2)} \right] e^{k_2 t} \qquad (12.31)$$

$$P_f(t) = \frac{\lambda_f \mu_p}{k_1 k_2} + \left[\frac{\lambda_f k_1 + \lambda_f \mu_p}{k_1(k_1 - k_2)} \right] e^{k_1 t} - \left[\frac{(\mu_p + k_2)\lambda_f}{k_2(k_1 - k_2)} \right] e^{k_2 t} \qquad (12.32)$$

where $\quad k_1 k_2 = \mu_p \mu_f + \lambda_p \mu_f + \lambda_f \mu_p$ $\qquad (12.33)$

$$k_1 + k_2 = -(\mu_p + \mu_f + \lambda_p + \lambda_f) \qquad (12.34)$$

The generator unit availability is given by Eq. (12.30). This availability expression is valid if and only if k_1 and k_2 are negative. As t becomes very large in Eq. (12.30), the steady-state generator availability, A_g, equation can be expressed as

$$A_g = \lim_{t \to \infty} P_0(t) = \frac{\mu_1 \mu_2}{k_1 k_2} \qquad (12.35)$$

12.3.4 Steam Turbine Overspeed Control System Model

This model is used to evaluate the unreliability of a high pressure steam turbine overspeed system [14]. Figure 12.4 shows the block diagram of steam turbine overspeed control system.

It can be seen from Fig. 12.4 that the valve and its controller form a series configuration. In other words if either controller or valve fails then the valve-controller fails. For the statistically independent components, the failure probability, F_{vc}, of the valve-controller assembly is

$$F_{vc} = f_v + f_c - f_v f_c \qquad (12.36)$$

where f_v is the valve failure probability
$\quad f_c$ is the valve controller failure probability

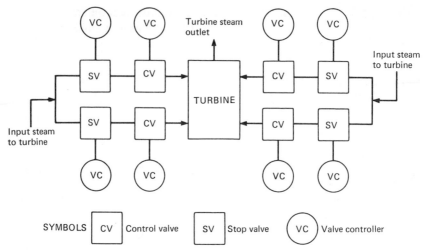

Figure 12.4. Represention of the overspeed control system.

Figure 12.4 indicates that if either the stop valve assembly or the control valve assembly fails to close the incoming steam then it means that a steam line control system has failed.

Each steam line failure probability, F_{sl}, is

$$F_{sl} = F_{sv}F_{cv} \tag{12.37}$$

where $F_{sv} = f_{sv} + f_{sc} - f_{sv}f_{sc}$
$\quad\quad F_{cv} = f_{cv} + f_{cc} - f_{cv}f_{cc}$
$\quad\quad F_{sv} \quad$ is the stop valve-controller unreliability
$\quad\quad F_{cv} \quad$ is the control valve-controller unreliability
$\quad\quad f_{sv} \quad$ is the stop valve unreliability
$\quad\quad f_{sc} \quad$ is the stop valve controller unreliability
$\quad\quad f_{cv} \quad$ is the control valve unreliability
$\quad\quad f_{cc} \quad$ is the control valve controller unreliability

Using Eq. (12.37), the reliability, R, of an n identical steam line turbine overspeed control system is

$$R = (1 - F_{sv}F_{cv})^n \tag{12.38}$$

EXAMPLE 2

If $F_{sv} = 0.1$, $F_{cv} = 0.2$ and $n = 4$ then calculate the turbine overspeed control system reliability, R, Eq. (12.38). Thus

$$R = (1 - 0.1 \times 0.2)^4 = 0.9224$$

12.3.5 Solar Domestic Hot-Water System Model

The reliability of this system is predicted by assuming that the statistically independent components of the solar domestic hot-water system [43] form a series configuration. The reliability block diagram of the solar domestic hot-water system is shown in Fig. 12.5. The reliability, R_{sd}, of the block diagram shown in Fig. 12.5 is

$$R_{sd} = \prod_{i=1}^{6} R_i \qquad (12.39)$$

where R_1 is the water heater/storage tank reliability
$\quad R_2$ is the check valve reliability
$\quad R_3$ is the pump reliability
$\quad R_4$ is the piping system reliability
$\quad R_5$ is the solar collector reliability
$\quad R_6$ is the differential controller reliability

For a constant failure rate, the solar hot water system's ith unit reliability is

$$R_i = e^{-\lambda_i t} \qquad (12.40)$$

where λ_i is the failure rate of the ith unit/subsystem of the solar hot water system
$\quad t \quad$ is time

By substituting Eq. (12.40) in Eq. (12.39) we get

$$R_{sd}(t) = \prod_{i=1}^{6} e^{-\lambda_i t} \qquad (12.41)$$

To obtain the solar hot water system mean time to failure, MTTF, integrate Eq. (12.41) over the time interval $[0, \infty]$.

$$R_{sd} = \int_0^\infty e^{-\sum\limits_{i=1}^{6} \lambda_i t} \, dt = \frac{1}{\sum\limits_{i=1}^{6} \lambda_i} \qquad (12.42)$$

Figure 12.5. Solar domestic hot-water system reliability block diagram.

12.3.6 Transmission System Two-Weather Model

Equipment installed indoors generally experiences a relatively constant environment [2, 4]. Environments vary for a transmission line and other outdoor transmission line equipment. These equipment failure rates may vary with the changes in environments. Figure 12.6 shows the state transition diagram of a single component or system operating under two types of environments. This is a repairable, single system Markovian model. The following assumptions are associated with the model:

1. Failures are statistically independent.
2. All failure and repair rates are constant.
3. The repaired system is as good as new.

In Fig. 12.6 the normal and stormy weather states are in dotted rectangles. The associated notations with the model are

i ith state of the system: $i = 0$ (operating-normal weather), $i = 1$ (failed-normal weather), $i = 2$ (operating-stormy weather), $i = 3$ (failed-stormy weather)

$P_i(t)$ probability of the system being in state i at time t.

μ_i constant ith repair rate of the system: $i = 1$ (normal weather), $i = 2$ (stormy weather)

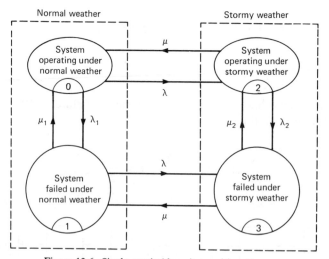

Figure 12.6. Single repairable unit transition diagram.

λ_i constant ith failure rate of the system: $i = 1$ (normal weather), $i = 2$ (stormy weather)

λ $1/k$ where k is the normal weather period expected duration

μ $1/g$ where g is the stormy weather period expected duration

The system of differential equations are

$$P_0'(t) + (\lambda_1 + \lambda)P_0(t) - \mu P_2(t) - \mu_1 P_1(t) = 0 \qquad (12.43)$$

$$P_1'(t) + (\mu_1 + \lambda)P_1(t) - \mu P_3(t) - \lambda_1 P_0(t) = 0 \qquad (12.44)$$

$$P_2'(t) + (\mu + \lambda_2)P_2(t) - \mu_2 P_3(t) - \lambda P_0(t) = 0 \qquad (12.45)$$

$$P_3'(t) + (\mu + \mu_2)P_3(t) - \lambda P_1(t) - \lambda_2 P_2(t) = 0 \qquad (12.46)$$

At $t = 0$, $P_0(0) = 1$, $P_1(0) = P_2$, $(0) = P_3(0) = 0$. The prime denotes differentiation with respect to time t. The steady-state probability solutions [2] of Eqs. (12.43) to (12.46) are

$$P_0 = \frac{\mu A_1}{\lambda(A_2 + A_3) + \mu(A_4 + A_1)} \qquad (12.47)$$

where $A_1 = \mu_2\lambda + \mu_1 A_5$
$A_2 = \mu_1\mu + \mu_2 A_6$
$A_3 = \lambda_1\mu + \lambda_2 A_6$
$A_4 = \lambda_2\lambda + \lambda_1 A_5$
$A_5 = \lambda_2 + \mu + \mu_2$
$A_6 = \lambda_1 + \lambda + \mu_1$

$$P_1 = A_4 P_0 / A_1 \qquad (12.48)$$

$$P_2 = \lambda P_0 A_2 / \mu A_1 \qquad (12.49)$$

$$P_3 = \lambda P_0 A_3 / \mu A_1 \qquad (12.50)$$

Note that Eqs. (12.47) to (12.50) are obtained by setting derivatives equal to zero in Eqs. (12.43) to (12.46) and using the relationship $\sum_{i=0}^{3} P_i = 1$.

12.3.7 Single-Phase Transformers with One Standby

This model [4] represents a mathematical model of three operating single-phase transformers with one standby unit. The following assumptions are used to develop the repairable unit Markovian model:

1. Transformers failures are statistically independent.
2. Transformers failure, repair, and replacement rates are constant.
3. When more than one transformer has failed then the whole transformer bank failure is assumed. In addition, it is assumed that no more transformer failures occur.
4. One repair person repairs the failed transformer. He or she starts repair on the second failed unit after the first failed unit is repaired and put back into service.
5. Repaired transformers are as good as new.
6. The standby unit can not fail in its standby mode.
7. All the system units are identical.

The system transition diagram is shown in Fig. 12.7. The following notation is associated with the model:

i denotes the ith state of the system: $i = 0$ (three transformers operating, one standby), $i = 1$ (two transformers operating, one standby), $i = 2$ (three transformers operating, no standby), $i = 3$ (two transformers operating, no standby)

$P_i(t)$ denotes the probability of the system being in ith state at time t

$\lambda_1 = 3\lambda$ where λ represents the single-phase transformer failure rate

μ_1 denotes the unit repair rate

α denotes the standby unit constant installation rate

The system of differential equations with associated Fig. 12.7 is

$$P_0'(t) + \lambda_1 P_0(t) - \mu_1 P_2(t) = 0 \tag{12.51}$$

$$P_1'(t) + \alpha P_1(t) - \lambda_1 P_0(t) - \mu_1 P_3(t) = 0 \tag{12.52}$$

$$P_2'(t) + (\lambda_1 + \mu_1) P_2(t) - \alpha P_1(t) = 0 \tag{12.53}$$

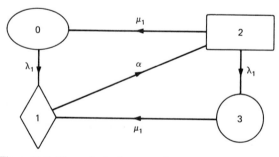

Figure 12.7. Three single-phase transformers with one standby.

$$P_3'(t) + \mu_1 P_3(t) - \lambda_1 P_2(t) = 0 \qquad (12.54)$$

$$\lambda_1 = 3\lambda \qquad (12.55)$$

At $t = 0$, $P_0(0) = 1$, $P_1(0) = P_2(0) = P_3(0) = 0$. The prime denotes differentiation with respect to time t. The steady-state probability solutions to Eqs. (12.51) to (12.55) are

$$P_0 = [1 + A_1(1 + A_2 + A_1)]^{-1} \qquad (12.56)$$

where $A_1 \equiv \lambda_1/\mu_1$
$A_2 \equiv (\lambda_1 + \mu_1)/\alpha$

$$P_1 = A_1 A_2 P_0 \qquad (12.57)$$

$$P_2 = A_1 P_0 \qquad (12.58)$$

$$P_3 = P_0 A_1^2 \qquad (12.59)$$

Note that the Eqs. (12.56) to (12.59) are obtained by setting derivatives equal to zero in Eqs. (12.51) to (12.54) and using the relationship $\sum_{i=0}^{3} P_i = 1$.

12.3.8 Pulverizer Model

At a coal-fired power station a generating unit may have a number of pulverizers to grind coal. Normally there is one standby pulverizer for each generating unit to improve the reliability of pulverizer system. In this model we assume

1. There are five pulverizers associated with one power generating unit. Only four are required to obtain full output capacity of the generating unit. Each pulverizer failure causes a 25 percent reduction in output capacity of the generating unit.
2. All five pulverizers form an active redundant network.
3. Pulverizers failures are statistically independent.
4. All pulverizers are identical.

Figure 12.8 shows the block diagram representation of the model. The four-out-of-five pulverizer operating reliability, $R_{4/5}(t)$, for a pulverizer constant failure rate λ is

$$R_{4/5}(t) = 5e^{-4\lambda t} - 4e^{-5\lambda t} \qquad (12.60)$$

where t is time.

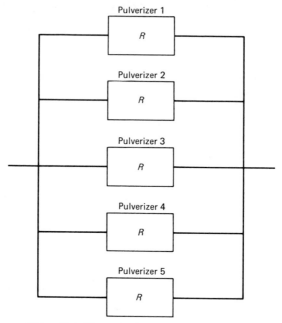

Figure 12.8. Five pulverizers in active redundancy.

Equation (12.60) gives the probability that at time t the power generating unit will be operating at its 100 percent capacity. One should note that pulverizer repair is not considered in Eq. (12.60).

Similarly the three-out-of-five nonrepairable pulverizer operating reliability, $R_{3/5}(t)$, for the pulverizer constant failure rate, λ, is

$$R_{3/5}(t) = 6e^{-5\lambda t} - 15e^{-4\lambda t} + 10e^{-3\lambda t} \tag{12.61}$$

Equation (12.61) will give the probability result that the power generating unit will be operating at least at 75 percent of its full generating capacity.

EXAMPLE 3

Suppose a power generating unit requires at least four out of five pulverizers operating for 100 percent output capacity and three out of five for operating at 75 percent of its full generating capacity. For a pulverizer failure rate of $\lambda = 0.001$ failures/hour, calculate the generating unit reliabilities for 1000 hours when the unit is to operate at 100 percent or 75 percent of its full capacity.

For 100 percent capacity. Using Eq. (12.60), for $\lambda = 0.0001$ failure/hour and $t = 1000$ hours

$$R_{4/5}(1000) = 5e^{-4(0.0001)(1000)} - 4e^{-5(0.0001)(1000)}$$

$$= 0.9255$$

For 75 percent capacity.   Using Eq. (12.61) for $\lambda = 0.0001$ failures/hour and $t = 1000$ hours

$$R_{3/5}(1000) = 6e^{-5(0.0001)(1000)} - 15e^{-4(0.0001)(1000)} + 10e^{-3(0.0001)(1000)}$$

$$= 0.9926$$

12.3.9 A Two-Unit System Model with Restricted Repair

This model represents a two nonidentical unit repairable system. This Markovian model has application to much power equipment. The model transition diagram is shown in Fig. 12.9

The assumptions associated with the model are:

1. Units failures are statistically independent.
2. All failure and repair rates are constant.
3. The repaired unit is as good as new.
4. Only one repairperson repairs a failed unit.

One unit is repaired at a time. In a situation in which one unit is under repair and the other unit fails, then the most recent failed unit waits for the repair.

The notation associated with the model is

i denotes the ith state of the system: $i = 0$ (both units operating), $i = 1$ (unit 2 operating, unit 1 under repair), $i = 2$ (unit 1 operating,

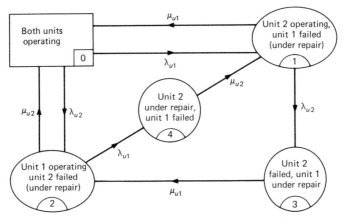

Figure 12.9. Two nonidentical unit system transition diagram.

unit 2 under repair), $i = 3$ (unit 2 failed, unit 1 under repair), $i = 4$ (unit 2 under repair, unit 1 failed)

$P_i(t)$ denotes probability that system is in state i at time t

λ_{ui} denotes the constant failure rate of ith unit: $i = 1$ (unit 1), $i = 2$ (unit 2)

μ_{ui} denotes the constant repair rate of ith unit: $i = 1$ (unit 1), $i = 2$ (unit 2)

The system of differential equations associated with Fig. 12.9 is

$$P'_0(t) + (\lambda_{u1} + \lambda_{u2})P_0(t) - \mu_{u1}P_1(t) - \mu_{u2}P_2(t) = 0 \qquad (12.62)$$

$$P'_1(t) + (\mu_{u1} + \lambda_{u2})P_1(t) - \lambda_{u1}P_0(t) - \mu_{u2}P_4(t) = 0 \qquad (12.63)$$

$$P'_2(t) + (\mu_{u2} + \lambda_{u1})P_2(t) - \lambda_{u2}P_0(t) - \mu_{u1}P_3(t) = 0 \qquad (12.64)$$

$$P'_3(t) + \mu_{u1}P_3(t) - \lambda_{u2}P_1(t) = 0 \qquad (12.65)$$

$$P'_4(t) + \mu_{u2}P_4(t) - \lambda_{u1}P_2(t) = 0 \qquad (12.66)$$

At $t = 0$, $P_0(0) = 1$, $P_1(0) = P_2(0) = P_3(0) = P_4(0) = 0$. The following steady-state probability solutions are obtained by setting derivatives equal to zero in Eqs. (12.62) to (12.66) and using the relationship $\sum_{i=0}^{4} P_i = 1$.

$$P_0 = A_1A_2/T \qquad (12.67)$$

where $A_1 \equiv \mu_{u1}\mu_{u2}$

$A_2 \equiv \lambda_{u1}\mu_{u1} + A_1 + \lambda_{u2}\mu_{u2}$

$T \equiv \mu_{u1}(\lambda_{u2} + \mu_{u2})A_2 + \lambda_{u1}(A_1 + \mu_{u2}\lambda_{u2} + \mu_{u1}\lambda_{u2})A_3$

$A_3 \equiv \lambda_{u1} + \lambda_{u2} + \mu_{u2}$

$$P_1 = A_1\lambda_{u1}A_3/T \qquad (12.68)$$

$$P_2 = A_1A_4\lambda_{u2}/T \qquad (12.69)$$

where $A_4 \equiv (\lambda_{u1} + \lambda_{u2} + \mu_{u1})$

$$P_3 = A_5A_3\mu_{u2}/T \qquad (12.70)$$

where $A_5 \equiv \lambda_{u1}\lambda_{u2}$

$$P_4 = A_5A_4\mu_{u1}/T \qquad (12.71)$$

12.4 TRANSMISSION LINE COMMON-CAUSE OUTAGES MODEL

Multiple transmission lines are quite commonly used on the same tower [8]. This model represents two nonidentical lines subject to common-cause failures. Common-cause failures can occur due to aircraft crash, flood, vehicles, and so on [3]. The state-space diagram of the model is shown in Fig. 12.10.

The notation associated with Fig. 12.10 is

i represents the ith state of the system: $i = 0$ (both transmission lines operating) $i = 1$ (line 1 failed, line 2 operating), $i = 2$ (line 2 failed, line 1 operating), $i = 3$ (both transmission lines failed)

$P_i(t)$ represents probability that system is in state i at time t

λ_{cc} represents the constant common-cause failure rate of transmission lines

λ_{tj} represents the jth transmission line constant failure rate: $j = 1$ (line 1), $j = 2$ (line 2)

μ_{tj} represents the jth transmission line constant repair rate: $j = 1$ (line 1), $j = 2$ (line 2)

The following assumptions are associated with the model:

1. Common-cause and other failures are statistically independent.
2. Transmission lines are nonidentical.
3. A repaired transmission line is as good as new.

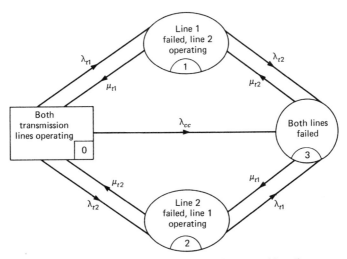

Figure 12.10. Two nonidentical transmission lines transition diagram.

The system of differential equations associated with Fig. 12.10 is

$$P_0'(t) + (\lambda_{t1} + \lambda_{t2} + \lambda_{cc})P_0(t) - \mu_{t1}P_1(t) - \mu_{t2}P_2(t) = 0 \qquad (12.72)$$

$$P_1'(t) + (\lambda_{t2} + \mu_{t1})P_1(t) - \mu_{t2}P_3(t) - \lambda_{t1}P_0(t) = 0 \qquad (12.73)$$

$$P_2'(t) + (\lambda_{t1} + \mu_{t2})P_2(t) - \mu_{t1}P_3(t) - \lambda_{t2}P_0(t) = 0 \qquad (12.74)$$

$$P_3'(t) + (\mu_{t1} + \mu_{t2})P_3(t) - \lambda_{t1}P_2(t) - \lambda_{t2}P_1(t) - \lambda_{cc}P_0(t) = 0 \qquad (12.75)$$

At $t = 0$, $P_0(0) = 1$, $P_1(0) = P_2(0) = P_3(0) = 0$. The prime denotes differentiation with respect to time t. The steady-state probability solutions to Eq. (12.72) to (12.75) are

$$P_0 = \mu_{t1}\mu_{t2}A/A_3 \qquad (12.76)$$

where $A \equiv A_1 + A_2$
$A_2 \equiv (\lambda_{t2} + \mu_{t2})$
$A_1 \equiv (\lambda_{t1} + \mu_{t1})$
$A_3 \equiv A_1 A_2 A + \lambda_{cc}[A_1(A_2 + \mu_{t1}) + \mu_{t2}A_2]$

$$P_1 = [A\lambda_{t1} + A_4\lambda_{cc}]\mu_{t2}/A_3 \qquad (12.77)$$

where $A_4 \equiv \lambda_{t1} + \mu_{t2}$

$$P_2 = [A\lambda_{t2} + A_5\lambda_{cc}]\mu_{t1}/A_3 \qquad (12.78)$$

where $A_5 \equiv \lambda_{t2} + \mu_{t1}$

$$P_3 = A\lambda_{t1}\lambda_{t2} + A_4 A_5 \lambda_{cc}/A_3 \qquad (12.79)$$

Equations (12.76) to (12.79) are obtained by setting derivatives equal to zero in Eq. (12.72) to (12.75) and using the relationship $\sum_{i=0}^{3} P_i = 1$.

12.5 LIFE CYCLE MAINTENANCE COST MODEL

This model is used to estimate the life cycle maintenance cost of transformers, switchgear, pumps, motors, generators, and so on. Furthermore, the model can be utilized when planning maintenance, purchasing new equipment, etc. When buying new equipment the life cycle maintenance cost must be added to the equipment acquisition and other costs to obtain the total life cycle cost of the equipment in question.

The following five-step approach [35] can be used to calculate the life cycle maintenance cost:

1. Estimate the expected useful life of the equipment. Generally for power equipment a 30-year life is assumed, because of most power generating stations are built to last at least 30 years. However, some of the power equipment may vary in useful life from this estimate.
2. Estimate the annual cost or interest rate of the money to be invested to purchase the equipment.
3. Estimate the annual hazard rate (failure rate) per unit.
4. For each failure, estimate the expected annual maintenance cost per unit.
5. Calculate the total present value of the annual maintenance costs. Use the present worth factor formula for a uniform series given in Chapter 9.

These steps are demonstrated by the following example:

EXAMPLE 4

A power corporation is considering purchasing a heavy duty power transformer. There are two manufacturers, A and B, who are bidding to sell such a transformer. To measure the field performance of A and B's transformers the following data are obtained:

	MANUFACTURER A	MANUFACTURER B
1. Constant failure rate, λ	0.01 failures/year	0.05 failures/year
2. Annual expected cost of each failure	$20,000	$15,000
3. Annual interest rate or cost of money	20 percent	20 percent
4. Expected useful life of the transformer	30 years	30 years

From the point of view of field performance find out the best purchase by assuming that the result of this exercise will provide an input to a life cycle cost model. The formula given as Eq. (9.15) in Chapter 9 can be used to obtain the present value.

$$PV = p \left[\frac{1 - (1 + i)^{-m}}{i} \right] \qquad (12.80)$$

Manufacturer A. Using Eq. (12.80) for $i = 0.2$, $p = \$20,000$, and $m = 30$ years, the present worth, PV, of the money is

$$PV = \$20,000 \left[\frac{1 - (1 + 0.2)^{-30}}{0.2} \right]$$

$$= \$99,578.73$$

For $\lambda = 0.01$ failures/year the present worth of the life cycle maintenance cost C_{la}, is

$$C_{la} = \lambda \times 99,578.73 = 0.001 \times 99,578.73$$

$$= \$995.79$$

Similarly for

Manufacturer B. Using Eq. (12.80) for $i = 0.2$, $p = \$15,000$, and $m = 30$ years, the present value, PV, of the money is

$$PV = \$15,000 \left[\frac{1 - (1 + 0.2)^{-30}}{0.2} \right]$$

$$= \$74,684.05$$

For $\lambda = 0.05$ failures/year, the present worth of the life cycle maintenance cost, C_{lb}, is

$$C_{lb} = 0.05 \times \$74,684.05 = \$3734.20$$

To maintain the manufacturer B's transformer it will cost \$2,734.20 more than to maintain manufacturer A's transformer over the life cycle years. Therefore, from the point of view of maintenance cost, to purchase manufacturer A's equipment will be a wise decision.

12.6 SUMMARY

The following areas of power equipment reliability are presented in this chapter:

1. Power system reliability definitions and indices
2. Equipment reliability models
3. Transmission lines common-cause outages
4. Life cycle maintenance cost

The author has attempted to present in this chapter some of the most useful models, definitions, and indices associated with power equipment reliability decision making. Like many other reliability areas the availability of getting good field data is still one of the drawbacks associated with power equipment reliability. A list of selective references is presented at the end of this chapter for further reading. References 7, 11, and 41 list most of the published literature up to 1977.

12.7 REFERENCES

Books

1. R. Billinton, R. J. Ringlee, and A. J. Woods, *Power-System Reliability Calculations*, MIT Press, Cambridge, MA, 1973.
2. R. Billinton, *Power System Reliability Evaluation*, Gorden and Breach, New York, 1970.
3. B. S. Dhillon and C. Singh, *Engineering Reliability: New Techniques and Applications*, Wiley, New York, 1981.
4. J. Endrenyi, *Reliability Modeling in Electric Power Systems*, Wiley, New York, 1978.
5. C. Singh and R. Billinton, *System Reliability Modelling and Evaluation*, Hutchinson, London, 1977.

Articles

6. K. Ayoub, D. Guy, and D. Patton, "Evaluation and Comparison of Some Methods for Calculating Generating System Reliability," *IEEE Trans. on Power Apparatus and Systems*, Vol. PAS-89, 1970, pp. 537–544.
7. "Bibliography on the Application of Probability Methods in Power System Reliability Evaluation," *IEEE Trans. on Power Apparatus and Systems*, Vol. PAS-97, 1978, pp. 2235–2242.
8. R. Billinton, T. L. P. Medicherla, and M. S. Sachdev, "Common Cause Outages in Multiple Circuit Transmission Lines," *IEEE Trans. on Reliability*, Vol. R-27, 1978, pp. 128–131.
9. R. Billinton and M. Alam, "Effect of Restricted Repair on System Reliability Indices," *IEEE Trans. on Reliability*, Vol. 27, 1978, pp. 376–379.
10. R. Billinton and K. E. Bollinger, "Transmission System Reliability Evaluation Using Markov Processes," *IEEE Trans. on Power Apparatus and Systems*, Vol. PAS-67, 1968, pp. 538–547.
11. R. Billinton, "Bibliography on the Application of Probability Methods in Power System Reliability Evaluation," *IEEE Trans. on Power Apparatus and Systems*, PAS-91, 1972, pp. 649–660.
12. R. Billinton and R. K. Ringlee, "Models and Techniques for Evaluation Power Plant Auxiliary Equipment Reliability," *1971 Power Generation Conference*, St. Louis, published as IEEE Paper No. 71, CP 702-PWR. Available from IEEE.
13. L. E. Booth, B. W. Logan, J. R. Frangola, and L. O. Hecht, "The Delphi Procedure as Applied to IEEE Project 500 (Reliability Data Manual for Nuclear Power Plants)," *Third Annual Reliability Engineering Conference for the Electric Power Industry*, 1976, pp. 49–61. Available from IEEE.
14. J. Burns, "Reliability of Nuclear Power Plant Steam Turbine Overspeed Control System, Failure Prevention and Reliability," *Design Engineering Technical Conference*, 1977, pp. 27–38. Available from ASME.
15. J. Cadwallader, P. Scaletta, N. Sharmahd, and R. J. Stokely, "Nuclear Reliability/Availabil-

ity Monte Carlo Analysis," *Proceedings Annual Reliability and Maintainability Symposium,* 1975, pp. 6–12.

16. V. M. Cook, R. K. Ringlee, and J. P. Whooley, "Suggested Definitions Associated With the Status of Generating Station Equipment and Useful in the Application of Probability Methods for System Planning and Operation," 1972, pp. 1–14. Available from IEEE.

17. S. Collard, E. Parascos, and A. Kressner, "Root Cause Failure Analysis in Electric Transmission and Distribution Equipment," *Annual Reliability and Maintainability Symposium,* 1980, pp. 428–432.

18. "Common Mode Forced Outages of Overhead Transmission Lines," IEEE Committee Report, *IEEE Trans. on Power Apparatus and Systems,* Vol. PAS-85, 1966, pp. 859–864.

19. C. F. DeSieno and L. L. Stine, "A Probability Method for Determining the Reliability of Electric Power Systems," *IEEE Trans. on Power Apparatus and Systems,* 1964, pp. 30–35.

20. J. F. Dapazo, D. E. Tuite, E. L. Stein, and H. M. Merill, "Objective Criteria for Power Plant Maintenance Scheduling," *Third Annual Reliability Engineering Conference for the Electric Power Industry,* 1976, pp. 97–107. Available from IEEE.

21. "Distribution Protection and Restoration Systems: Design Verification by Reliability Indices," IEEE Committee Report, *IEEE Trans. on Power Apparatus and Systems,* Vol. PAS-93, 1974, pp. 564–570.

22. "Definitions of Customer and Load Reliability Indices for Evaluating Electric Power System Performance," IEEE Committee Report, IEEE Paper A 75 588–4. Available from the IEEE.

23. L. Garver, "Reserve Planning Using Outage Probabilities and Load Uncertainties," *IEEE Trans. on Power Apparatus and Systems,* 1970, pp. 514–520.

24. D. P. Gaver, F. E. Mountmeat, and A. D. Patton, "Power System Reliability I—Measures of Reliability and Methods of Calculation," *IEEE Trans. (Power Apparatus and Systems),* Vol. 83, 1964, pp. 727–736. Available from the IEEE.

25. P. Gall and S. Musick, "Experience with Data Collection on Operation of General Electric Large Steam Turbine-Generator Sets, *Third Annual Reliability Engineering Conference for the Electric Power Industry,* 1976, pp. 117–124. Available from the IEEE.

26. M. S. Grover and M. T. G. Gilleapie, "A Cost-Oriented Reliability and Maintainability Assessment of Circuit Breakers in the Design of Auxiliary Electrical Power Systems of a Generating Station," *Third Annual Reliability Engineering Conference for the Electric Power Industry,* 1976, pp. 125–134. Available from the IEEE.

27. M. S. Grover and R. Billinton, "Quantitative Evaluation of Maintenance Policies in Distribution Systems," IEEE Paper C 75 112–8, 1975. Available from the IEEE.

28. H. Karr and L. Wagner, "R&M—Today's Heating and Cooling vs Solar Energy," *Annual Reliability and Maintainability Symposium,* 1976, pp. 491–499.

29. B. W. Logan and W. R. Riggs, "Economic Assessment of a Spare Step-up Transformer for a System of Generating Stations," *Third Annual Reliability Engineering Conference for the Electric Power Industry,* 1976, pp. 135–138. Available from the IEEE.

30. A. J. Marino and R. S. Ullman, "Use of Aerospace Industry Failure Analysis Techniques in Electrical Utility Operation," *Annual Reliability and Maintainability Symposium,* 1980, pp. 433–441.

31. M. E. Nelson, C. C. Richard, and P. D. Sierer, "An Application of a Reliability Analysis Method to Actual Unit Failure Data," *Century 2 Potpourri Conference,* San Francisco, 1980, pp. 1–4. Available from the ASME.

32. V. I. Nitu, H. Albert, S. Iomescu, and H. Cimpeanu, "Power System Reliability Indicators," *Proceedings Fourth Power System Computation Conference,* Grenoble, France, 1972, Vol. 1, Paper 1.2/4.

33. A. D. Patton, "Report on Reliability Survey on Industrial Plants," *Industrial and Commercial*

Power Systems Technical Conference, IEEE Industry Application Society, Atlanta, 1973. Available from IEEE.

34. A. D. Patton, "Short-Term Reliability Calculation," *IEEE Trans. on Power Apparatus and Systems,* Vol. PAS-89, 1970, pp. 509–513.

35. T. Parascos and J. A. Arceri, "Reliability Engineering and Underground Equipment Failure, Cost and Manufacturer's Analysis," *Third Annual Reliability Conference for the Electric Power Industry,* 1976, pp. 25–30. Available from the IEEE.

36. D. Raheja, J. Muench, and J. McNulty, "Developing High Reliability Transformer Components A Manufacturer's Viewpoint," *Annual Reliability and Maintainability Symposium,* 1980, pp. 442–447.

37. M. Reusterholz, Arceri, and T. Parascos, "Computerized Inspection and Maintenance of Underground Equipment A Construction Man's Tool," *3rd Annual Reliability Engineering Conference for the Electric Power Industry,* 1976, pp. 31–37. Available from the IEEE.

38. M. Ramamoorty and Balgopal, "Block Diagram Approach to Power System Reliability," *IEEE Trans. on Power Apparatus and Systems,* Vol. PAS-89, 1970, pp. 802–811.

39. J. Ringlee and D. Goode, "On Procedures for Reliability Evaluations of Transmission Systems," *IEEE Trans. on Power Apparatus and Systems,* Vol. PAS-89, 1970, pp. 527–536.

40. H. T. Spears, K. L. Hicks, and S. T. Y. Lee, "Probability of Loss of Load for Three Areas," *IEEE Trans. on Power Apparatus and Systems,* Vol. PAS-89, 1970, pp. 521–526.

41. S. Vemuri, "An Annotated Bibliography of Power System Reliability Literature—1972–1977," *IEEE PES Summer Meeting,* Los Angeles, July 16–21, 1978. Available from IEEE.

42. J. E. Winkler, D. A. Rickett, W. P. Gould, "Determining the Optimum Number of Pulverizer Mill-Mill Feeder Sets for a Pulverized Coal Fired Generating Unit," *The Winter Annual Meeting,* Houston, December 1975, Paper No. 75-WA/PWR-12. Available from ASME.

43. R. M. Wolosewicz and P. S. Chopra, "Application of Reliability, Maintainability and Availability Engineering to Solar Heating and Cooling Systems," *Proceedings at the Annual Reliability and Maintainability Symposium,* 1980, pp. 248–253.

Index